Undergraduate Lecture Notes in Physics

Undergraduate Lecture Notes in Physics (ULNP) publishes authoritative texts covering topics throughout pure and applied physics. Each title in the series is suitable as a basis for undergraduate instruction, typically containing practice problems, worked examples, chapter summaries, and suggestions for further reading.

ULNP titles must provide at least one of the following:

- An exceptionally clear and concise treatment of a standard undergraduate subject.
- A solid undergraduate-level introduction to a graduate, advanced, or non-standard subject.
- A novel perspective or an unusual approach to teaching a subject.

ULNP especially encourages new, original, and idiosyncratic approaches to physics teaching at the undergraduate level.

The purpose of ULNP is to provide intriguing, absorbing books that will continue to be the reader's preferred reference throughout their academic career.

Series editors

More information about this series at http://www.springer.com/series/8917

Rosario Bartiromo · Mario De Vincenzi

Electrical Measurements in the Laboratory Practice

Springer

Rosario Bartiromo
Università Roma Tre
Rome
Italy

Mario De Vincenzi
Università Roma Tre
Rome
Italy

ISSN 2192-4791 ISSN 2192-4805 (electronic)
Undergraduate Lecture Notes in Physics
ISBN 978-3-319-31100-5 ISBN 978-3-319-31102-9 (eBook)
DOI 10.1007/978-3-319-31102-9

Library of Congress Control Number: 2016934203

Printed on acid-free paper

This Springer imprint is published by Springer Nature
The registered company is Springer International Publishing AG Switzerland

Foreword

This book is intended for students in physics or engineering attending laboratory classes in electrical measurements. It covers the basic theory of electrical circuits, describes analog and digital instrumentation, and applies modern methods to evaluate uncertainties in electrical measurements. Its structure and content are the result of a specific teaching experience lasted for more than 5 years.

Indeed when we first faced the task of offering laboratory classes in electricity to physics students, we soon realized that available textbooks had two major shortcomings. First, the evaluation of measurement uncertainties they propose is often inconsistent with modern methodology, as elaborated in the review of the expert panel appointed by the Bureau International des Poids et Measures (BIPM). Second, digital measuring devices, now ubiquitous in class and research laboratories, are not described or marginally mentioned. Moreover, textbooks available in the published literature are often very specialized and do not pay enough attention to the links that the matter dealt with has with the general framework and methods of physics and engineering.

The material presented in these pages can be taught during the second semester of the second year of the course of studies. At this stage, students should have completed their basic physics studies and achieved a level of maturity enabling them to generalize to other subjects the concepts they learn in the study of a particular topic. Therefore, we were encouraged to use the matter subject of the course, which is electrical circuits with the relevant instrumentation and experiments, as an opportunity for students to:

- Consolidate the understanding of the concepts needed for the evaluation of the uncertainties in experimental measurements and complete the learning of necessary formal tools
- Illustrate with examples the methods to construct models of physical systems, a very important task in the professional life, emphasizing their potentialities and limitations

- Familiarize with a modern electronic equipment where the understanding of the functional relationships between different parts of the measuring device allows for an effective exploitation even without the knowledge of the construction details
- Learn to use the tools of linear system analysis, such as harmonic decomposition, symbolic method, the duality between frequency and time domain, and the concept of normal modes
- Learn to design and perform simple laboratory experiments to verify physical laws and to measure parameters used to characterize components and materials
- Take advantage of the availability of data in digital form to learn how to write simple software for their analysis
- Learn how to report on laboratory activities in a manner suitable for a scientific presentation of experimental results.

We are convinced that the availability of a valuable written text is a necessary condition for students to learn and internalize the arguments dealt with during the lectures and the laboratory sessions and to reach a level of knowledge significant at the professional level. Since we did not find in the available literature a book, or a reasonable combination of books, that was suitable to support the type of course we had in mind to offer, we decided the writing of these notes.

This book consists of ten chapters, each followed by an adequate number of exercises, which are briefly outlined below. In addition, in the first of its three appendices we propose a series of nine experiments to carry out in the laboratory, with an illustration mainly intended for the use of student tutors.

In line with the experimental approach we have chosen, we start in Chap. 1 with the definition of electrical circuits and the description of components required for their construction. We then introduce the discrete component model and discuss its validity limits. Next, we discuss the characteristics of ideal components by introducing the concepts of impedance and admittance. Construction methods of resistors, inductors, and capacitors are then briefly outlined. A discussion of the skin effect is presented at this stage, and completed in a separate appendix, to illustrate the limitations of the discrete component model with a concrete example. This chapter assumes a certain degree of familiarity with the content of General Physics courses, but where we deem it useful, or necessary, the relevant concepts are recalled in special notes.

In Chap. 2, we introduce the two Kirchhoff laws and we illustrate the topological properties of electric networks. The principal methods used under steady-state conditions to compute the distribution of voltages and currents in a circuit are then presented. After a brief description of problems posed by the possible nonlinearity in the characteristics of components, we discuss the properties of linear networks: superposition and reciprocity theorems and the concept of equivalent circuit with related theorems (Norton and Thévenin).

Chapter 3 examines the uncertainties associated with electrical measurements adopting as a reference the guide developed by the International Organization for Standardization (ISO) [1] that has standardized the treatment of measurement uncertainties in science and industry. We first discuss the different types of uncertainties and their causes, and introduce the concept of standard uncertainty. Then we show how to evaluate the combined effect of various sources of uncertainty. Particular emphasis is given to the possibility of correlation between the measurements, a concept that can be usefully introduced at this stage of undergraduate studies program. All these concepts are applied in a number of examples in order to achieve an appropriate level of clarity.

In Chap. 4, we first discuss methods to measure steady-state currents and voltages. After recalling the important features of measuring instruments, the use of a current meter is illustrated and the principles of operation of the moving coil galvanometer are presented in detail. This is followed by a discussion of voltage measurements and an illustration of the different types of voltmeter and their characteristics. Throughout this chapter, a special attention is given to explain to students the influence of the instrument used to perform the measurement on its result and the importance of systematic uncertainties that may come with it. The chapter ends with a detailed discussion of methods for measuring the resistance of the circuit components.

Chapter 5 begins with the definition and characterization of time-dependent signals. We recall again the properties of the inductors and capacitors and then proceed to the derivation of the differential equation that governs the evolution of current and voltage in a series *RLC* circuit. This equation is solved for a sinusoidal signal after illustrating the specificity and the importance of this type of signal. It is then explained how the generalization to the space of complex numbers allows a more easy solution of linear circuits under sinusoidal excitation by reducing the differential equations into algebraic equations. This result is finally formalized in the symbolic method with the introduction of the complex representation of voltages and currents and the concept of complex impedance of circuit components.

In Chap. 6, the frequency response of important alternating current circuits is obtained. It begins with resonant circuits, series and parallel, and then it deals with bridge circuits that allow the measurement of unknown impedances. The series–parallel transformation is used for the assessment of the effect due to nonideal components in resonant circuits. We then introduce the notion of two-port circuits and discuss Bode diagrams for their amplitude and phase response. The chapter continues with a detailed solution of the *RC* and *RL* circuits in both of the high-pass and low-pass filter configuration and a discussion of their possible use as differentiators and integrators of waveforms. Then we recall the phenomenon of mutual induction and its use in electrical transformers whose ideal behavior is evaluated in detail. The chapter ends with a discussion of the problem of impedance matching and a presentation of the main methods used in this regard.

Chapter 7 is devoted to the illustration of alternating signals other than sinusoidal and to the experimental methods useful to characterize them. The concept of effective value of an electrical quantity is introduced and methods for its measurement are presented. Finally, phase measurements are briefly discussed.

In Chap. 8, both analog and digital oscilloscopes are introduced and their operation is analyzed in detail. We first illustrate the principles of operation of a cathode ray tube and the block diagram of an analog oscilloscope. We deal in detail with the need to synchronize this instrument with the phenomena whose time evolution we want to study and we discuss the methods to achieve it. The modern version of the instrument, the digital oscilloscope, is then introduced and its functional structure is described. The benefits obtained are illustrated, and emphasis is given to the possibility of reading the instrument memory with a personal computer to process the data numerically in a personalized way. Finally, we describe the methods for measuring the phase shift between two electrical signals with an oscilloscope.

Chapter 9 is devoted to the analysis of the behavior of non-sinusoidal waveforms with particular regard to circuit excitation by pulses. We first discuss the techniques for solving electrical circuits in the time domain and the special role played by the step function. Then we proceed to obtain the step response of the two-port circuits already studied in Chap. 6 in the frequency domain, with a particular emphasis on the initial conditions to be imposed to the solution of the relevant differential equations. Again, we find the conditions under which they behave as integrators or differentiators of input signals. The response of *RLC* circuits is then studied in detail and it is used, as an example, to evaluate the parasitic inductance in a RC high-pass filter. The chapter ends with an exhaustive evaluation of the properties of a compensated voltage divider that we analyze in both the time and the frequency domains. In this chapter we did not use Laplace methods to be consistent with the undergraduate level of the book and to avoid the risk of hiding behind a formal approach, the basic physics of circuit operation.

Finally, Chap. 10 addresses the problem of the behavior of components that cannot be treated as discrete and in particular discusses the properties of transmission lines. After introducing an appropriate representation, we derive the so-called telegrapher equation. Its solution in the time domain shows that, in the absence of losses, signals are transmitted without distortion with a propagation speed given by line parameters. The general solution in the frequency domain is worked out to introduce the concept of line impedance. Then, the issue of the termination of the line is addressed and the expression for the reflection coefficient is obtained. After discussing the attenuation due to resistive losses, we describe the electrical behavior of terminated transmission lines and obtain the expression for their apparent impedance. The chapter ends with some practical considerations on the cables used for connections in laboratory experiments.

The book contains three appendices. Appendix A is an important complement to the main text. There we illustrate a series of nine experiments that we have used extensively in our laboratory classes. For each of them we give a plan of action that students should follow and a note that can be useful to tutors both in the preparation and in the illustration of the experiment to students. Appendix B and C deal, respectively, with the skin effect and the Fourier analysis. They are meant to give more information on their subject to the interested readers.

Most chapters in this book are equipped with a number of problems that students can solve using concepts and methods discussed therein. For these problems, we tried to maintain a level of difficulty in line with the preparation of students in the second semester of undergraduate course of studies. Problems that require much effort are marked with an asterisk.

Reference

1. BIPM, IEC, IFCC, ILAC, ISO, IUPAC, IUPAP and OIML, Evaluation of measurement data - Guide to expression of uncertainty in measurement. JCGM 100:2008 (2008)

Contents

Chapter 1
The Electrical Circuit and Its Components

1.1 Introduction

An electrical circuit, or electrical network, consists of a set of physical devices interconnected so as to allow the flow of the electrical current. The devices used in electrical circuits are called *electrical components* or *circuit elements*. Note that the term *circuit* highlights a key feature of an electrical network, consisting in the presence of one or more closed paths through which the current can circulate.

Electrical circuits are used to transfer, distribute, or process energy or information in electromagnetic form. This requires the manipulation of electrical currents and/or voltages in a wide range of values and with very different time dependencies. We can find countless examples of use of these circuits around us in everyday life.

In our smartphone, a low voltage battery converts chemical into electrical energy and supplies a constant voltage. The circuits in the device fulfill the job to convert it into the signals needed by logical elements to perform their functions. An example is shown in Fig. 1.1 representing the output of a digital clock, a high frequency square wave with carefully stabilized 5 MHz frequency, and a standard voltage swing of 1 V. When the battery needs recharging, electrical power is drawn from the main distribution network and accumulated as chemical energy in the battery.

However, the voltage available from the network is not constant but has sinusoidal time dependence and a rather high maximum value. Sinusoidal waveforms offer a convenient opportunity to transmit power over long distances. They play a special role in circuit analysis since their shape does not change when propagating in a linear circuit.

Figure 1.2 shows the voltage available in Europe from the main distribution network. This waveform needs to be transformed in a time-independent voltage of low value, a few volts, to power most of the consumer electronics available on the market. This is done in electrical circuits using nonlinear components, such as semiconductor diodes.

R. Bartiromo and M. De Vincenzi, *Electrical Measurements in the Laboratory Practice*, Undergraduate Lecture Notes in Physics,
DOI 10.1007/978-3-319-31102-9_1

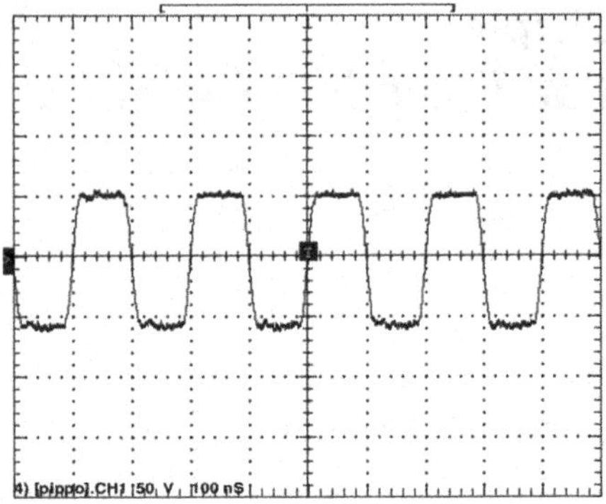

Fig. 1.1 Typical output of a digital clock. It consists of a square wave voltage signal with 5 MHz frequency and 1 V amplitude. Plot obtained on the screen of a digital oscilloscope with vertical sensitivity 1 V per division and horizontal sensitivity 100 ns per division

Waveforms that are even more complicated are produced by a host of sensors that convert physical quantities into electrical signals. For example, an ionization chamber detects the passage of a fast particle by the charge it leaves behind. The chamber electrodes detect a very fast current signal that must be integrated to recover a measurement of the particle energy. Therefore, integration is one of the many operations that we must be able to perform with an electric circuit.

In many cases, it is necessary to use techniques that require concepts that go beyond the physics of electricity and rely upon the use of devices controlling the behavior of electrons at the microscopic level. These are subjects of the discipline of electronics that is beyond the scope of this book. However, it is not possible nowadays to deal in a meaningful way with electrical measurements without accounting for the opportunities offered by electronic devices. Therefore, in this book we introduce the reader to the most important examples of these kinds of instruments, limiting ourselves to a functional description of their behavior with a minimum of details on their practical implementation.

In this chapter, we describe the most common electrical components. In Sect. 1.2, we first establish the basic framework allowing a useful definition and classification of circuit elements. Then we introduce the ideal version of discrete components of most common use. In Sect. 1.3, we illustrate the properties of the practical implementation

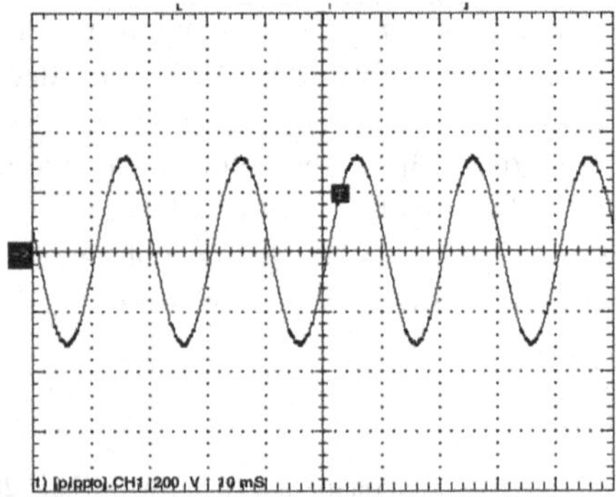

Fig. 1.2 The voltage supplied by the public distribution network in Europe. It has a sinusoidal waveform with a frequency of 50 Hz and a root mean square voltage of 230 V. The root mean square is the standard way to characterize the amplitude of a time-dependent signal, as we will learn later in this book

of each of them, the materials and the fabrication techniques involved. We also give an introductory discussion of the skin effect to illustrate the limits of the discrete component approximation. Finally, in Sect. 1.4 we briefly address the limits of use we face when deploying a component in a practical circuit.

1.2 Electrical Components

The electrical components (also referred to as circuit elements) are devices designed to control the relation between the voltage difference at their terminals and the electrical current flowing through them. This definition of electrical or electronic[1] component covers a wide spectrum of devices ranging from a simple resistor to electromechanical devices such as the relays, from semiconductor devices such as transistors, to the latest chips with the functionality of millions of transistors in a single component.

Most of this book will deal with *lumped parameters* circuits, those in which the signal wavelength[2] is much larger than the physical dimension of the circuit and of its components. This allows us to assume that electric and magnetic fields are uniform within each component and possibly dependent only upon time.

When this approximation does not hold, that is, when the wavelength of the signals propagating in the circuit is comparable to components dimensions, we say that the circuit has *distributed parameters* and we cannot neglect the nonuniformity of the electric and magnetic fields within its components. A noteworthy distributed parameters circuit is the transmission line and we will deal in detail with it in the last chapter of this book.

Electrical components have at least two terminals and for each of them we must define a positive direction for the *electrical current* flowing through it. Moreover, for each couple of terminals we must define the positive direction for *voltage difference* or *electrical tension* across them.

The simplest components have just two terminals. They are characterized by the voltage difference between their terminals and by the current flowing through them. Figure 1.3 shows the symbol of a generic component whose terminals are identified by the letters A and B. Denoting with $v(t) = v_A(t) - v_B(t)$ the voltage drop across it and with $i(t)$ the current flowing through it, in general we can express them as

[1] Nowadays it is customary to use the term electronic for all electrical components even if electronics itself was founded in the early 1900s with the invention, by A. Fleming and L. De Forest, of vacuum valves, the first devices able to control the motion of electrons.

[2] The electromagnetic field propagates in electrical circuits with a speed of the order of light speed in vacuum c. Denoting with ν the frequency of the signal, its wavelength is $\lambda \simeq c/\nu$. To get an idea of the order of magnitude of lengths, consider that for $\nu = 1\,\mathrm{kHz}$, $\lambda = 3.010^5\,\mathrm{m}$, for $\nu = 1\,\mathrm{MHz}$, $\lambda = 3.010^2\,\mathrm{m}$ and for $\nu = 1\,\mathrm{GHz}$, $\lambda = 0.3\,\mathrm{m}$. These lengths are larger than typical dimensions of components $\simeq 10^{-2} \div 10^{-3}\,\mathrm{m}$ and of circuits $\simeq 10^{-1}\,\mathrm{m}$.

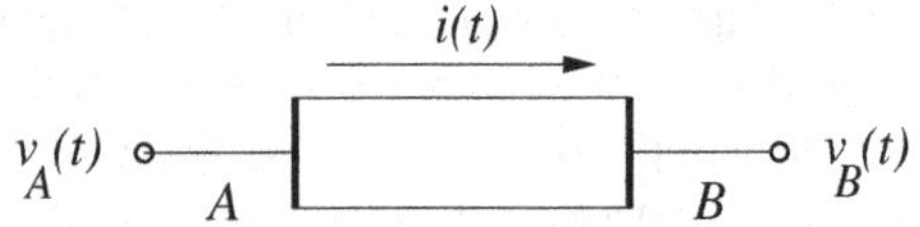

Fig. 1.3 Symbolic representation of a generic two-terminal electrical component

$$\begin{cases} v(t) = \hat{Z}i(t) \\ i(t) = \hat{Y}v(t) \end{cases} \tag{1.1}$$

where $\hat{Z}$ and $\hat{Y}$ are two *operators* named, respectively, *impedance* and *admittance*. The operator *impedance* $\hat{Z}$ acts on the mathematical representation of the electrical current $i(t)$ flowing through the component and yields the potential drop $v(t)$ across it. The operator *admittance* $\hat{Y}$ acts on the mathematical representation of the potential drop $v(t)$ across the component and yields the electrical current $i(t)$ flowing through it.

The operators *impedance* and *admittance* have a *physical dimension*: the unit of measure of the impedance is the *Ohm* (symbol: Ω) while the unit of measure of the admittance is the *Siemens* or *Mho* (symbol: Ω^{-1} or ℧). Moreover, they are the inverse of each other and the following operational relationship holds: $\hat{Z} = \hat{Y}^{-1}$. The mathematical properties of the impedance and admittance operators are a useful starting point to classify the circuit elements they describe. The first important classification distinguishes *linear* from *nonlinear* components.

Linear components. Electrical components are linear if operators $\hat{Z}$ and $\hat{Y}$ are linear, meaning that *their mathematical expression does not contain explicitly the voltage drop across the component or the current flowing through it*. Examples of linear components are the ideal resistor, the ideal capacitor, the ideal inductor, the ideal amplifier. Examples of nonlinear components are the diode, whose impedance is a function of the voltage drop at its terminals, the ferromagnetic inductor, whose inductance is a function of the current flowing through the component. Also nonlinear are all circuits used in digital electronics since they are stable only for a finite number of output voltage levels.

The second important classification distinguishes *active* from *passive* components.

Active and passive components. *Active* electrical components are devices able to transfer net electrical power. On the contrary, *passive* components can only store or dissipate energy. An active component is able to supply or transfer energy to other components in the circuit. Examples of active components are the voltage generator (battery), the transistor, the operational amplifier; examples of passive components are the resistor, the capacitor, the inductor, the diode.

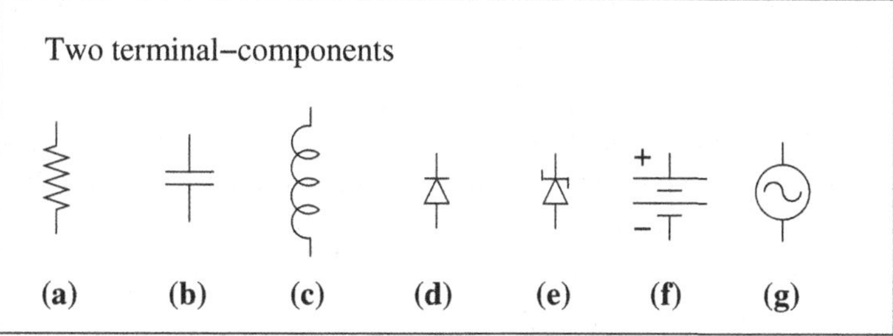

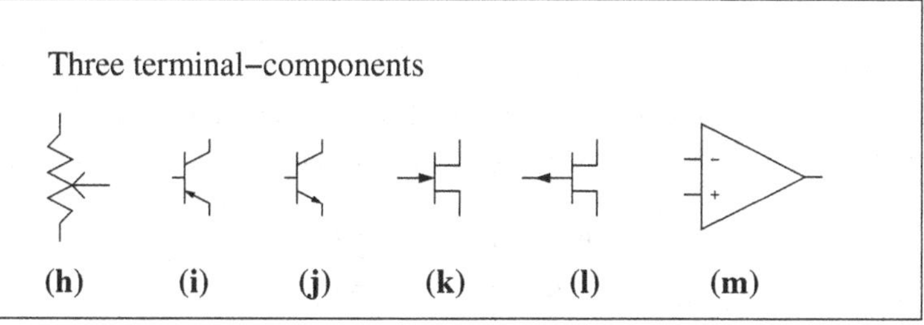

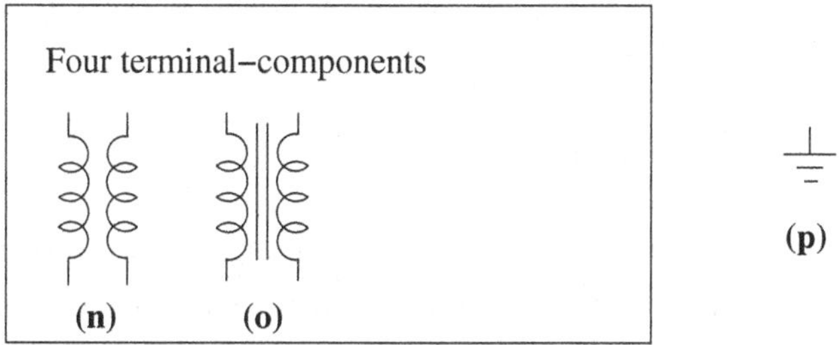

(p)

Fig. 1.4 Examples of component symbols used in diagrams of electrical circuits. **a** resistance, **b** capacitance, **c** inductance, **d** diode, **e** Zener diode, **f** DC voltage generator, **g** alternating voltage generator, **h** potentiometer, **i** pnp transistor, **j** npn transistor, **k** field effect transistor (FET) channel n, **l** p-channel FET, **m** operational amplifier, **n** air core transformer, **o** ferrite core transformer, **p** ground connection

1.2.1 Electrical Symbols

It is customary to use specific graphic symbols to represent components in circuit diagrams. The symbol specifies the qualitative properties of a component while a number representing the value of the relevant parameter, with its physical units, supplies quantitative information nearby the symbol. In Fig. 1.4 we show the symbols of the most common electrical components.

For the description of real circuit elements, it is useful first to introduce ideal components, which are described by simple expression of the operators $\hat{Z}$ and $\hat{Y}$. We shall see that we can describe satisfactorily real components as a properly connected collection of ideal elements.

1.2.2 Resistor

The *ideal resistor* is a two-terminal component whose behavior is determined uniquely by the first Ohm's law[3]; if $v(t)$ is the voltage drop across the resistor terminals and $i(t)$ the current flowing through it, using notations in Fig. 1.3 we can write

$$v(t) = v_A(t) - v_B(t) = Ri(t). \tag{1.2}$$

where R is the resistor resistance. The expression (1.2) can be inverted yielding

$$i(t) = G[v_A(t) - v_B(t)] = Gv(t). \tag{1.3}$$

where $G = 1/R$ is the resistor *conductance*. The operators, impedance and admittance, of the ideal resistor are simple multiplicative constants

$$\text{resistor impedance}: \quad \hat{Z} = R \tag{1.4}$$

$$\text{resistor admittance}: \quad \hat{Y} = G = 1/R \tag{1.5}$$

The ideal resistor is a dissipative component. The energy dissipated per unit of time, i.e., the dissipated power, is given by

$$W(t) = v(t)i(t) = Ri^2(t) = \frac{v^2(t)}{R} \tag{1.6}$$

Dissipation of electric as heat in a resistor is universally known as the *Joule effect*.

Resistors in series. Two electrical components are connected in series when the current flowing through them is identical (see Fig. 1.5). Therefore, if n resistors, each with resistance R_i, are connected in series, the additive property of the electric potential and a simple calculation show that the equivalent resistance of the n resistors in series is equal to

$$R = \sum_{i=1}^{n} R_i$$

[3]A simple model of the electrical conduction due to *Drude* provides the behavior of ohmic conductors. The model is based on the assumption that, inside the conductor, the electrons (point-like particles with electrical charge e) are subject to two forces. The first is due to the external (constant) electric field ($F_D = eE$). The second is a viscous force proportional to the electron speed and takes into account the diffusion effects suffered by the electrons in the solid ($F_V = -\gamma v$). In one spatial dimension we can write $m_e\dot{v} = eE - \gamma v$. In stationary conditions the electron drift velocity (v_D) is $v_D = eE/\gamma$, proportional to the electric field. Since the current density J is given by $J = nev_D$, where n is the electron particle density, we have $J = ne^2E/\gamma$. If A is the conductor cross sectional area and l its length, the electric field is $E = V/l$ and the current is $I = JA = (ne^2/\gamma)(A/l)V$. The first Ohm's law is thus recovered with the resistance $R = (\gamma/ne^2)(l/A)$. Furthermore, we note that the quantity γ/ne^2 represents the material resistivity ρ as defined by the second Ohm's law $R = \rho l/A$.

Fig. 1.5 Series (*left*) and parallel (*right*) connections of two resistors

Resistors in parallel. Two electrical components are connected in parallel when the voltage drop across each of them is the same (see Fig. 1.5). If n resistors, each with resistance R_i, are connected in parallel, the current flowing through the shared connections is the sum of the individual currents flowing in each resistor. A simple calculation shows that the equivalent resistance of the n resistors in parallel is given by

$$\frac{1}{R} = \sum_{i=1}^{n} \frac{1}{R_i}$$

1.2.3 Capacitor

A capacitor is made of two conductors, isolated from each other and in condition of complete electrostatic induction, and exploits the proportionality between the electric charge $q(t)$ accumulated on each conductor and their voltage difference $v(t)$. Therefore, for an ideal capacitor the following relation holds:

$$q(t) = Cv(t) \tag{1.7}$$

where C is the capacitor capacitance. The quantity $S = 1/C$ is the capacitor elastance, a somewhat obsolete term. Using the definition of electrical current, $i(t) = dq/dt$, in Eq. (1.7) we get

$$i(t) = C\frac{dv(t)}{dt} \quad \text{and, after integration,} \quad v(t) = \frac{1}{C}\int i(t)dt$$

Therefore, for the ideal capacitor the operators, impedance and admittance, are

$$\text{capacitor impedance}: \quad \hat{Z} = \frac{1}{C}\int dt \tag{1.8}$$

$$\text{capacitor admittance}: \quad \hat{Y} = C\frac{d}{dt} \tag{1.9}$$

It is well known that the capacitor is a device able to store electrostatic energy in the space where the electric field is not null. The electrostatic energy $\mathcal{E}_C$, stored by a capacitor of capacitance C, is given by

$$\mathscr{E}_C(t) = \frac{1}{2}q(t)v(t) = \frac{1}{2}Cv^2(t) = \frac{1}{2}\frac{q^2(t)}{C}. \tag{1.10}$$

Capacitors in series. When n capacitors, each with capacitance C_i, are connected in series, *the charge on each of them is the same*. Therefore, the additive property of the electric potential and a simple calculation show that the equivalent capacitance of n capacitors in series is given by

$$\frac{1}{C} = \sum_{i=1}^{n} \frac{1}{C_i}$$

Capacitors in parallel. When n capacitors, each with capacitance C_i, are connected in parallel, the voltage drop across each of them is the same while the charge on the shared connection is the sum of the individual charges on each capacitor. A simple calculation shows that the equivalent capacitance of the n capacitors in parallel is equal to

$$C = \sum_{i=1}^{n} C_i$$

1.2.4 Inductor

The ideal inductor is an electrical component where the electric field is entirely due to the variation of the magnetic flux generated by the current flowing through it. Denoting by Φ the magnetic induction flux through the inductor and by L its *self-inductance coefficient* or, in short, *inductance*, in an ideal and isolated inductor the following relations holds:

$$\Phi = Li(t) \tag{1.11}$$

$$i(t) = \frac{1}{L}\Phi = \Gamma\Phi \tag{1.12}$$

The quantity $\Gamma = 1/L$ is the *inductive elastance*, a somewhat obsolete term. The law of *Faraday-Neumann* yields

$$v(t) = \frac{d\Phi}{dt} = L\frac{d}{dt}i(t) \quad \text{and, after integration} \quad i(t) = \frac{1}{L}\int v(t)dt$$

These relations yield the expressions for the operators, impedance and admittance, of the ideal inductor

$$\text{inductor impedance}: \quad \hat{Z} = L\frac{d}{dt} \tag{1.13}$$

$$\text{inductor admittance}: \quad \hat{Y} = \frac{1}{L}\int dt \tag{1.14}$$

While a capacitor stores electrostatic energy, the inductor stores *magnetic energy*, $\mathscr{E}_L$, in the space region where magnetic induction B is not null. For an inductor with inductance equal to L, we have

$$\mathscr{E}_L(t) = \frac{1}{2}Li^2(t)$$

where $i(t)$ is the current flowing through the inductor.

Mutual induction. When the magnetic field, generated by the current flowing through an inductor, links to a different inductor, it can induce an electric field there, and the two inductors are not anymore isolated. This phenomenon known as *mutual induction* is at the heart of the operation of the voltage transformer, an electrical component we will discuss in Chap. 6 when dealing with time-dependent currents.

Here we want to show how it is possible to take it into account the simultaneous presence of different inductors in the same circuit. For this purpose let us consider n inductors connected in a network; denoting with $L_{jk}(=L_{kj})$[4] the induction coefficient between inductors j and k, the voltage induced in the inductor j due to the current flowing through the inductor k is[5]:

$$V_{jk} = L_{jk}\frac{d}{dt}i_k(t) \tag{1.15}$$

where $i_k(t)$ is the current flowing through the kth inductor. The total voltage drop across inductor j is then easily obtained as

$$V_j(t) = \sum_k V_{jk}(t) = \sum_k L_{jk}\frac{d}{dt}i_k(t) \tag{1.16}$$

Denoting with $\Phi_j(t)$ the magnetic induction flux linked to the jth inductor we have

$$V_j(t)dt = \sum_k L_{jk}di_k(t) = d\Phi_j(t) \tag{1.17}$$

$$\Phi_j(t) = \int_0^t V_j(t')dt' + \Phi_j(0) = \sum_k L_{jk}i_k(t) + \Phi_j(0) \tag{1.18}$$

---

[4]The equality $L_{jk} = L_{kj}$ is recovered rather easily when the coefficients L_{jk} are derived using the vector potential A. Details can be found in general physics textbooks, as for example *The Physics of Feynmann* Vol. II-17-11 [1].

[5]In this notation, L_{jj} is the self-inductance of the jth inductor.

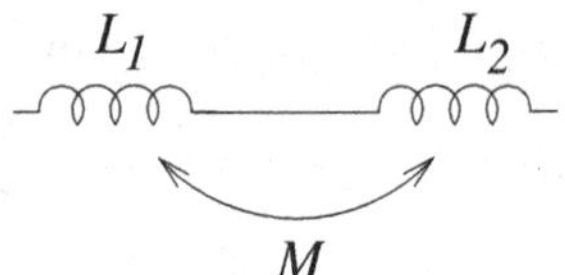

Fig. 1.6 Two inductors connected in series; M is the coefficient of mutual inductance between the two inductors

Solving the linear equation system obtained from the last equation, we get the relation

$$i_k(t) = \sum_j \Gamma_{jk}[\Phi_j(t) - \Phi_j(0)] \tag{1.19}$$

where the coefficients Γ_{jk}, whose physical dimension is the inverse of the inductance, are obtained from the matrix $L^\dagger_{jk}$, adjoint of L_{jk} through the equation

$$\Gamma_{jk} = \frac{1}{\|L_{jk}\|} L^\dagger_{jk} \tag{1.20}$$

where $\|L_{jk}\|$ denotes the determinant of matrix L_{jk}.

As an example of application of these results, we consider now the case of two inductors connected in series or in parallel.

Inductors in series. Consider two inductors connected in series. Denoting with L_1 and L_2 their self-inductances and with M their mutual induction coefficient[6] (see Fig. 1.6) and using Eq. (1.16) we get:

$$\begin{aligned} v(t) = v_1(t) + v_2(t) &= (L_{11} + L_{12} + L_{21} + L_{22})\frac{d}{dt}i(t) \\ &= (L_1 + L_2 + 2M)\frac{d}{dt}i(t) \end{aligned}$$

Therefore, the equivalent inductance of the two inductors in series is

$$L = L_1 + L_2 + 2M \tag{1.21}$$

Inductors in parallel. Consider two inductors connected in parallel. Denoting with L_1 and L_2 their self-inductances, with M their mutual induction coefficient and considering that the total current flowing through the parallel connection is the sum of the currents i_1 and i_2 flowing through the individual inductors, we use relations (1.19) to obtain

$$i(t) = i_1(t) + i_2(t) = (\Gamma_{11} + \Gamma_{12} + \Gamma_{21} + \Gamma_{22})\Phi(t) \tag{1.22}$$

[6] The correspondence between the notation used in this example and the one used previously is as follows: $L_1 = L_{11}$, $L_2 = L_{22}$ and $M = L_{12} = L_{21}$.

where we have taken into account that, since the inductors are connected in parallel, relation (1.17) yields $\Phi_1(t) = \Phi_2(t) \equiv \Phi(t)$. Applying relation (1.20), we get

$$\Gamma_{jk} = \frac{1}{\|L_{jk}\|} L_{jk}^{\dagger} = \frac{1}{L_1 L_2 - M^2} \begin{vmatrix} L_2 & M \\ M & L_1 \end{vmatrix}$$

Taking the time derivative of relation (1.22) and using $\Gamma_1 = \Gamma_{11}$, $\Gamma_2 = \Gamma_{22}$, and $\Gamma_M = \Gamma_{12} = \Gamma_{21}$ we finally obtain:

$$v(t) = \frac{1}{\Gamma_1 + \Gamma_2 + 2\Gamma_M} \frac{d}{dt} i(t)$$

Therefore, the equivalent inductance of the two inductors connected in parallel is

$$L = \frac{L_1 L_2 - M^2}{L_1 + L_2 + 2M}$$

Generalization to more than two inductors in series or in parallel only requires the handling of more complex mathematical computations.

Note that in case the inductors are magnetically isolated, $M = 0$, the expressions for the equivalent inductance are similar to those obtained for resistors in both series and parallel connection case. Finally, we recall that M is an algebraic quantity, its sign depending on the geometry of the inductor. More details will be given in Sect. 6.11 dealing with the voltage transformer.

1.2.5 Generators

A generator is an *active* electrical component supplying the energy that other components can either store or dissipate as heat through the *Joule* effect. Both voltage and current generators can be used to power electric circuits.

Voltage generator. The voltage difference v at the terminals of an ideal voltage generator does not depend upon the current i flowing through it (although it can possibly be time dependent). In formulas, this is expressed as

$$\frac{dv}{di} = 0$$

implying that the ideal voltage generator has *null impedance* and *infinite admittance*. Two circuit symbols are used for ideal voltage generators, one for the case of a constant voltage, or battery (see Fig. 1.7a) and one for the case of a time depending voltage (see Fig. 1.7b).

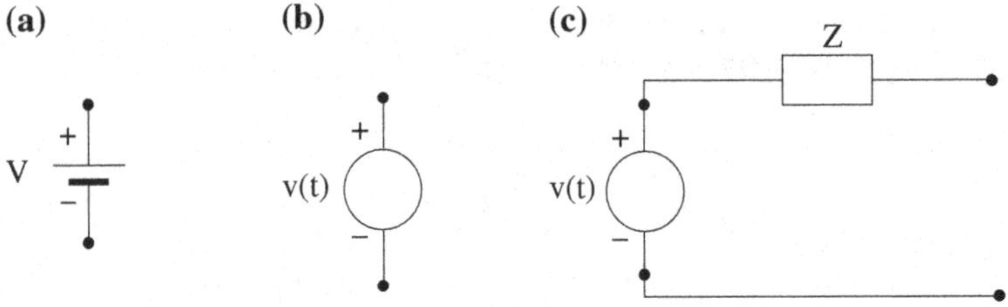

Fig. 1.7 Circuit symbols of voltage generators. **a** the battery symbol; **b** symbol of a variable generator; **c** symbol of a real voltage generator

Real voltage generator. The behavior of a real voltage generator can be represented with an ideal generator in series with an impedance that is referred to as the internal generator impedance (see Fig. 1.7c).

Current generator. The current i flowing through an ideal current generator does not depend upon the voltage drop v across its terminals (although it can possibly be time dependent). In formulas, this is expressed as

$$\frac{di}{dv} = 0$$

implying that the ideal current generator has *infinite impedance* and *null admittance*.

Real current generator. The behavior of a real current generator can be represented with an ideal generator in parallel with an impedance that is referred to as the internal generator impedance (see Fig. 1.8).

Special ideal generators. Of particular importance are two generators frequently used in the analysis of circuits, whose formal definitions are

- **Short Circuit**: a voltage generator with $v(t) = 0$ for any $i(t)$
- **Open Circuit**: a current generator with $i(t) = 0$ for any $v(t)$

1.2.6 Controlled Generators

The operation of many active circuits, such as for example the amplifiers, can be better described introducing special components behaving as voltage (or current)

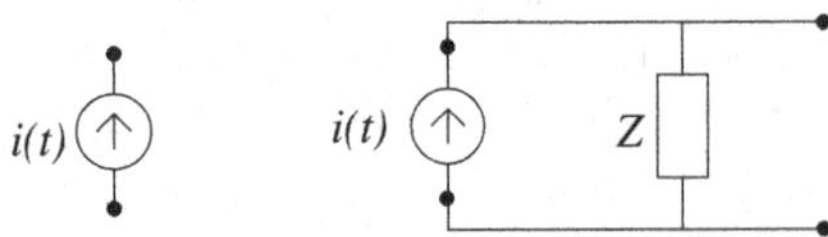

Fig. 1.8 On the *left*, the circuit symbol of the ideal current generator, on the *right* the representation of a real current generator. Conventionally, the *arrow* indicates the positive direction of the current

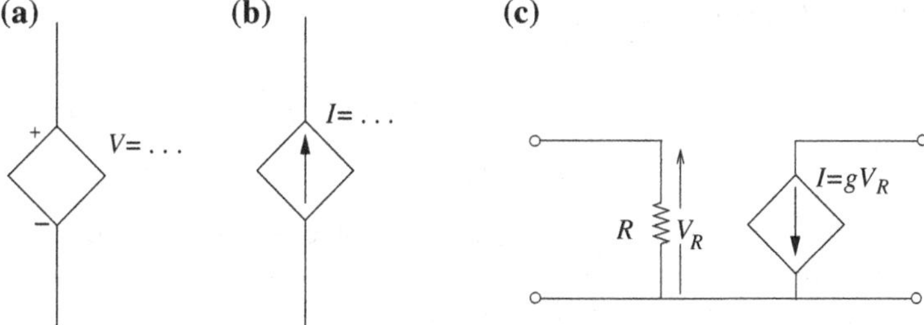

Fig. 1.9 Circuit symbols of controlled voltage (**a**) and current (**b**) generators. With **c** we show an example of a circuit that uses a current controlled generator. The current $I = gV_R$, supplied by the generator "depends" upon the voltage drop across the resistor R (in this case proportionally). If the voltage drop on R changes, the current circulating in the output loop will change according to the formula $I = gV_R$

Table 1.1 Elementary Circuit Components

Ideal element	Circuit symbol	Impedance (applies to I)	Admittance (applies to V)
Resistor		R	$G = 1/R$
Capacitor		$\frac{1}{C}\int dt$	$C\frac{d}{dt}$
Inductor		$L\frac{d}{dt}$	$\frac{1}{L}\int dt$
Voltage generator		0	∞
Current generator		∞	0

generators whose output value *depends upon the voltage or the current in a different component of the circuit.* These components are referred to as *controlled (or dependent) generators*. Figure 1.9 shows the circuit symbols used for such kind of generators.

In conclusion of this section, we summarize in Table 1.1 symbols and properties of the ideal components introduced so far.

1.3 Real Components

When designing an electrical circuit we may need, for example, to use a *resistance* in a given branch to obtain the effect desired; in practice, however, the component eventually inserted in that location is a *resistor* having, among its many parameters, the required resistance value. Therefore, the resistance is a property of the physical object that we are going to use whose correct name is *resistor*. The *resistor* is ideal when it is characterized uniquely by the value of its *electrical resistance*. Likewise, the *capacitance* is sufficient to characterize an ideal *capacitor* and the *inductance* is sufficient to characterize an ideal *inductor*. However, it is customary to identify a real component with the name of its main property and, as a consequence, a resistor is normally identified as a resistance. This practice, unfortunately, is likely to lead us to forget that a real component, in particular at high frequencies, must be characterized with more than one property, as we will discuss later in this chapter.

Electromagnetic properties of materials. It is adequate for most purposes to characterize the electromagnetic behavior of materials in terms of the following parameters[7]:

- **Electrical conductivity** σ. This parameter is the inverse of resistivity ρ ($\sigma = 1/\rho$) and is used to quantify the capability of the material to support the passage of an electrical current. Its physical units are $(ohm \cdot meter)^{-1}$.
- **Electrical permittivity** ε. This parameter quantifies the dielectric response of the material to an external electric field. Its physical units are *farad·meter*$^{-1}$
- **Magnetic permeability** μ. This parameter quantifies the material magnetization when immersed in an external magnetic field. Its physical units are *henry·meter*$^{-1}$

With the help of these parameters, we can reconsider ideal electrical components from a different perspective that helps to understand the problems posed by their physical realization. We will therefore say that we can build an ideal resistor only with an ideal material having an electrical conductivity $\sigma \neq 0$ and null dielectric and magnetic permittivity: $\varepsilon = \mu = 0$. Likewise, for an ideal capacitor we need an ideal material with dielectric permittivity $\varepsilon \neq 0$ and $\sigma = \mu = 0$. Finally, for an ideal inductor we need an ideal material with magnetic permittivity $\mu \neq 0$ and $\sigma = \varepsilon = 0$. Obviously, real components are built with real materials and show parasitic effects that, although manufacturers strive to minimize them, are unavoidable. For example, the material of a real resistor in general will have $\varepsilon \neq 0, \sigma \neq 0$, and $\mu \neq 0$. Therefore, a real resistor will also display a non-null capacitance and a non-null inductance, which in general will both depend on the geometry and dimensions of the component. These parasitic properties will affect the performance of the component,

[7] On this argument, the reader can consult any general physics textbook dealing with electromagnetic properties of materials, such as for example [2] or [3].

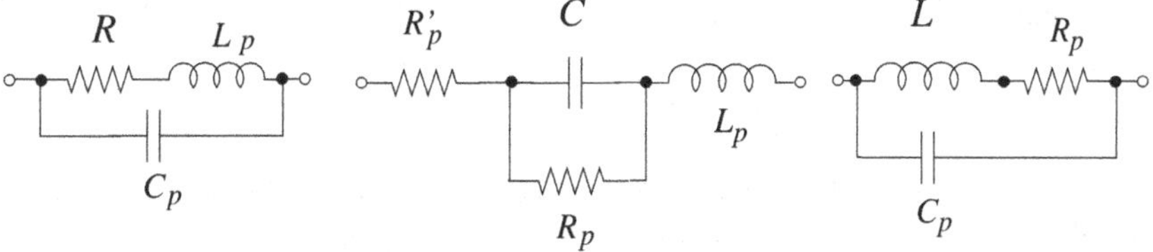

Fig. 1.10 Possible representations of real components: **a** Real resistor, **b** Real capacitor, **c** Real inductor. Parasitic characteristics are denoted by the suffix p

especially for high frequency signals. A possible representation of a real resistor in terms of discrete[8] ideal components is depicted in Fig. 1.10a.

A real capacitor has always a stray resistance connecting its two terminals, due to the finite conductivity of the dielectric separating its plates (dielectric losses). In addition, it has an inductance because of its finite dimensions and a series resistance due to the resistivity of its terminals (see Fig. 1.10b). Similarly, for a real inductor it is necessary to take into account the resistance of its coil and the capacitance between turns (see Fig. 1.10c).

We now give some details about the construction techniques and the characteristics of passive electrical components most commonly used: resistors, capacitors, and inductors.

1.3.1 *Real Resistors*

We have introduced the concept of *resistance* in Sect. 1.2.2 as the impedance of an electrical component, the *ideal resistor*. In this section, we discuss the practical realization of such components and their properties and limitations.

Resistors are components with a well-defined electrical resistance whose value in practice is obtained exploiting the second Ohm's law, which links it to the geometrical dimension of the component and to the resistivity of its material:

$$R = \rho \frac{l}{A} \tag{1.23}$$

where ρ is the component resistivity, l its length, and A its cross-sectional area, assumed constant.

Materials resistivity is temperature dependent. Therefore, manufacturers quote the value of a component resistance at a reference temperature T_o (usually 20 °C)

[8]Note that parasitic effects in real component depend on their geometry and extension. Therefore, a representation in terms of discrete elements is always an incomplete approximation of the component behavior that could be better described in terms of distributed constants.

and its value $R(T)$ at a generic temperature[9] T is given by

$$R(T) = R_o[1 + \alpha(T - T_o)]$$

where the parameter α, the temperature coefficient, is a characteristic of the resistor material. In Table 1.2, we list values of resistivity and temperature coefficient for materials of common use in resistor production.

The resistors available on the market range from resistance value of fractions of Ohm up to several hundreds of megaohms ($M\Omega$). Manufacturers adopt different production techniques depending on the resistance value required and on the parasitic effects to be reduced.

Wire wound resistors. Wrapping a wire of length l and cross section area A around an insulating and refractory support, we can obtain a resistance value given by Eq. (1.23). Selecting the material and the geometrical parameters of the wire a wide resistance range can be covered. In practice, however, the choice of the wire material is limited to those with a small temperature coefficient. Metallic alloys, like *Constantan*, *Manganin* and *Argentan*, offer the best opportunity to obtain resistance values with very little temperature dependence, see Table 1.2.[10]

Thin film resistors. These resistors are made of a thin film of conducting material with thickness $\sim 10^{-2}\mu m$ deposited on an isolating substrate (typically silicon, alumina or gallium arsenide). The required value of the resistance is obtained selecting the thickness of the conducting material that is typically made of tantalum nitride (TaN) or nichrome.

Carbon composite resistors. These resistors are obtained compressing a mixture of carbon granules, ceramic powder and epoxy resin in a tubular insulating container. The relative amount of carbon in the mixture controls the resistance value in this kind of resistors.

SMD resistors. The techniques of miniaturization in electronics have led to the development of resistors (and other components) of small size which are called "Surface Mounted Device" or SMD.

Color code. Closing this section, we point to the reader's attention that it is customary to give the resistance value and its tolerance with color bars printed directly on the resistor case. Detailed description of the codes adopted are available on the Internet from manufacturers and a host of other sources.

---

[9]The temperature is an *influence variable* for the value of R. As we will see in detail in the chapter dealing with the handling of measurement uncertainties, the *influence variables* are physical quantities that, although not directly involved in the definition of the physical variable of primary interest, in this case the resistance, can have an influence on their value.

[10]Data taken from: "https://en.wikipedia.org/wiki/Electrical_resistivity_and_conductivity." This site contains references to many texts and other web sites on resistivity values.

Table 1.2 Resistivity and temperature coefficient of conducting materials

Material	Alloy	Resistivity (μΩ cm)	Temperature coefficient ($10^{-3}K^{-1}$)
Aluminum	Al	2.828	3.6
Antimony	An	42	3.8
Stainless steel	Fe-C	10–25	4.5–5.0
Argentan	~(Cu 55-Zn 22-Ni 23)	35–41	0.07
Bismuth	Bi	120	4
Carbon (Grafite)	C	800	−0.5
Cobalt	Co	5.7	5.5
Constantan	~(Cu 55-Ni45)	52.0	±0.02
Iron	Fe	9.8	5
Manganin	~(Cu84-Mn12-Ni 4)	43–48	0.01
Mercury	Hg	96	0.89
Nickel	Ni	7.2	5.4
Gold	Au	2.42	3.6
Nichrome		90–104	0.11–0.19
Platinum	Pl	10.5	3.7
Lead	Pb	20.6	4.0
Copper	Cu	1.72	4.0
Tin	Sn	11.4	4.4
Tungsten	Wl	5.5	5.2
Soil (average value)		10^9–10^{11}	

1.3.2 Real Capacitors

Real capacitors are manufactured by coupling two conducting surfaces, usually called plates, separated by a dielectric with thickness, as far as possible, constant. The characteristics of a capacitor, as a circuit component, depend on the technique used for its construction. Besides the relative dielectric constant ε_r of the insulation material, multiple *influence variables* determine the characteristics of a real capacitor. Their effects are not always easily quantified. Among them:

- *Temperature*. The geometrical dimensions of the dielectric material are temperature dependent because of its finite thermal expansion coefficient, and consequently, the component capacitance becomes a function of its temperature. The *temperature coefficient* α describes this dependence as $C(T) = C_o(T_o)[1 + \alpha(T - T_o)]$, where T is the operating temperature and $C_o(T_o)$ is the capacitance value at the reference temperature T_o. When nonlinearity is important, i.e., when α depends upon T, manufacturers may prefer to give the plot of the capacitance relative change as a function of the operating temperature.

- *Dielectric conductivity*. Although very low, the conductivity of dielectrics used in capacitors is finite and, often, frequency dependent. This adds a resistive component to the capacitor impedance that can be described as a resistance R_p connected in parallel to the capacitor (see Fig. 1.10b). Often, manufacturers quote the amount of this effect through a parameter δ, named *loss angle*, which we will define in the following section.
- *Aging*. The dielectric *aging*, particularly relevant for electrolytic capacitors, can cause changes in its chemical–physical characteristics that make the capacitor noncompliant with the characteristics declared by the manufacturer, and then often unusable. This phenomenon depends on the conditions of use and in particular, on the temperature; for example, an electrolytic capacitor used at a temperature of 45 °C shows a useful life span of about 15 years that is reduced to a few months in case the component is used, or even stored, at 100 °C.
- *Parasitic inductance*. Capacitors are built with nonmagnetic materials; nevertheless, they have a parasitic inductance mainly depending on their geometry. Capacitors whose plates are rolled to shrink their size present relatively high values of inductance. In addition, all real capacitors have some inductance (and resistance) associated with their leads.
- *Microphone effects*. Materials are deformed when subject to the pressure of acoustic waves. Some ceramics used as dielectric in capacitors are slightly piezoelectric and their deformation causes a variation in their dielectric constant. Consequently, the capacitance value can change depending on the amplitude and frequency of the acoustic noise present in the near environment. This effect can be particularly relevant in electronic amplifiers placed in proximity of loudspeakers.

Dielectric materials more commonly used in capacitors are:

- **Plastic film**. Various kinds of plastic films are well suited as capacitor dielectric. Among them, *polyester* (PET) and *polytetrafluoroethylene* (PTFE) are common choices. Film capacitors can cover a wide range of capacitances, ranging from below 1 nF up to 30 μF. They can be made with breakdown voltage from 50 V up to a few kV. Film capacitors can be made with high precision capacitance values, and are less affected by aging than other kinds of capacitors.
- **Ceramic**. Capacitors made with a ceramic as dielectric are used in high frequency circuits and in high voltage devices. These capacitors have a low temperature coefficient, typically ranging in the interval ±30 ppm/°C for a temperature excursion in the interval +25 °C, +85 °C. Unfortunately, ceramic capacitors tend to show a capacitance that depends on the applied voltage across their terminals and their use is not advisable in analog circuits.
- **Glass**. Glass is used in applications requiring stable and reliable components able to operate in extreme conditions as required for example by the aerospace and nuclear industries.
- **Air**. Air is often used as dielectric in variable capacitors with maximum capacitance in the range of a few hundred pF.
- **Paper**. Paper has been widely used in the past in radio-frequency devices. Nowadays, it is used only for special applications such as high voltage capacitors.

Electrolytic capacitors. High capacitance values (in the range $1 \div 10^4 \mu F$) can be obtained with electrolytic capacitors. In these components, the dielectric is a very thin layer of metal oxide, typically aluminum or tantalum oxide, obtained electrolytically through a chemical reaction from a liquid or gel contained between the capacitor plates. In this case the two plates are not equivalent by construction and the lead marked with a "+" must be connected to a voltage higher than the other lead, marked with a "−". In this way, the capacitor is properly polarized and the electrochemical reaction can proceed in the right direction.[11] Therefore, the use of electrolytic capacitors is limited to circuit locations where the voltage difference does not change sign. Another limitation of these components is their slow response time that can be compensated by a plastic or ceramic capacitor connected in parallel. Finally, it must be remarked that aging in electrolytic capacitors proceeds through the evaporation of the electrolyte and can be very relevant, as discussed above.

Modeling a real capacitor. A real capacitor can be represented in terms of ideal discrete components by the diagram shown in Fig. 1.10b. The resistance R_p represents the effect of dielectric losses; R'_p and L represent, respectively, the resistance and the inductance due to the plates and their leads, as previously described. Often the inductance can be neglected up to the highest frequencies used in discrete components circuits; in these circumstances, the loss angle is defined by

$$\tan\delta = \frac{1}{\omega R_p C}$$

where $\omega = 2\pi\nu$, ν being the frequency of the sinusoidal voltage difference applied to the capacitor terminals. The meaning of this formula will be clarified once the behavior of circuit for alternating currents has been discussed. Here we just note that the loss angle is frequency dependent and usually manufacturers quote its value for discrete frequency values. Obviously, a capacitor approximates the ideal behavior better when the loss angle is small.

An alternative to account for capacitor losses is to measure its equivalent series resistance, ESR in short. As we will discuss later in this book, this parameter is frequency dependent and could be computed with a parallel to series transformation if we happen to know the values of R_p, R'_p and L.

1.3.3 Real Inductors

The procedure to build a real inductor consists in winding an adequate number of turns of a conducting wire along a cylindrical or toroidal core. In practice, this technique is nowadays in use only for radio-frequency circuits, up to a few hundred

[11] An electrolytic capacitor incorrectly connected, with the wrong polarity, is permanently damaged and can create a dangerous situation with the emission of toxic fumes and with the possibility of component explosion.

MHz. For higher frequencies, inductances are obtained with spiral tracks on printed circuits or even directly on the silicon wafer of integrated circuits. A large inductance at relatively low frequencies can be usefully replaced by using a capacitor and an operational amplifier performing as a *gyrator*. The gyrator is an active circuit that converts an impedance in its inverse.[12]
Inductors made of a wire-winding present an inductance that depends on its geometry[13] and on the magnetic permeability of the material used for the core. Note that saturation effects in this material can generate a nonlinear dependence of inductance on the current flowing through it.

Besides its inductance, a real inductor has both a resistive component of the impedance, mainly due to electrical resistance of the winding (usually denoted by R_{DC} to underline that this parameter is easily measured at low frequency), and a capacitive component, due to the distributed capacitance between the different turns of the winding. Other mechanisms responsible for power dissipation in an inductor are parasitic Foucault's currents in the core, due to its finite resistivity, *skin effect* in the wire at high frequency, see the following section, and, at very high frequency, irradiation of electromagnetic waves.

Total dissipation in an inductor can be usefully accounted by its quality factor Q, a parameter that can be easily measured using a time-dependent current.[14] Q is given by the ratio of the energy stored in the inductor to the dissipated energy

$$Q = 2\pi \frac{P_L}{P_R} = \frac{\omega L I^2}{R I^2} = \frac{\omega L}{R}$$

Since dissipation in an inductor is frequency dependent, the technical specifications supplied by the manufacturer usually report the Q value for some frequency values or a plot of its value as a function of the frequency. Obviously, the higher its quality factor, the better the inductor approximates the ideal behavior.

1.3.4 Skin Effect

As an example of the limits of discrete component approximation, in this section we give a simplified discussion of the physics of current diffusion in a conductor of finite dimensions. When the electrical current flowing through a conductor is time dependent, the phenomenon of self-induction modifies its space distribution. As a result, increasing the frequency, the current is progressively confined in the conductor periphery, hence the name *skin effect*. This amounts to a reduction of the useful cross

---

[12]The practical implementation of such a device requires familiarity with working principles of operational amplifiers and cannot be discussed here.

[13]Only for a few geometries we can easily calculate the inductance. Among these: the toroidal winding and, neglecting edge effects, the winding on a cylinder.

[14]Here it is supposed to use the inductor with a sinusoidal generator of frequency $\nu = \frac{\omega}{2\pi}$. Periodic currents will be treated extensively in Chap. 5.

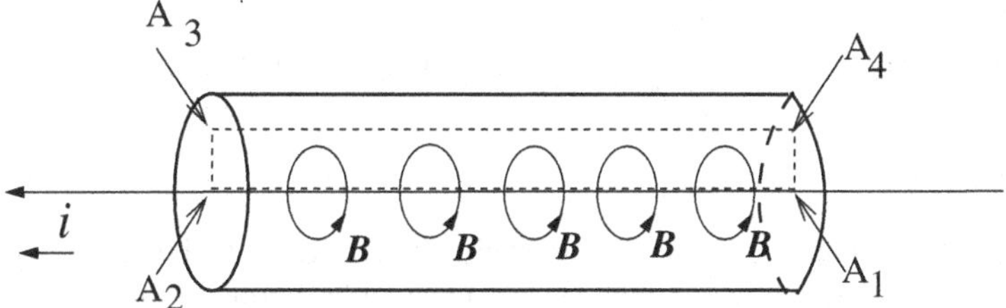

Fig. 1.11 Qualitative deduction of the skin effect

section for the current flow, causing an increase in its resistance. In the following, we give a qualitative understanding of the physics of skin effect in a simple geometry.

Consider an alternating current $i(t)$ flowing through a cylindrical wire as shown in Fig. 1.11. The rectangular path A_1, A_2, A_3, A_4 inside the wire has one of its edges, A_1A_2, along the cylindrical axis. When the current $i(t)$ is directed along the arrow in the figure the field lines of the magnetic induction B are circles perpendicular to the axis as shown in the figure. The magnetic flux across the path A_1, A_2, A_3, A_4 increases with the current intensity and the electric field resulting from self-induction along the path has the same direction of the current at the periphery and becomes opposite to it on the axis. This field opposes the current variation in the conductor core but reinforces it at the edges. A similar conclusion can be reached when the current decreases; in this case, the induced electric field in the core is directed in the same direction of the current opposing the current variation while at the periphery it becomes opposite to the current helping to reduce it. The overall effect is that *the alternating current becomes confined in a circular ring at the edge of the cylinder* whose depth is therefore referred to as the *skin-depth*. This causes the increase of the wire resistance and the reduction of its inductance. Moreover, since, as we shall see, this reduction is stronger the higher the oscillation frequency, both these parameters become frequency dependent. A quantitative evaluation of the skin effect can be very complex and simple results are obtained only for special cases. For a current flowing parallel to the surface of a plane conductor of infinite thickness, the current density $J(x)$ has its maximum right at the surface and decreases with the distance x from it according to the law:

$$J(x) = J_0 e^{-\frac{x}{\delta}}$$

where we have denoted with δ the *skin depth* given by

$$\delta = \sqrt{\frac{2\rho}{\omega\mu}} \tag{1.24}$$

Table 1.3 Skin depth in copper as a function of the signal frequency

Frequency	50 Hz	10 kHz	100 kHz	1 MHz	10 MHz
δ (mm)	7.82	0.66	0.21	66×10^{-3}	21×10^{-3}

In this equation, ρ and μ denote, respectively, the resistivity and the magnetic permeability of the conducting material and ω is the angular frequency of the oscillating current. Table 1.3 gives the extension of the *skin-depth*, as derived from Eq. (1.24), for a copper conductor as a function of the current oscillation frequency.

In Appendix A we illustrate a more formal approach to the skin effect.[15]

1.4 Usage Limits for Real Electrical Components

All electrical components suffer of usage limits due to the effects that the passage of an electric current and/or the application of an electric field can cause in the materials used for their construction. The most important limit comes from the heat production caused by the resistivity of the material; indeed, when this heat is not transferred easily to the environment, the temperature of the component increases eventually leading to its damage. For this reason manufacturers of resistors and inductors usually give, in their technical specifications, the maximum electrical power input that the component can safely withstand.

Inductors in addition suffer from a current limitation related to the mechanical stress due to magnetic forces generated by current–current interaction that can cause the rupture of the conductor or of its support.

Capacitors usually do not present an important limitation due to thermal dissipation but suffer from a specific problem due to dielectric strength. Indeed, every insulating material can withstand up to a maximum electric field without losing its insulation properties. Therefore, for every capacitor, depending on its geometry and the dielectric used, the manufacturer gives the maximum voltage that can be applied without damaging the component.

Problems

Problem 1 A 220 Ω resistor can dissipate up to 0.5 W without damage. Compute the maximum voltage and the maximum current that it can withstand. [A. $V_{max} = 10.5$ V, $I_{max} = 47.7$ mA.]

[15]More detailed discussion on the skin effect can be found in electrodynamics textbook such as, for example, [4].

Problem 2 Calculate the amount of energy that can be accumulated in a capacitor of 12 nF whose dielectric can sustain a maximum voltage of 50 V. [A. 1.5×10^{-5} J.]

Problem 3 Calculate the amount of energy that can be accumulated in an inductor of 15 mH that can sustain a maximum current of 0.5 A. [A. 1.87×10^{-3} J.]

Problem 4 The battery of your car has 0.4 MJ of stored energy, 12 V of open circuit voltage and an internal resistance of 1.0 Ω. If you forget the lights on and the lamps have an equivalent resistance of 5 Ω, after how long will the battery be fully discharged? Use the simplifying assumption that the internal resistance of the battery remains unchanged until the discharge completes. [A. 4 h38 m.]

Problem 5 Calculate the resistance of a 1000 W electric heater, recalling that the electricity network delivers an effective voltage of 220 V. [A. 48.4 Ω.]

Problem 6 A voltage generator with output V connects to the series of two resistors R_1 and R_2 forming a voltage divider. Compute the voltage drop across each resistor and their ratio. [A. $\Delta V_1 = VR_1/(R_1 + R_2)$, $\Delta V_2 = VR_2/(R_1 + R_2)$, $\Delta V_1/\Delta V_2 = R_1/R_2$.]

Problem 7 N resistors with resistance values R_n are connected in series to form a chain; calculate the voltage drop ΔV_n across each resistor R_n when voltage V is applied to the chain. [A. $\Delta V_n = VR_n/\sum_i R_i$.]

Problem 8 A current generator with output I connects to the parallel of two resistors R_1 and R_2 that form a current divider. Compute the current value in each resistor and their ratio. [A. $I_1 = IR_2/(R_1 + R_2)$, $I_2 = IR_1/(R_1 + R_2)$, $I_1/I_2 = R_2/R_1$.]

Problem 9 N resistors with resistance values R_n are connected in parallel. Compute the current value in the generic n-*th* resistor as a function of the total current I supplied by the generator powering the circuit. [A.$I_n = I/R_n/\sum_i(1/R_i)$.]

Problem 10 Compute the voltage V_{AB} in the circuit shown in the figure using the expressions deduced for the voltage divider in problem 6. Assume $V = 100$ V, $R_1 = 100\,\Omega$, $R_2 = R = 200\,\Omega$. [A. $V_{AB} = 50$ V.]

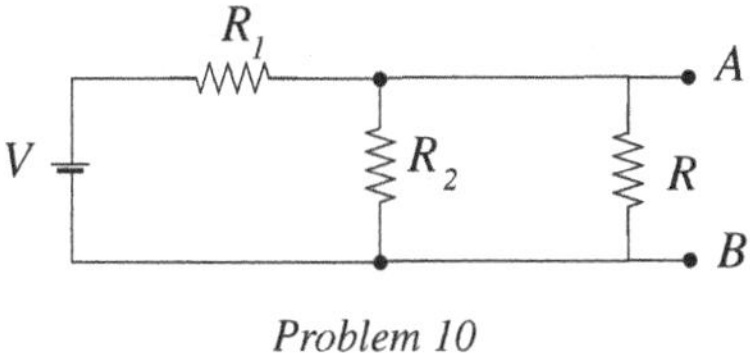

Problem 10

Problem 11 Assume to have a 7 A current generator and three resistors of values 0.25, 0.5, 1 Ω respectively. How would you arrange them to build a current generator with output equal to 1 A? [A. Connect the three resistor in parallel and take the output in series with the 1 Ω resistor.]

Problem 12 Compute the power dissipated by each of the three resistors in the previous problem. [A. $P_{0.25\Omega} = 4\text{W}$; $P_{0.5\Omega} = 2\text{W}$; $P_{1\Omega} = 1\,\text{W}$.]

Problem 13 Compute the current in the resistor R_3 in the circuit of the figure using the expression deduced for the current divider in problem 8 assuming $R_1 = 100\,\Omega$, $R_2 = R_3 = 50\,\Omega$. [A. $I_{R_3} = 2.5\,\text{A}$.]

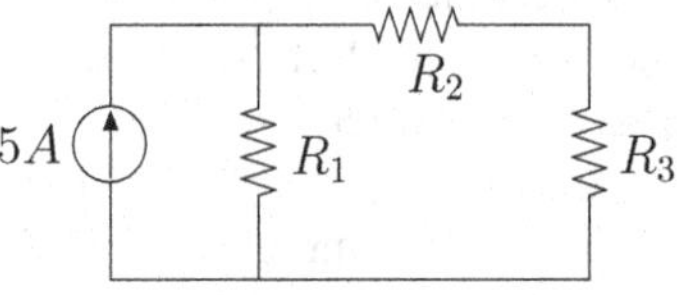

Problem 13

Problem 14 Compute the power delivered by the generator and the power dissipated in the resistor R_1 of the circuit in the figure. [A. $W_g = V^2/R_{eq}$, $W_{R_1} = V^2 R_1/R_{eq}^2$ con $R_{eq} = R_1 + \frac{1}{\frac{1}{R_2}+\frac{1}{R_3+R_4}}$.]

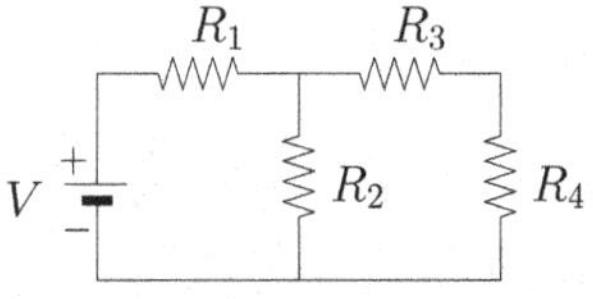

Problem 14

Problem 15 Compute the power delivered by the current generator and the power dissipated in the resistor R_2 of the circuit in the figure. [A. $W_g = I^2(R_1+R_2)(R_3+R_4)/(R_1+R_2+R_3+R_4)$, $W_{R_2} = I^2 R_2 (R_3+R_4)^2/(R_1+R_2+R_3+R_4)^2)$.]

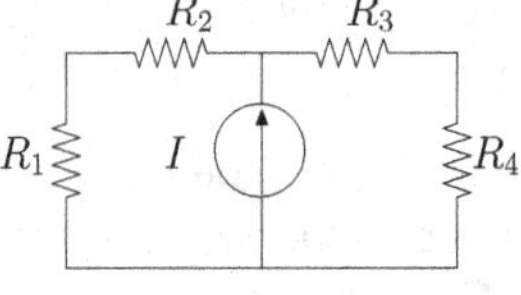

Problem 15

Problem 16 Compute the resistance between the two points A and B and the resistance between the two points C and D of the circuit in the Figure. [A. $R_{AB} = R_1 + R_3 + \frac{1}{\frac{1}{R_2}+\frac{1}{R_6}+\frac{1}{R_4+R_5}}$, $R_{CD} = \frac{1}{\frac{1}{R_5}+\frac{1}{R_4+\frac{1}{\frac{1}{R_2}+\frac{1}{R_6}}}}$.]

Problem 16

Problem 17 Two capacitors with capacitance C_1 and C_2 are connected in series and the series is connected to a voltage generator of output V. Compute the charge accumulated on each of them. [A. $Q = VC_1C_2/(C_1 + C_2$.]

Problem 18 Two capacitors with capacitance C_1 and C_2 are connected in series and the series is connected to a voltage generator of output V. Compute the voltage drop across each capacitor and their ratio. [A. $\Delta V_1 = VC_2/(C_1 + C_2), \Delta V_2 = VC_1/(C_1 + C_2), \Delta V_1/\Delta V_2 = C_2/C_1$.]

Problem 19 Two inductors with inductance L_1 and L_2 are connected in series and the series is connected to a voltage generator of output v(t). Neglecting mutual inductance, compute the voltage drop across each inductor and their ratio. [A. $\Delta V_1 = v(t)L_1/(L_1 + L_2), \Delta V_2 = v(t)L_2/(L_1 + L_2), \Delta V_1/\Delta V_2 = L_1/L_2$.]

Problem 20 * The second Ohm's law holds for resistors with a constant cross section. Consider how to solve the problem of computing the resistance R of a resistor having the form of a truncated cone with length l and a cross section radius ranging from r_1 to r_2. Assume that the resistor material is homogeneous with resistivity ρ. Many textbooks give the answer: $R = \rho l/\pi r_1 r_2$; recover the procedure leading to this answer and show that it is wrong since it violates charge conservation. The full response can be found in Ref. [5].

References

1. R.P. Feynmann, R.B. Leighton, M. Sands, *The Feynmann Lectures on Physics*, vol. II (Dover Publications, New York, 1967)
2. Munir H. Nayfeh, Morton K. Brussel, *Classical Electricity and Magnetism*, 2nd edn. (Dover Publications, Inc., Mineols New York, 2015)
3. K.H. Wolfgang, *Panofsky and Melba Phillips*, 2nd edn. (Melba Phillips, Classical Electricity and Magnetism 2005)
4. J.D. Jackson, *Classical Electrodynamics*, 3rd edn. (Academic Press, New York, 1998)
5. J.D. Romano, R.H. Price, Am. J. Phys **64**, 1150 (1996)

Chapter 2
Direct Current Circuits

2.1 Introduction

Nowadays, our lives are increasingly dependent upon the availability of devices that make extensive use of electric circuits. The knowledge of the electrical current flowing in each component of a circuit, and the corresponding voltage drop, is essential to obtain the desired function from the device exploiting it. When we obtain these values, we say that the circuit is solved.

Circuits where voltages and currents are not time dependent are called direct current circuits or, using an acronym, DC circuits. They are a useful starting point to confronting the task of constructing procedures useful to obtain circuit solutions.

This chapter deals with methods to obtain the solution of a DC circuit. First, we discuss in Sect. 2.2 the basic physical principles to be used and we recall the two Kirchhoff's laws. Then we introduce some basic concepts of graphs theory useful to analyze the topological properties of electrical circuits. Using these results, we move on in Sect. 2.3 to describe the implementation of the two principal approaches to circuit solution, the method of nodes and the method of meshes. As an example of the application of both methods, we give the detailed solution of the Wheatstone's bridge, a circuit used to obtain accurate measurement of a resistance. Next in Sect. 2.4, we turn our attention to circuits made only of components whose current is linearly dependent upon the applied voltage and we prove a number of theorems that simplify considerably the task of solving linear circuits. Finally, in Sect. 2.5, we discuss briefly the issue of power transfer among different parts of an electric circuit.

2.2 Kirchhoff's Laws

The tools adopted to solve electrical circuits are based on the two *Kirchhoff's laws*. The first, known as the *Kirchhoff's law of currents*, uses the concept of *node*, i.e., a point connecting three or more electrical components, and is based on the principle of

R. Bartiromo and M. De Vincenzi, *Electrical Measurements in the Laboratory Practice*, Undergraduate Lecture Notes in Physics,
DOI 10.1007/978-3-319-31102-9_2

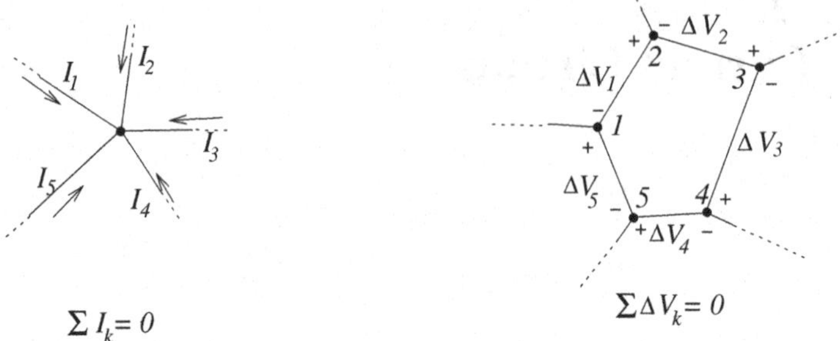

Fig. 2.1 Schematic illustration of the two Kirchhoff's laws: the diagram on the *left* refers to the law of the nodes, while diagram on the *right* illustrates the loops law. The two figures show, respectively, the positive direction of the current (indicated by the *arrows*) and the positive voltage (indicated by + and −)

conservation of electric charge. The second, known as the *Kirchhoff's law of voltages*, uses of the concept of *loop*, a closed path in a circuit, and is based on the consideration that the electrostatic field is conservative. The formulation of Kirchhoff's laws is the following:

1. **Kirchhoff's law of currents**, also known as the *law of the nodes* or the *first Kirchhoff's law*: the sum of all currents converging in a node is zero: $\sum_k I_k = 0$ (see Fig. 2.1).
2. **Kirchhoff's law of voltages**, also known as the *law of the loops* or the *second Kirchhoff's law*: the sum of the potential differences across components forming a closed loop is zero: $\sum_k \Delta V_k = 0$ (see Fig. 2.1).

The two Kirchhoff's laws are based on the general principles of conservation of charge and energy, respectively, and are valid independently of the nature of the components in the circuit.

In the two Kirchhoff's laws both I_k and ΔV_k are algebraic quantities. By convention, the I_k are positive when directed toward the node. Likewise, the ΔV_k are positive when the voltage increases along the arbitrarily chosen direction of the loop.

2.2.1 Network Geometry

Consider a circuit with more than one loop, such as, for example, the circuit shown in Fig. 2.2. One can easily verify that the Kirchhoff's laws allow writing a number of equations that exceed the number of unknown currents required for the circuit solution. In the circuit, three closed loops are present and to each of them we can apply the law of voltages. Denoting with i_1, i_2, and i_3 the currents flowing, respectively, in resistors R_1, R_2, and R_3 (in the figure the positive directions of the currents are indicated by the arrows), we obtain

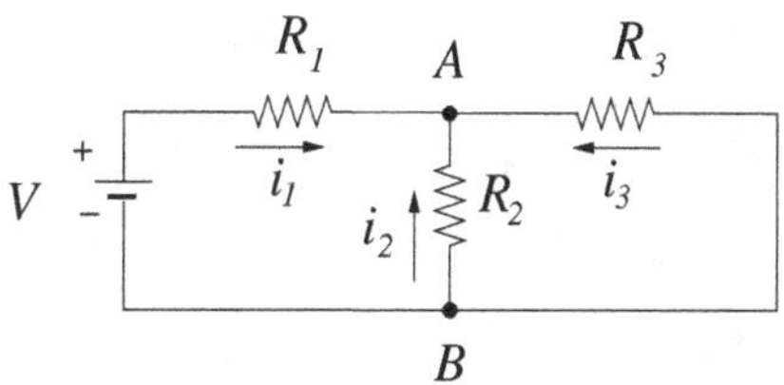

Fig. 2.2 Example of a circuit with two independent loops

$$\begin{cases} V = R_1 i_1 - R_2 i_2 \\ V = R_1 i_1 - R_3 i_3 \\ 0 = R_2 i_2 - R_3 i_3 \end{cases} \tag{2.1}$$

However, these three equations are not enough to obtain the values of the three unknown currents since they are linearly dependent. In fact, the third equation can be obtained by difference from the first two and it is easy to recognize that this is only a consequence of the geometrical properties of the network. We say, therefore, that the three loops in the circuit are not independent. To solve this circuit we must also take into account the Kirchhoff's law of currents applied to the two nodes in the circuit under test. In fact, it is easy to verify that for both nodes the same equation is obtained: $i_1 + i_2 + i_3 = 0$ which means that the two nodes are not independent. We leave to the reader the solution of the circuit with the first two Eq. (2.1) and the relation among the currents given above.

Our next task is to consider how to determine in a generic circuit the number of independent loops and the number of independent nodes. To answer this question, we need to exploit some simple and intuitive concepts borrowed from graph theory.

The graphs. In mathematics, a graph consists of a set of points called *vertices* or *nodes* of the graph, and of a set of *edges* or *branches* joining pairs of *nodes*. Graph theory is devoted to the study of the topological properties[1] of these structures. Consider a graph with n nodes and r branches such as, for example, the one shown in Fig. 2.3. By definition, a *tree* of the graph is a set of its branches connecting all its nodes without making a closed path. It is easy to verify that the number of branches in a tree must be $n - 1$; the set of branches that do not belong to the tree will form a so-called *co-tree* that consists of $r - (n - 1)$ branches.

Adding to the tree a branch of the corresponding *co-tree* will result in a closed loop. Since $r - n + 1$ is the number of branches of the *co-tree*, you can then get in this way $r - n + 1$ closed loops that are referred to as the *fundamental loops* of the graph. Obviously, the number of fundamental loops is independent from the tree initially selected.

Another operation defined on graphs is the *cut*, which consists of isolating some nodes of the graph with a surface (a line in the plane). Eliminating all branches intersected by the *cut*, the graph becomes divided into two parts *non-connected* between them. A cut that intersects only one branch of the tree (the other intersected

[1]The topological properties of a geometrical entity are those invariant for elastic deformations.

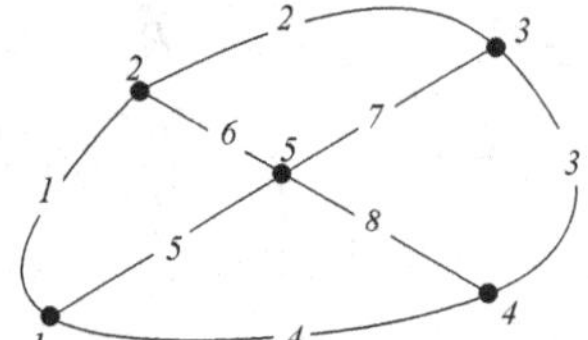

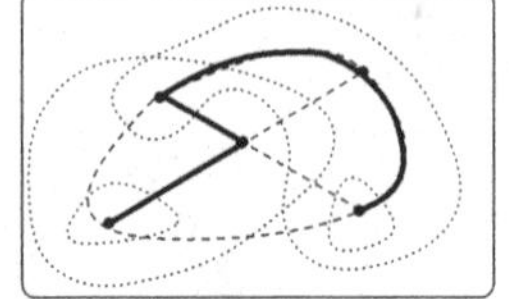

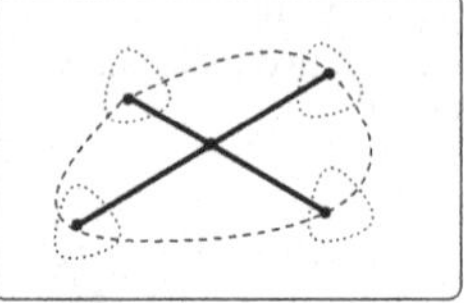

Fig. 2.3 Graph with five nodes and eight branches. In the two inserts, two trees of the graph (in *bold*) with their co-trees (*dashed branches*) are shown. The *dotted* closed curves represent the fundamental cuts intersecting only one branch belonging to the tree

branches belonging to the co-tree) is called *fundamental* or *independent*; the number of fundamental cuts is obviously equal to the number of branches in a tree, i.e., $n-1$.

The parallel between graphs and electrical circuits is evident: in fact, nodes of the graphs can be identified with the nodes of the electrical circuits and the branches of the graph can be identified with the bipolar components (possibly in series) of electrical circuits. In particular, the number of fundamental loops in the graphs coincides with the number $m = r - n + 1$ of independent loops to which apply Kirchhoff's voltage law, and the number of independent cuts $(n - 1)$ coincides with the number of independent nodes to which apply Kirchhoff's current law.

Considering again the circuit of Fig. 2.2, we note that the network represented there consists of three branches and two nodes. Therefore, the number of independent loops $r - (n - 1)$ is equal to two, while the number of independent nodes $n - 1$ is equal to one, as obtained previously from a direct analysis of equations resulting from Kirchhoff's laws.

2.3 Solution Methods for Electric Circuits

As previously mentioned, the solution of an electrical circuit consists in the determination of the voltage at each node and the current in each branch. The principles underlying the methods of solution are the two Kirchhoff's laws to be used together with relationships linking current and voltage in each component of the circuit. For resistive components, the *Ohm's law* provides these relations. In next paragraphs, we will work out two methods, the first said method *of the nodes* and the second said method *of the loops* that, using the two Kirchhoff's law, give the solution of a generic DC circuit.

2.3.1 The Method of Nodes

The method of nodes addresses the solution of a circuit by applying Kirchhoff's law of currents ($\sum I_k = 0$) to its independent nodes. Its name should be, more correctly, *the method of the voltage of the nodes*. In fact, the application of this method starts with the arbitrary choice of a node of reference against which to measure the potential differences of all other nodes. The voltages of these $n-1$ nodes are the unknowns of our problem, while the current in each branch can be expressed in terms of the node voltages and of the impedances of the electrical components in the branch. Application of Kirchhoff's law of currents to each independent node yields the $n-1$ equations needed to obtain the circuit solution. Note that assigning a value of voltage to each node, with respect to the reference node, implies *the automatic verification of the Kirchhoff's law of voltages*; in fact, with this assignment, in a closed path in the circuit the sum of the voltage drops in each branch will be always identically zero. In other words, the value of the electric potential at a given point does not depend on the path we follow to reach it but only on the point itself.

The steps for solving electrical circuits with this method are as follows:

1. Among the n nodes of a circuit, choose a reference node against which to measure the voltages of the other $n-1$ nodes.
2. Assign unknown voltage values $(V_1, V_2, \ldots, V_k, \ldots, V_{n-1})$ to these $n-1$ nodes.
3. Evaluate the currents that converge in each node using the values of the voltages V_k and the impedances of the circuit elements through which they flow.
4. Write the $n-1$ equations expressing Kirchhoff's current law.
5. Solve the system to obtain the $n-1$ node voltages.
6. Compute the currents flowing in various circuit components using the known values of node voltages and the relationship between voltage and current for the circuit components connecting them. When the circuit consists only of resistors, this relation is the first Ohm's law.

As an application, we solve by this method the circuit of Fig. 2.2 that we already addressed in the previous section. In this circuit there are $n = 2$ nodes, and then $n - 1 = 1$ independent node. We choose the node B as the reference for the voltage of the other node (in this particular case only for the voltage in A) and denote by i_1, i_2, and i_3 currents that, flowing in the resistors with the same index, converge in the node A. We denote by V_A the unknown voltage of node A, and we write the Kirchhoff's current law for this node:

$$\sum_{k=1}^{3} i_k = i_1 + i_2 + i_3 = \frac{V - V_A}{R_1} + \frac{0 - V_A}{R_2} + \frac{0 - V_A}{R_3} = 0$$

To simplify this expression, we introduce the conductance $G_k = 1/R_k$ $(k = 1, 2, 3)$ and the solution of the previous equation becomes

$$V_A = \frac{G_1}{G_1 + G_2 + G_3} V$$

With a simple algebraic manipulation, it can be shown that this expression of V_A coincides with the one that can be derived from the equations given in the previous Sect. 2.2.1.

2.3.2 The Method of Meshes

The method of the meshes is based on the Kirchhoff's law of voltages requiring that the sum of the voltage drops across components forming a closed path is zero. This method introduces a fictitious current, said the *mesh current*, for each of the independent loops in the circuit. Expressing the physical currents, those that actually flow in the various components, as the algebraic sum of the mesh currents that affect them, *the Kirchhoff's law of currents is everywhere automatically satisfied* because all currents are now sums of closed currents (the mesh currents).

The mesh currents are now the unknowns of the problem and it is sufficient to use the Kirchhoff's law of the voltages applied to the independent meshes to obtain their value. The steps for solving electrical circuits with this method are as follows:

1. Count the number n of nodes and r of branches in the circuit and compute the number m of independent loops: $m = r - n + 1$. This is the number of equations needed to solve the circuit.
2. Select m between all possible independent loops[2] by arbitrarily choosing a tree on the graph of the circuit and proceeding as described in Sect. 2.2.1. The m loops selected in this way are those to which the Kirchhoff's law of voltages applies.
3. For each of the loops identified in the previous paragraph indicate a mesh current (unknown) with an arbitrary direction.
4. Compute the currents in the individual components by adding up the mesh currents flowing in them taking into account their mutual direction.
5. Apply the law of voltages to each loop and obtain the m equations of the form $\sum_k V_k = 0$ using the relationships between voltage and current for the components in the circuit. In the case that only resistors form the circuit, this relationship consists of the simple Ohm's law. Note that the e.m.f. of each generator in the mesh has to be taken positive if it generates a current flowing contrary to the positive direction chosen for the mesh and negative if it generates a current along that direction.
6. Solve the resulting system of equations to obtain the value of the m mesh currents. The currents flowing in single elements are obtained by adding up the mesh currents that flow in the element taking into account their mutual direction.
7. The voltages at the nodes of the circuit are obtained by selecting a node as reference and assigning an arbitrary voltage to it, typically the null value. The voltage of any other node is then obtained by applying sequentially the Ohm's law[3] to the branches of a path that connects it to the reference node.

[2] The importance of choosing independent loops can be understood from the example in the following Sect. 2.3.3 on the solution of the Wheatstone's bridge shown in Fig. 2.5. If we had chosen for the solution of the circuit with the method of meshes, instead of the mesh supporting the current I_a, the one formed by the resistances R_1, R_2, R_3, R_4, the determinant of the resulting system would have been zero, because the mesh R_1, R_2, R_3, R_4 is linearly dependent (it is the sum) of those supporting the currents I_b and I_c.

[3] The voltage drop across a branch consisting of resistors and generators is simply equal to the sum of voltage drops across each individual element in the branch.

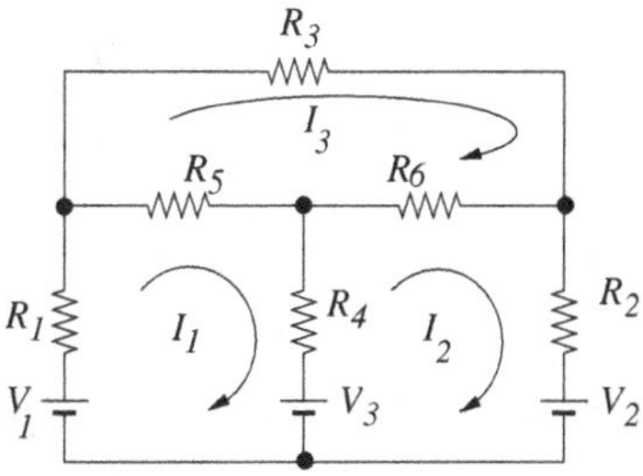

Fig. 2.4 Network with three independent loops

We now apply the method of the loops described above to the solution of the circuit of Fig. 2.4. The circuit has $n = 4$ nodes and $r = 6$ branches and then the number of independent loops is $m = r - n + 1 = 3$. We choose the three meshes[4] and their positive direction as shown in the figure; we can write the following equations that follow from the application of the Kirchhoff's law of voltages to the loops with currents I_1, I_2, and I_3, respectively:

$$\begin{cases} R_1I_1 + R_5(I_1 - I_3) + R_4(I_1 - I_2) - V_1 + V_3 = 0 \\ R_4(I_2 - I_1) + R_6(I_2 - I_3) + R_2I_2 - V_3 + V_2 = 0 \\ R_3I_3 + R_6(I_3 - I_2) + R_5(I_3 - I_1) = 0 \end{cases}$$

Rearranging this system and isolating the unknowns I_1, I_2, and I_3:

$$\begin{cases} (R_1 + R_4 + R_5)I_1 & -R_4I_2 & -R_5I_3 = V_1 - V_3 \\ -R_4I_1 & +(R_2 + R_4 + R_6)I_2 & -R_5I_3 = V_3 - V_2 \\ -R_5I_1 & -R_6I_2 & +(R_3 + R_5 + R_6)I_3 = 0 \end{cases}$$

The solution of this system is obtained with a long calculation of determinants and minors that is left as an exercise to the interested reader.[5] In case all the resistors are equal, $R_k = R$, one can easily get the following solution:

[4]The three meshes used for the solution are generated from T-shaped tree obtained by the branches that contain resistors R_5, R_6, and R_4.

[5]As a hint to the solution we give the value of the determinant Δ of the system and the expression for the current I_1:

$$\Delta = R_1R_2R_3 + R_1R_3R_4 + R_2R_3R_4 + R_1R_2R_5 + R_2R_3R_5 + R_1R_4R_5 + R_2R_4R_5 + R_3R_4R_5 + \\ + R_1R_2R_6 + R_1R_3R_6 + R_1R_4R_6 + R_2R_4R_6 + R_3R_4R_6 + R_1R_5R_6 + R_2R_5R_6 + R_3R_5R_6$$

$$I_1 = \frac{1}{\Delta}\{[R_4R_5 + R_6(R_4 + R_5)](V_1 - V_2) + R_2(R_3 + R_5 + R_6)(V_1 - V_3) \\ + R_3(R_3V_1 + R_6V_1 - R_4V_2 - R_6V_3)\}$$

$$\begin{cases} I_1 = \dfrac{2V_1 - V_2 - V_3}{4R} \\ I_2 = -\dfrac{2V_2 - V_1 - V_3}{4R} \\ I_3 = \dfrac{V_1 - V_2}{4R} \end{cases}$$

A Note on the Best Method Selection

In the preceding paragraphs, we illustrated two ways to deal with the challenge of solving electrical circuits: the method of nodes and that of meshes. The two methods are equivalent and the choice among them is largely subjective. A possible criterion for this choice, based on the economy of the calculations, is the comparison between the number of independent nodes $n - 1$ and the number of independent meshes $m = r - n + 1$; if $n - 1 < m$ it is convenient to use the method of nodes; otherwise it is more convenient to apply the meshes method. It can be shown that if $n < 4$ and the circuit has only voltage generators, the method of the nodes is the most convenient. See in this respect the Problem 2 at the end of this chapter.

2.3.3 *Example of Circuit Solution: The Wheatstone's Bridge*

A circuit particularly important for the accurate measurement of resistance values is the Wheatstone's bridge that is shown in Fig. 2.5. We first apply the method of the meshes for the solution of this circuit that has four nodes and six branches. The number of independent loops is therefore 3; the three loops selected for the solution are indicated in the figure by the currents I_a, I_b, and I_c, which circulate in them. The reader can easily verify that the three meshes have been identified from the tree that starts from node A and contains the three branches with resistors R_1, R_5, and R_3.

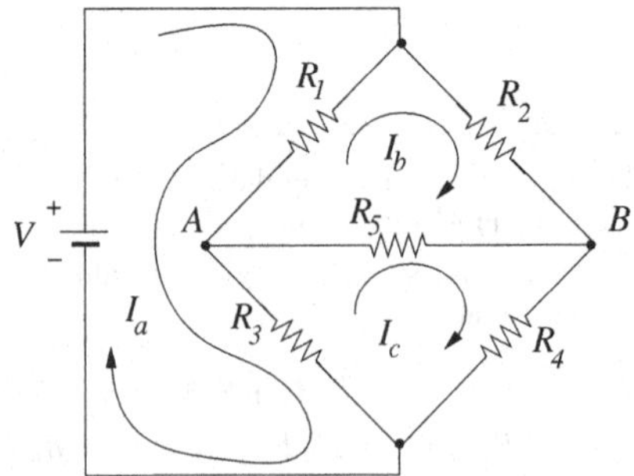

Fig. 2.5 Solution of the Wheatstone's bridge by the method of the meshes. The circuit has three independent meshes for which we have defined the three mesh currents I_a, I_b, I_c

Applying Kirchhoff's law of voltages to these three meshes, we have

$$\begin{cases} R_1(I_a - I_b) + R_3(I_a - I_c) - V = 0 \\ R_2 I_b + R_5(I_b - I_c) + R_1(I_b - I_a) = 0 \\ R_3(I_c - I_a) + R_5(I_c - I_b) + R_4 I_c = 0 \end{cases}$$

(you need to pay close attention to how you add up the currents in the individual resistors). Rearranging the above equations, we have

$$\begin{cases} (R_1 + R_3)I_a - R_1 I_b - R_3 I_c = V \\ -R_1 I_a + (R_1 + R_2 + R_5)I_b - R_5 I_c = 0 \\ -R_3 I_a - R_5 I_b + (R_3 + R_4 + R_5)I_c = 0 \end{cases} \tag{2.2}$$

Denoting with Δ the determinant of the matrix of the coefficients of the currents,

$$\begin{aligned} \Delta &= \left\| \begin{matrix} R_1 + R_3 & -R_1 & -R_3 \\ -R_1 & R_1 + R_2 + R_5 & -R_5 \\ R_3 & -R_5 & R_3 + R_4 + R_5 \end{matrix} \right\| \\ &= R_3[R_4 R_5 + R_2(R_4 + R_5)] + R_1[R_4(R_3 + R_5) + R_2(R_3 + R_4 + R_5)] \end{aligned}$$

we obtain the expression for the current I_b:

$$I_b = \frac{1}{\Delta} \begin{vmatrix} R_1 + R_3 & V & -R_3 \\ -R_1 & 0 & -R_5 \\ -R_3 & 0 & R_3 + R_4 + R_5 \end{vmatrix} = \frac{V}{\Delta}[R_5 R_3 + R_1(R_3 + R_4 + R_5)]$$

Similarly, we get for I_c:

$$I_c = \frac{1}{\Delta} \begin{vmatrix} R_1 + R_3 & -R_1 & V \\ -R_1 & R_1 + R_2 + R_5 & 0 \\ -R_3 & -R_5 & 0 \end{vmatrix} = \frac{V}{\Delta}[R_1 R_5 + R_3(R_1 + R_2 + R_5)]$$

Now we calculate the current I_5 passing in R_5, taking as the positive direction that of I_b:

$$I_5 = I_b - I_c = \frac{V}{\Delta}(R_3 R_5 + R_1 R_3 + R_1 R_4 + R_1 R_5 - R_1 R_5 - R_1 R_3 - R_2 R_3 - R_3 R_5)$$

$$I_5 = \frac{V}{\Delta}(R_1 R_4 - R_2 R_3) \tag{2.3}$$

The relation (2.3) characterizes the Wheatstone bridge. In particular, the Wheatstone bridge is *balanced* when it has $I_5 = 0$ which implies that between the four resistors R_1, R_2, R_3, and R_4, the following relation holds:

$$R_1 R_4 = R_2 R_3 \quad \text{or equivalently} \quad \frac{R_1}{R_2} = \frac{R_3}{R_4} \tag{2.4}$$

The value of an unknown resistance (R_1 for example) can be obtained by balancing the Wheatstone's bridge using for R_2, R_3, and R_4 precision resistors that can be varied in a known manner (resistance box).

To complete the solution of the circuit of the Wheatstone bridge we calculate the voltages of nodes A and B. Putting to zero the value of the potential of the negative terminal of the battery, we get for the node B

$$V_B = R_4 I_c = \frac{VR_4}{\Delta}[R_1 R_5 + R_3(R_1 + R_2 + R_5)] \tag{2.5}$$

The voltage of node A can be obtained subtracting to the voltage in B (given the sign of I_5) the voltage drop on R_5:

$$\begin{aligned} V_A = V_B - R_5 I_5 &= \frac{V}{\Delta}\{R_4[R_1 R_5 + R_3(R_1 + R_2 + R_5)] - R_5(R_1 R_4 - R_2 R_3)\} \\ &= \frac{V}{\Delta}(R_1 R_3 R_4 + R_2 R_3 R_4 + R_3 R_4 R_5 + R_2 R_3 R_5) \end{aligned} \tag{2.6}$$

Of course when the bridge is balanced ($I_5 = 0$) we have $V_A = V_B$.

Having solved the circuit of the Wheatstone's bridge with the meshes method, we do it again with the method of the nodes. The nodes are in this case $n = 4$, and then we need to apply the first Kirchhoff's law to $n - 1 = 3$ nodes to obtain the equations for the solution of the circuit. We select the nodes A, B and the positive terminal of the generator but we note that the voltage of this last node is fixed by its connection to the battery. This observation reduces the unknowns, and therefore the equations we need, from three to two.[6] Therefore, we choose the nodes A and B for which to write Kirchhoff's law of currents. We get

$$\begin{cases} -\dfrac{V_A}{R_3} + \dfrac{V - V_A}{R_1} + \dfrac{V_B - V_A}{R_5} = 0 \\ -\dfrac{V_B}{R_4} + \dfrac{V - V_B}{R_2} + \dfrac{V_A - V_B}{R_5} = 0 \end{cases}$$

To simplify the notation we introduce the conductance $G_k = 1/R_k$ and the previous system becomes

$$\begin{cases} (G_1 + G_3 + G_5)V_A & -G_5 V_B = G_1 V \\ G_5 V_A & -(G_2 + G_4 + G_5)V_B = -G_2 V \end{cases}$$

[6] In general, it can be stated that if two nodes of a circuit are connected only by an ideal voltage generator, the number of equations necessary for its solution by the method of the nodes is reduced by one unit.

whose solution is

$$V_A = \frac{[G_2G_5 + G_1(G_2 + G_4 + G_5)]}{(G_1 + G_3)(G_2 + G_4) + (G_1 + G_2 + G_3 + G_4)G_5} V$$
$$V_B = \frac{G_1(G_2 + G_5) + G_2(G_3 + G_5)}{(G_1 + G_3)(G_2 + G_4) + (G_1 + G_2 + G_3 + G_4)G_5} V$$

It is a simple algebraic exercise to verify that the formulas of V_A and V_B are the same as in Eqs. (2.6) and (2.5). The equilibrium condition of the bridge, i.e., the vanishing of the current that flows in R_5 ($V_A = V_B$), leads to the following condition on the conductance of the resistors in the circuit:

$$G_1G_4 = G_2G_3$$

which, as we have expected, is identical to relation (2.4) we found by solving the circuit with the method of the meshes.

2.3.4 Nonlinear Circuits

The methods for the solution of circuits, illustrated in the previous paragraphs, are based on the Kirchhoff's laws that, in turn, are based on the general principles of conservation of energy and electric charge and therefore can be applied, in principle, to circuits with nonlinear components. The difficulties encountered when trying to solve this kind of circuits are of a mathematical nature and derive from the nonlinear relationship between current and voltage in the nonlinear element, described by the so-called *characteristic*. This is the reason why in the examples of solution of DC circuits discussed above we considered only resistors that are linear components. However, Kirchhoff's laws apply to *all circuits*, even those with nonlinear components.

As an example of application of the Kirchhoff's laws to a nonlinear circuit, we analyze the circuit shown in Fig. 2.6, which makes use of a diode, a nonlinear component. For the diode, whose junction we assume of the type *pn*, the relationship between the voltage across it, v_D, and the current flowing through it, i_D, is $i_D = i_s(e^{v_D/v_0} - 1) \simeq i_s e^{v_D/v_0}$, where i_s and v_0 are constants (see Fig. 2.6). Applying the Kirchhoff's law for voltages to the unique mesh in the circuit and reversing the relationship between i_D and v_D we obtain

$$V_0 = Ri + v_D = Ri + v_0 \log \frac{i}{i_s} \tag{2.7}$$

Formula (2.7) allows calculating, in principle, the current i that flows in the circuit. However, this equation is *transcendent*, and does not have an explicit solution making it necessary to resort to numerical or geometrical methods to solve it. The plot in

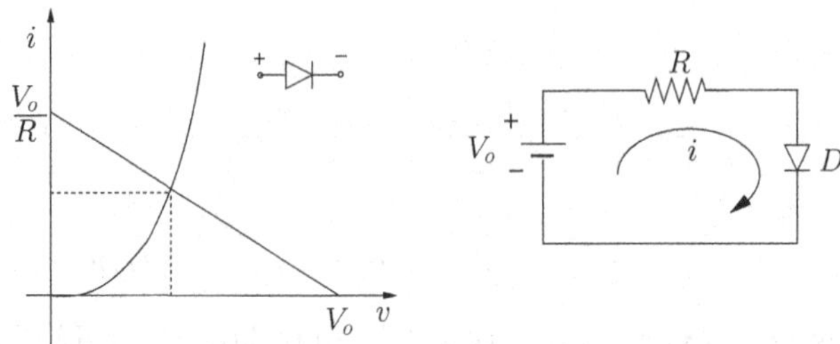

Fig. 2.6 Example of nonlinear circuit. On the *left* side of the figure the so-called "characteristic voltage–current" of a junction diode is represented while the *right* side shows the scheme of a simple circuit with a diode. The straight line in the plane $v = V_0 - Ri$ is referred to as the load line and gives the current in the resistor for a given voltage drop across the diode. Since the current flowing in the two components must be the same, the intersection of the load line with the characteristic of the diode gives the solution of the circuit

Fig. 2.6 shows how it is possible to obtain graphically the solution required. In the diagram, the two curves are the exponential *characteristic* of the diode and the current flowing in the resistance R as a function of the voltage across the diode. The point where the two curves intersect determines the solution of the circuit: the current flowing in R is equal to that flowing in the diode as required by charge conservation.

2.4 Analysis of Linear Networks

Circuits whose components are *all* linear are referred to as *linear circuits*. In practice, in DC circuits only resistors are linear passive components. In fact, we recall here that the impedance operator of an ideal resistor is given by a simple multiplicative constant (the resistance), and it is therefore obviously linear. The linearity of a circuit has important consequences on its properties. The relationships implied by linearity facilitate the understanding of the behavior of the circuit and simplify considerably their solution. Below we give the detailed statements and the proofs of the most important theorems on linear circuits

2.4.1 Superposition Theorem

As it is known from elementary mathematical analysis, the solution of a system of n linear equations can always be expressed as a linear combination of the known terms. Therefore, in a linear circuit the currents in all branches and the voltage of each node can always be expressed as a linear superposition of the intensities of source generators (the "known terms" in the equations). In particular if there are only n voltage generators V_k, $(k = 1, \ldots, n)$ the current in the generic branch j can always be expressed as

$$I_j = \sum_{k=1,n} A_k V_k \tag{2.8}$$

where A_k are constants determined only by the characteristics of the circuit components. Likewise, the voltage at a generic node can be obtained as a linear combination of the n V_k values. Similarly, if in the circuit only n current generators I_k are present, the voltage in the generic node j can be expressed as

$$V_j = \sum_{k=1,n} B_k I_k \tag{2.9}$$

where once again the B_k are constants determined only by the characteristics of the circuit components. A similar expression also gives the current in a generic branch. The generalization to the case where they are simultaneously present in both voltage and current generators is trivial. From this simple mathematical property, we derive the superposition theorem that can be stated in the two following forms.

For voltage generators. In any linear circuit with only n voltage generators $V_1, V_2, \ldots, V_n$, the current flowing in a generic branch (or the voltage of a generic node) of the circuit is equal to the sum of the currents (or the voltages) produced by each voltage generator, taken individually, assuming that all the others have been short-circuited.

For current generators. In any linear circuit with only n current generators $I_1, I_2, \ldots I_n$, the current flowing in a generic branch (or the voltage of a generic node) of the circuit is equal to the sum of the currents (or of the voltages) produced by each current generator considered individually, assuming that all the others have been disconnected.

2.4.2 Thévenin's Theorem

The practical exploitation of electrical circuits is greatly simplified by the existence of equivalence theorems. They allow for an easy evaluation of circuits behavior when they are connected among themselves or to a single component, the load, whose parameters can depend on the specific application. Remarkably, using equivalence theorems we can avoid the tedious task of solving the circuit when the load parameters are changed.

Thévenin's theorem, in its formulation for DC circuits, states the equivalence between a generic linear circuit and a voltage generator with a resistor in series; the formulation of the theorem is the following:

Any linear circuit "seen" between two of its nodes, A and B, is equivalent to an ideal voltage generator V_{oc} *whose tension is equal to the voltage measured between nodes A and B (open-circuit voltage), in series with the resistance "seen" between the two nodes which is called the equivalent resistance of Thévenin (see Fig. 2.7).*

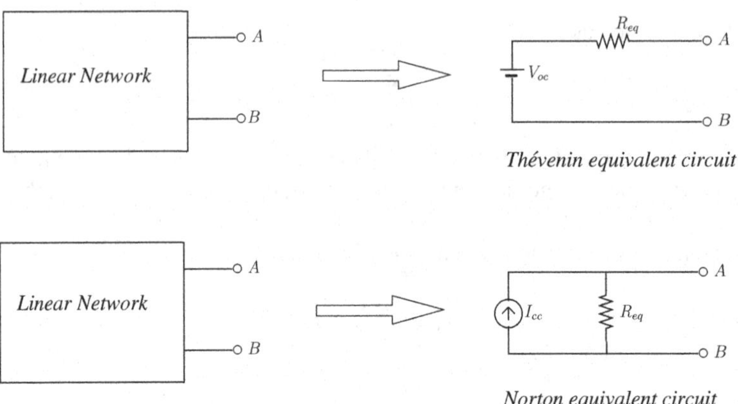

Fig. 2.7 Illustration of the theorems of Thévenin and Norton. On the *left*, we depict a generic linear network viewed from terminals A and B. On the *right*, the corresponding equivalent circuits according to Thévenin (*top*) and Norton (*bottom*)

The equivalent resistance R_{eq} can be obtained by analyzing the circuit between the two terminals A and B, combining resistors in series and in parallel according to their connections when all voltage generators present in the circuit have been *short-circuited* and all current generators have been *opened*. A method to obtain the value of R_{eq} when the layout of the circuit is not available will be given below.

The use of the equivalent circuit of Thévenin is particularly convenient when a resistance of the circuit, say R_L, can be considered as a variable load, i.e., a parameter that can vary from time to time, while all the other parameters of the circuit remain unchanged. This is elucidated in the following example.

Example. Consider the circuit of Fig. 2.8a whose solution can be obtained with the techniques already described (method of the nodes or of the meshes). Alternatively, you can deal with the solution of the circuit in question by calculating the equivalent of Thévenin considering R_L as the load of the circuit. Removing the load R_L, we apply Thévenin's theorem to the remaining circuit, see Fig. 2.8b. The resistance seen between A and B (Thévenin's equivalent resistance) amounts to $R_{eq} = R_1 \parallel R_2 = R_1R_2/(R_1+R_2)$. With a simple calculation[7] one gets the open-circuit voltage between terminals A and B as $V_{oc} = V_A - V_B = (R_1V_2+R_2V_1)/(R_1+R_2)$. The current flowing in R_L will then be equal to $V_{oc}/(R_{eq} + R_L)$.

Proof of Thévenin's Theorem.

According to the theorem of Thévenin, if V_{oc} is the open-circuit voltage between terminals A and B of a circuit and R_{eq} is the equivalent resistance seen between A and B, the current flowing in an additional load resistance r connected between A and B is

[7] The calculation of V_{oc} is immediate when using the superposition theorem.

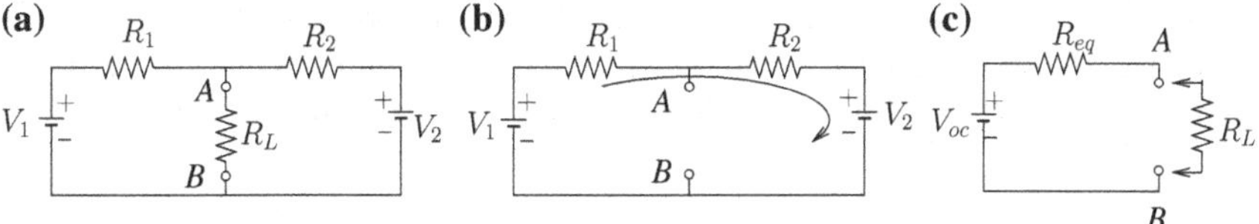

Fig. 2.8 Applying Thévenin's theorem: **a** original circuit; **b** circuit with the "load" R_L removed; **c** equivalent circuit of Thévenin with the load R_L connected

$$i = \frac{V_{oc}}{R_{eq} + r} \tag{2.10}$$

The proof of (2.10) is equivalent to the proof of Thévenin's theorem. Suppose to insert in series with the load r an additional generator whose e.m.f. is V_0. Since the circuit is linear by assumption, denoting by V_k, $k = 1, \ldots, n$ the generators belonging to the circuit, the current i' flowing in the resistance r, in the presence of the additional generator V_0, is given by the linear combination:

$$i' = \sum_{k=0}^{n} A_k V_k \tag{2.11}$$

where A_k are appropriate constants with dimensions of a conductance. Let us now reduce to zero the e.m.f. of the generators pertaining to the circuit, i.e., those with $k \neq 0$. This procedure is equivalent to replacing each generator with a *short circuit.* Equation (2.11) becomes

$$i' = A_0 V_0$$

Under these conditions, the constant $1/A_0$ can be identified by construction with the sum of r and R_{eq}:

$$\frac{1}{A_0} = r + R_{eq}$$

We now bring back the circuit to its original condition by setting the V_k to their values. In addition, we set the e.m.f. of the generator inserted in the load branch to a value of $-V$ such that the current flowing in r is reduced to zero. From Eq. (2.11) we get

$$0 = -VA_0 + \sum_{k=1}^{n} A_k V_k$$

that gives

$$\sum_{k=1}^{n} A_k V_k = VA_0$$

In this circumstance the load branch added, the one including r, does not absorb current and therefore V must coincide with the open-circuit voltage between the points A and B (V_{oc}). Finally, setting $V_0 = 0$ we obtain the current in the load as

$$i = \sum_{k=0}^{n} A_k V_k = \sum_{k=1}^{n} A_k V_k = VA_0 = \frac{V_{oc}}{r + R_0}$$

which proves Thévenin's theorem.

2.4.3 *Norton's theorem*

Norton's theorem for DC circuits establishes the equivalence between a generic linear circuit and a current generator with a resistor in parallel; the formulation of the theorem is the following:
Any linear circuit "seen" between two of its nodes, A and B, is equivalent to an ideal current generator whose output I_{sc} is equal to the short-circuit current between A and B in parallel with the resistance "seen" between the two nodes which is said (Norton's) equivalent resistance (see Fig. 2.7).

The proof of Norton's theorem is similar to that of Thévenin's theorem and is left to the reader as a useful exercise.

The same observation made for Thévenin's theorem about the impedance seen between two terminals, also applies to Norton's theorem. In fact, the equivalent resistances of Thévenin and Norton are identical and can be obtained in the same way.

Theorem of Open Circuit and Short Circuit

Thévenin's and Norton's theorems have the following remarkable corollary:
the impedance R_{eq} between two terminals of a linear circuit is equal to the ratio between the open circuit voltage V_{oc} and the short circuit current I_{sc}:

$$R_{eq} = \frac{V_{oc}}{I_{sc}}$$

It should be remarked that the above expression, extremely useful for the evaluation of the *output impedance* of various circuits, should not be confused with Ohm's law, with which it only shares the mathematical expression.

Example. Consider the circuit of Fig. 2.9 where $V_s = 6.0\,\text{V}, R_1 = 1.5\,\text{k}\Omega, R_2 = 2.2\,\text{k}\Omega, R_3 = 3.3\,\text{k}\Omega$ with A and B nodes for which we need to calculate the Norton equivalent circuit.

We start with the calculation of the resistance R_{eq} "seen" between A and B. To get R_{eq} we must short-circuit the voltage generator appearing in the circuit. Therefore, R_{eq} is the parallel of R_1, R_2, and R_3:

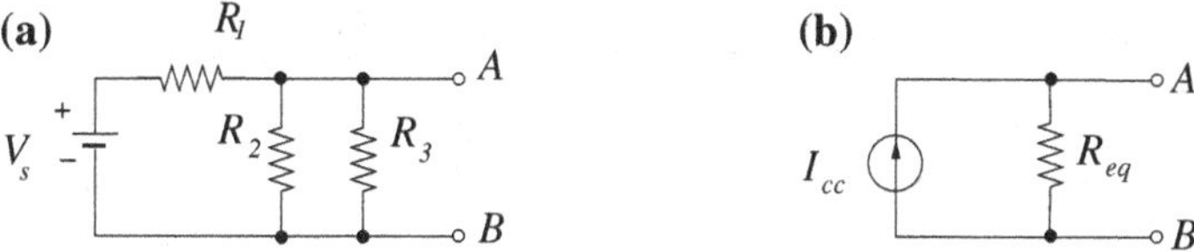

Fig. 2.9 Example of application of Norton's theorem

$$\frac{1}{R_{eq}} = \frac{1}{R_1} + \frac{1}{R_2} + \frac{1}{R_3} = \left(\frac{1}{1.5} + \frac{1}{2.2} + \frac{1}{3.3}\right) \times 10^{-3}\,\Omega^{-1} = 1.42 \times 10^{-3}\,\Omega^{-1}$$

that yields $R_{eq} = 437\,\Omega$

To complete the Norton equivalent circuit we must determine the value of I_{sc}, the short-circuit current between A and B. We get very easily

$$I_{sc} = \frac{V_s}{R_1} = \frac{6.0}{1.5 \times 10^3}\,A = 4.0\,\mathrm{mA}$$

2.4.4 *Reciprocity Theorem*

The proof of this important theorem starts with the analysis of the power balance in a given circuit. In a generic branch connecting two nodes i and j of a network, the product of the voltage drop on the branch $V_{ij} = V_i - V_j$ times the current I_{ij} flowing through it (from node i to node j) yields the power P_{ij} that is generated, if negative, or dissipated, if positive, in the branch (see Fig. 2.10). Since energy is conserved, the total balance of power must be zero and the following equality must hold

$$\sum_{branches} V_{ij} I_{ij} = 0 \tag{2.12}$$

where the sum is extended to all the branches in the network. We now prove that Kirchhoff's laws imply the power balance. In order to proceed, we note that being $V_{ij} = V_i - V_j$, the first member of (2.12) becomes

Fig. 2.10 Notation and conventions for the power balance

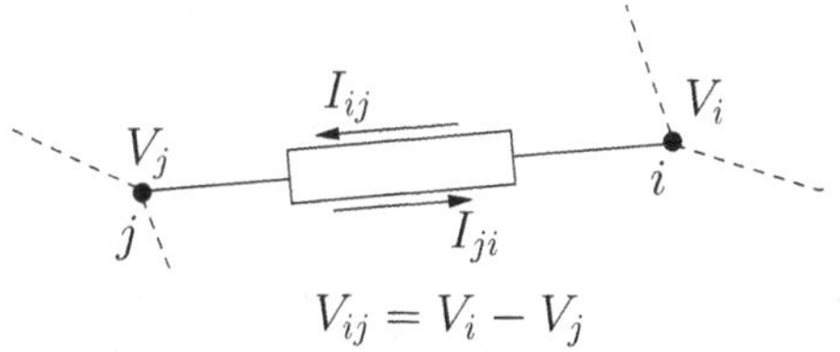

$$\sum_{branches} V_{ij}I_{ij} = \sum_{branches} V_iI_{ij} - \sum_{branches} V_jI_{ij} = \sum_{branches} V_iI_{ij} + \sum_{branches} V_jI_{ji} \quad (2.13)$$

where we used the relation $I_{ji} = -I_{ij}$. Rearranging the terms, the previous expression becomes

$$\sum_i V_i \cdot \sum_j I_{ij} \quad (2.14)$$

where the first sum extends to all nodes and second only to the nodes connected to i and expresses just the sum of all the currents entering the node. Kirchhoff's law of the nodes states that the second sum is null for all i, and hence the equality (2.12) follows.

Suppose now to change the arrangement of the components on the network, without changing its topology, in order to build a new distribution of tensions V'_{ij} and currents I'_{ij} that obviously still respect Eq. (2.12). However, it is easy to see that, retracing the previous proof, we have also

$$\sum V'_{ij} \cdot I_{ij} = 0 \quad \text{e} \quad \sum V_{ij} \cdot I'_{ij} = 0 \quad (2.15)$$

In other words, the power balance remains in force even if the distribution of voltages and currents pertain to different distributions of components (*principle of conservation of virtual power*).

An important application of this result is the so-called reciprocity principle. To this aim, given a network having in the branch a the generator V_a we build a second network by removing the generator from the branch a and replacing it with a new generator V'_b now in the branch b different from a. Taking voltages of the first network and currents of the second we can write

$$V_aI'_a + \sum R_{ij}I_{ij}I'_{ij} = 0$$

where $R_{ij}I_{ij}$ are the voltage drops on the network resistances.
Taking instead currents of the first network and voltages of the second, we get

$$V'_bI_b + \sum R_{ij}I'_{ij}I_{ij} = 0$$

From these two relations we can easily deduce that

$$V_aI'_a = V'_bI_b$$

Now if we make $V_a = V'_b$ we get $I'_a = I_b$. This proofs the following theorem:

Reciprocity Theorem
In a reciprocal network if a voltage generator V in the branch AA′ produces a current I in the branch BB′, the same generator V in the branch BB′ produces the same current I in the branch AA′. Similarly, in a reciprocal network if a current generator I in

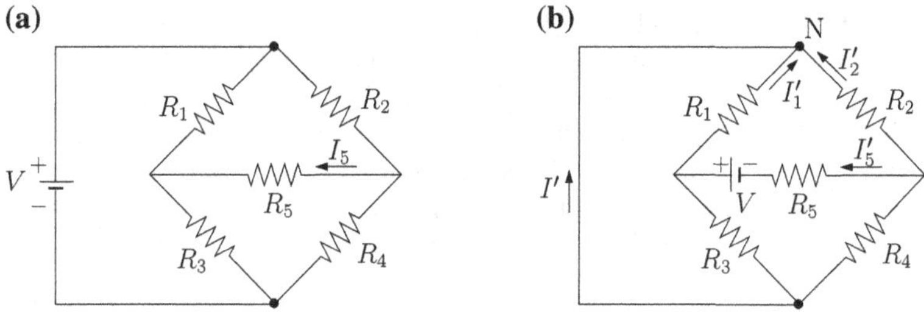

Fig. 2.11 Example of application of the reciprocity theorem

the branch AA′ produces a voltage drop V between nodes B and B′, the same the generator I in the branch BB′ produces the same voltage drop V between nodes A and A′.

It is easily shown that all linear networks are reciprocal while, if a network includes nonlinear elements or controlled generators, then in general the network is not reciprocal.

Example of Application of the Reciprocity Theorem

As an example, we apply the reciprocity theorem to the Wheatstone's bridge already studied in detail in Sect. 2.3.3 and shown again in Fig. 2.11a. The application of the reciprocity theorem allows calculating the current I_5 quickly and elegantly. Figure 2.11b shows the circuit in which the generator V has been removed from its original location (and replaced by a short circuit) and inserted into the branch where we need to calculate the current. The reciprocity theorem ensures that the current I_5 of Fig. 2.11 a is equal to the current I' of Fig. 2.11b.

Consider this last figure and apply to the node N the second Kirchhoff's law to obtain $I' = -(I'_1 + I'_2)$. Both resistances R_1 and R_3 and resistances R_2 and R_4 are connected in parallel, and this leads us to

$$I'_1 = \frac{R_3}{R_1 + R_3} I'_5 \quad \text{e} \quad I'_2 = -\frac{R_4}{R_2 + R_4} I'_5$$

where we have used the properties of the current divider (see Problem 8 of Chap. 1) and I'_5 is the current flowing through R_5. It is easy to get I'_5 as

$$I'_5 = \frac{V}{R_5 + R_1 \| R_3 + R_2 \| R_4}$$

Finally, we get the desired result:

$$I_5 = I' = \frac{V}{R_5 + R_1 \| R_3 + R_2 \| R_4} \left(\frac{R_4}{R_2 + R_4} - \frac{R_3}{R_1 + R_3} \right) \tag{2.16}$$

As a check of the reciprocity theorem, we leave to the reader the exercise to show that formula (2.16) is equal to expression (2.3), obtained for I_5 by the method of the meshes.

The reciprocity theorem has another interesting application to the solution of the circuit, known with the name of $R - 2R$ ladder, used in devices that convert digital signals into analog signals, see Problem 18 of Chap. 4.

2.5 Maximum Power Transfer Theorem

The simple circuit shown in Fig. 2.12 allows us to illustrate an important theorem, which is known as the theorem of *maximum power transfer*. In the diagram R_g represents the internal resistance of the voltage generator V_o and R the external load applied to it. Obviously, the current I that circulates in the circuit is given by $I = V_o/(R_g + R)$, and the voltage at the terminals of the load R is $V_R = RI$. The power P dissipated by the resistor R will be

$$P = V_R I = RI^2 = \frac{V_o^2 R}{(R_g + R)^2} \tag{2.17}$$

We now calculate the value of the load resistance R yielding the maximum power transfer from the generator to the load. By calculating the derivative of expression (2.17) with respect to R, we get

$$\frac{d}{dR}\frac{V_o^2 R}{(R_g + R)^2} = V_o^2 \frac{1}{(R_g + R)^2}\left(1 - \frac{2R}{R_g + R}\right)$$

that vanishes[8] for $R = R_g$.

We can therefore state the following maximum power transfer theorem:

If a generator of internal resistance R_g is closed on a resistive load R, the power transferred from the generator to the load reaches a maximum if the value of the load resistance is equal to the internal resistance of the generator.

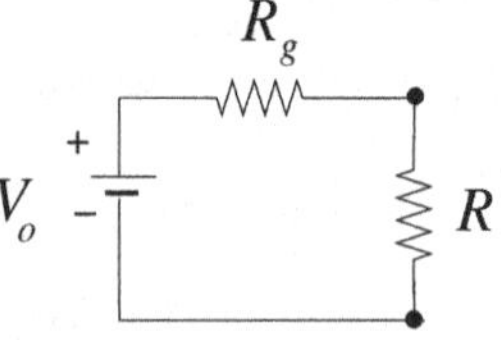

Fig. 2.12 Circuit for the proof of the theorem of maximum power transfer

[8]The second derivative of expression (2.17) is easily computed. Its value for $R = R_g$ is $-V_o^2/8\,R_g^3 < 0$, and confirms that in $R = R_g$ the expression (2.17) shows a maximum.

The previous discussion is related to the concept of impedance matching that will be dealt with in detail in Chap. 6.

Problems

Problem 1 Prove Norton's theorem following the layout of the proof of Thévenin's theorem illustrated in the text, adapting it where appropriate.

Problem 2 Recalling that at least three branches must converge in a node, show that a circuit with no more than three nodes is solved with the smallest number of equations using the method of the nodes when only voltage generators are present. Discuss why the same conclusion cannot be guaranteed in the presence of current generators.

Problem 3 What is the most effective method, i.e., the one using the smallest number of equations, to solve the circuit shown in the figure? [A. The method of the nodes.]

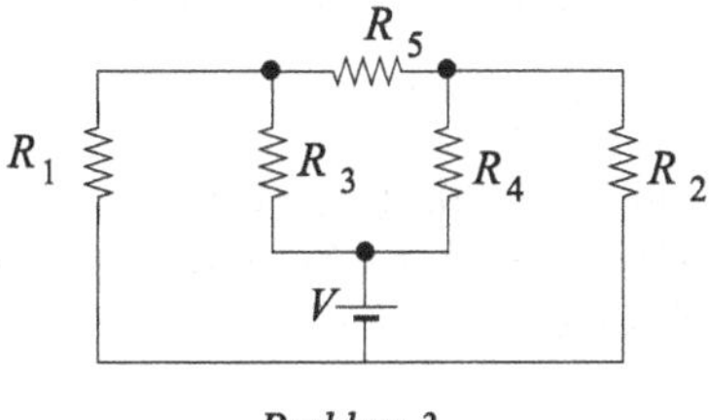

Problem 3

Problem 4 Which of the two circuits in the figure is solved more effectively with the method of the meshes? [A. Circuit B.]

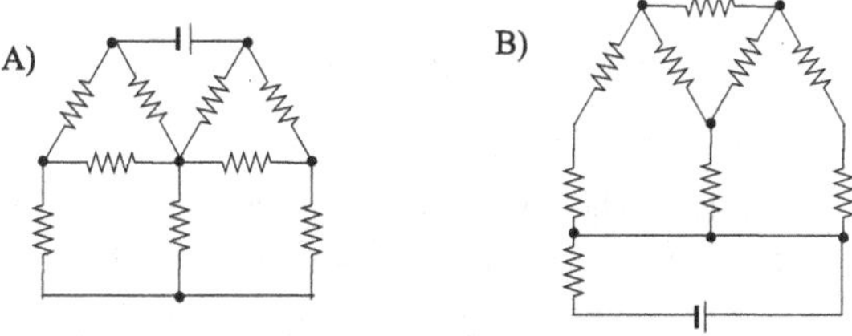

Problem 4

Problem 5 Solve the circuit shown in the figure with the most effective method and compute the voltage drop across R_5, assuming that the values of the six resistances in the circuit are equal. [A. $V_C = V/4$, the two methods being equivalent.]

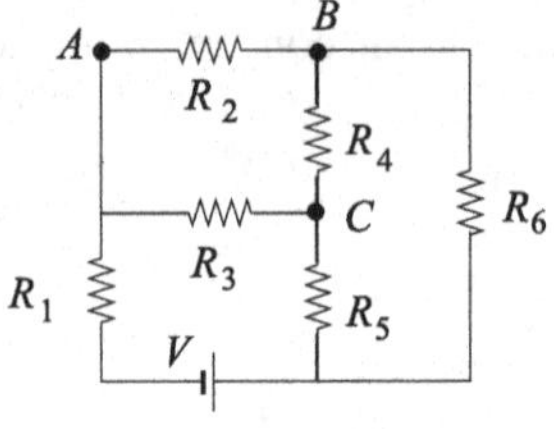

Problem 5

Problem 6 Solve the circuit shown in the figure with the most effective method and compute the voltage drop across R_1, assuming that the values of the seven resistances in the circuit are equal. [A. $\Delta V_{R_1} = V/3$, with the mesh method.]

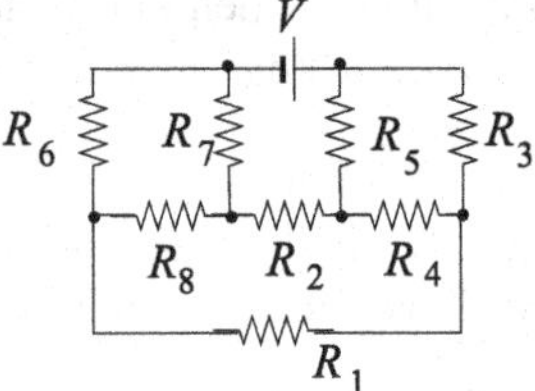

Problem 6

Problem 7 Solve the circuit shown in the figure with the most effective method and compute the current flowing in R_1 and R_5, assuming that the values of the four resistances in the circuit are equal. [A. $I_{R_1} = I/2, I_{R_5} = I/2$ with the mesh method.]

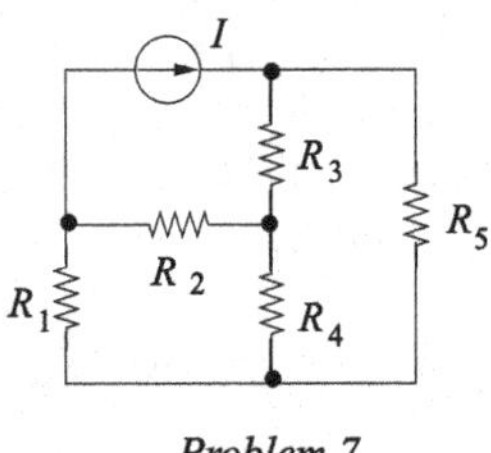

Problem 7

Problem 8 Solve the circuit shown in the figure with the most effective method and compute the current flowing in R_1 and R_5, assuming that the values of the seven resistances in the circuit are equal. [A. $I_{R_1} = 4(I_1 + I_2)/11$, $I_{R_5} = 6(I_1 + I_2)/11$ with the mesh method.]

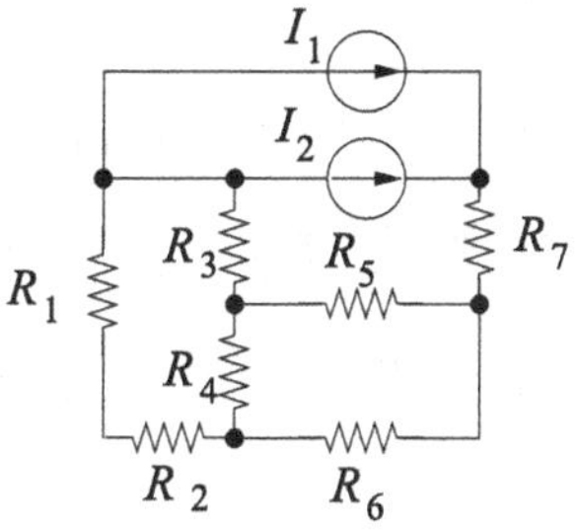

Problem 8

Problem 9 Solve the circuit shown in the figure with the most effective method and compute the current flowing in R_1, R_2, and R_3, assuming that the values of the nine resistances in the circuit are equal. [A. $I_{R_1} = 2I/11, I_{R_2} = I/11, I_{R_3} = I/11$ with the mesh method.]

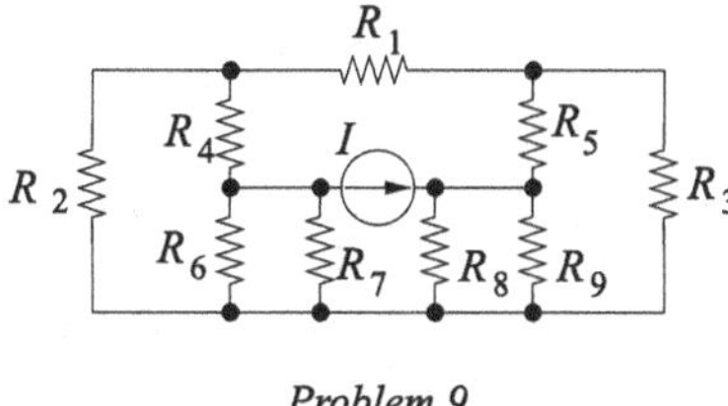

Problem 9

Problem 10 Solve the circuit shown in the figure with the most effective method and compute the current flowing in R_4, R_9, and R_{10}, assuming that the values of the nine resistances in the circuit are equal. [A. $I_{R_4} = \frac{9I}{21} - \frac{5V}{21R}$, $I_{R_9} = \frac{11I}{21} + \frac{5V}{63R}$, $I_{R_{10}} = \frac{I}{3} - \frac{2V}{9R}$ with the mesh method.]

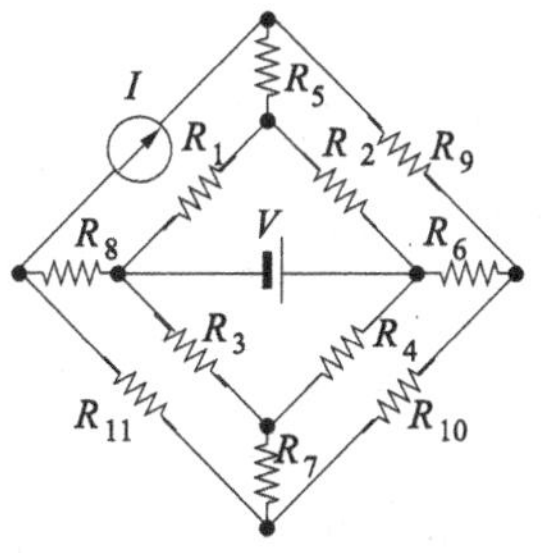

Problem 10

Problem 11 Compute the Norton equivalent with reference to the nodes A and B for the circuit in the figure. Use the following values for its parameters: $V = 5.0\,V$, $R_1 = R_2 = 2.0\,\Omega$, $R_3 = R_4 = 4.0\,\Omega$. [A. $I_{eq} = -2.5\,A$ $R_{eq} = 1.33\,\Omega$.]

Problem 12 Compute the Norton equivalent with reference to the nodes A and B for the circuit in the figure. Use the following values for its parameters: $I = 10\,A$, $R_1 = R_2 = 2.0\,\Omega$, $R_3 = R_4 = 4.0\,\Omega$. [A. $I_{eq} = 5\,A$; $R_{eq} = 1.33\,\Omega$.]

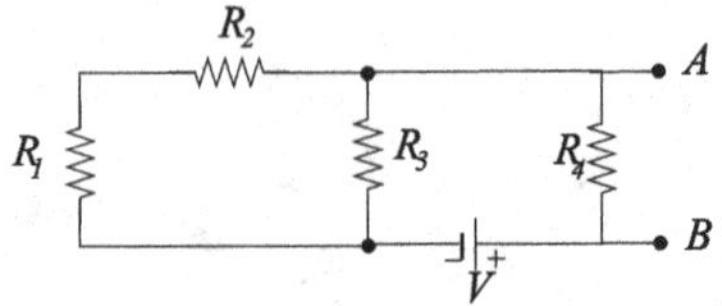

Problem 11

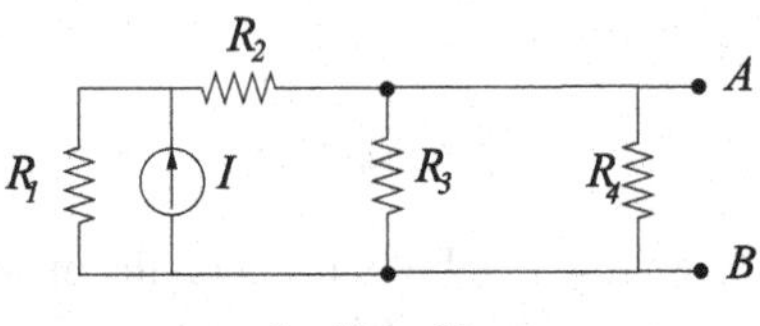

Problem 12

Problem 13 Compute the Norton equivalent with reference to the nodes A and B for the circuit in the figure. Use the following values for its parameters: $I = 10\,A,\ V = 5.0\,V,\ R_1 = R_2 = 2.0\,\Omega,\ R_3 = R_4 = 4.0\,\Omega$. [A. $I_{eq} = 2.5\,A,\ R_{eq} = 1.33\,\Omega$.]

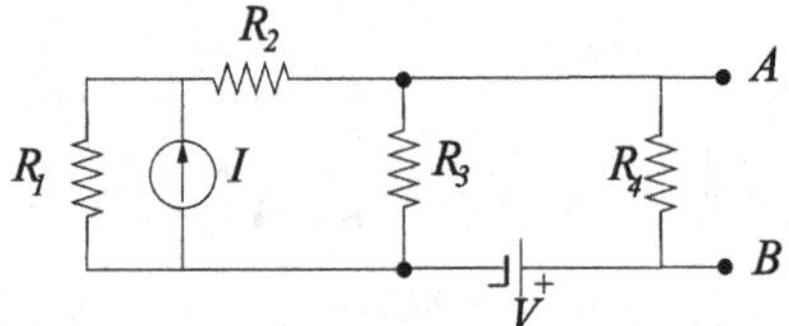

Problem 13

Problem 14 Make use of Thévenin's theorem to compute the current flowing in the resistance R_5 for the circuit in the figure with the following values for its parameters: $V = 10\,V, R_1 = R_3 = R_5 = 200\,\Omega, R_2 = 600\,\Omega,\ R_4 = 400\,\Omega$. [A. 23 mA.]

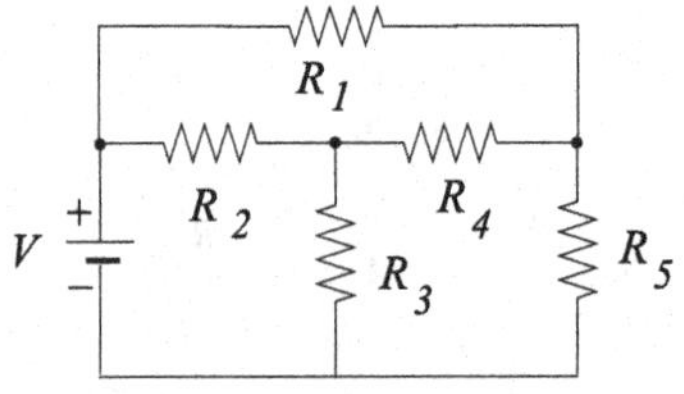

Problem 14

Problem 15 Make use of Thévenin's theorem to compute the power dissipated by the resistance R_4 for the circuit in the figure. Use the following values for its parameters: $V = 2.0\,\text{V},\ I = 4\,\text{mA},\ R_1 = 2.0\,\text{k}\Omega,\ R_2 = 1.0\,\text{k}\Omega, R_3 = 5.0\,\text{k}\Omega, R_4 = 1.0\,\text{k}\Omega$. [A. 3.4 mW.]

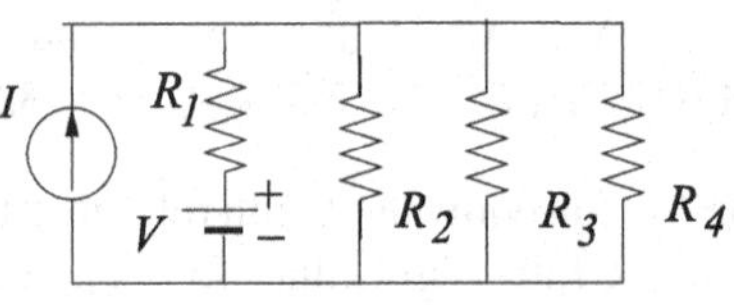

Problem 15

Problem 16 Solve the circuit indicated by (A) in the figure by an appropriate method obtaining the currents in its independent meshes and compute the current delivered by the generator V_1. Then modify the circuit as shown in (B) by inserting a voltage generator V_2. Use the theorems of reciprocity and superposition to evaluate the change in the output current from V_1. Compute the value of V_2 needed to reduce this current to zero. Modify again the circuit by inserting a generator V_3 as shown in (C) and compute the new value of the current delivered by V_1. [A. $I_1 = V_1(R_1 + R_3)/R_1R_3$; $I_1' = (V_1 - V_2)(R_1 + R_3)/R_1R_3$; $V_1 = V_2$; $I_1'' = -V_3/R_1$.]

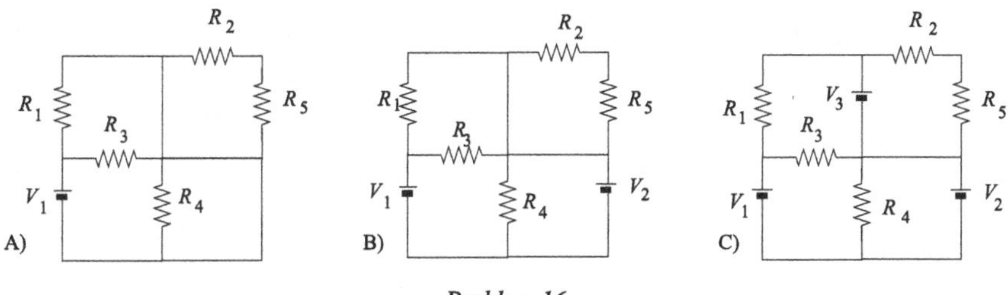

Problem 16

Problem 17 Compute the Thévenin equivalent with reference to the nodes A and B for the circuit in the figure and evaluate the voltage drop across R_3 assuming $V_1 = 25\,V$, $V_2 = -10\,V$, $R_1 = 10\,\text{k}\Omega$, $R_2 = 5\,\text{k}\Omega$, $R_3 = 20\,\text{k}\Omega$. [A. $V_A - V_B = 1.43\,\text{V}$.]

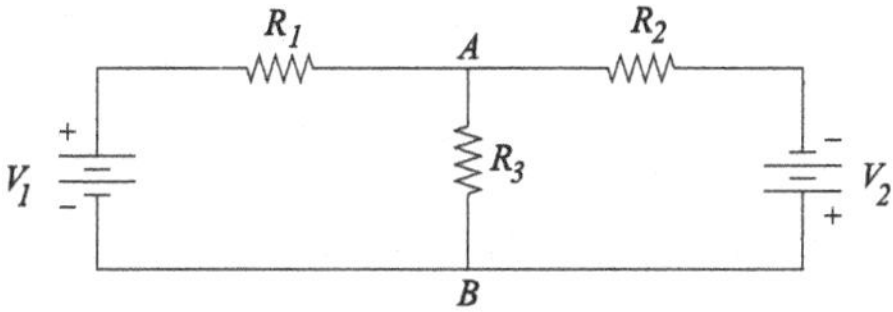

Problem 17

Chapter 3
Uncertainty in Electrical Measurements

3.1 Introduction

In this chapter, we deal with the problem of evaluating the uncertainty involved in the measurement of physical quantities with particular emphasis to the analysis of those of interest in electricity studies. To this purpose, in the following, we will refer to an important document, first published in 1993, that in the years has established the accepted standards for the evaluation of experimental uncertainties in metrology and in scientific research in general. This document, entitled "*Guide to the Expression of Uncertainty in Measurement*" [1], has been published by the *International Organization for Standardization* (ISO)[1] in order to standardize in a rational way the handling of uncertainties in Science and Engineering. In this book, we will refer to it with the acronym *GUM* (which stands for Guide, Uncertainty, and Measurements) adopting a convention nowadays widespread in metrology literature.

In the next Sect. 3.2, we give a short review of the basic principles of measurement theory. In Sect. 3.3, we introduce the important concept of *uncertainty* as the expected standard deviation of obtained results and we discuss the classification of uncertainties based on their evaluation method. In Sect. 3.4, we show the standard procedure to combine uncertainties of different measured variables in the evaluation of derived quantities and we illustrate the important concept of correlation among different measurement uncertainties. After introducing extended uncertainty in Sect. 3.5, we

[1] This guide has been elaborated by an expert group including scientists selected by the most important international organizations responsible of establishing metrology standards: the *Bureau International de Poids et Mesures* (BIPM), the *International Electrotechnical Commission* (IEC), the *International Federation of Clinical Chemistry* (IFCC), the *International Organization for Standardization* (ISO), the *International Union of Pure and Applied Chemistry* (IUPAC), the *International Union of Pure and Applied Physics* (IUPAP), and the *International Organization of Legal Metrology* (OIML). An introduction to the principal findings of this *Guide* has been published in 2002 by Les Kirkup [2]. Another text often quoted together with the *Guide* is the dictionary of the terms used in metrology in English and French language [3]. This text is referred to as "*VIM*", an acronym obtained from *Vocabulaire International de Mètrologie*.

R. Bartiromo and M. De Vincenzi, *Electrical Measurements in the Laboratory Practice*, Undergraduate Lecture Notes in Physics,
DOI 10.1007/978-3-319-31102-9_3

present in Sect. 3.6 a brief discussion on how to handle the problem of the compatibility among results of different measurements of the same quantity. After a brief summary on uncertainty evaluation in Sect. 3.7, in the last section of this chapter, we illustrate in detail a number of examples that make use of all the concepts and methods discussed.

For the purposes of this chapter, we assume that the reader is familiar with elementary principles of probability theory as treated, for example, in reference [4] or [5].

3.2 Notes on Measurement Theory

The aim of a measurement is to obtain the *value*[2] of a physical quantity identified as the measurand. In general, the result of a measurement is only an estimate of the measurand value and therefore it is complete only when the uncertainty associated to it is provided. Indeed, when the result of the measurement of a physical quantity is reported, it is *mandatory* to give a quantitative indication of the quality of the result so that users could judge its reliability. When uncertainties are unknown, one cannot compare different measurements among themselves or with reference data. Moreover, it is necessary that the procedure to characterize the quality of a measurement through its *uncertainty* is easily understandable, easily implementable and generally accepted by the scientific community.

The concept of *uncertainty* as a quantifiable attribute is relatively new in the theory of measurements and replaces the concepts of *error* and *error analysis* that have characterized for a long time the science of measure, or *metrology*.

We point to the reader's attention that the term "*uncertainty*", as defined above, takes a precise meaning in the theory of measurement as a *quantitative parameter* that we will better specified in the following.

One more point worth attention is that the authors of the *GUM* with the choice of the word uncertainty underlined the necessity to operate in a probabilistic framework. The fact that a measurement value is associated with an *uncertainty* means that one can only use it to draw probabilistic conclusions and that a coherent theory of measurement can only be constructed using probabilistic concepts.

Measurement errors Any measurement process is unavoidably influenced by many physical phenomena that cannot be controlled and that affects for an unknown amount its result. Therefore, we need to admit that *the true value of a physical quantity is in principle unknown.*

---

[2]Some textbooks use the words *true value* in place of *value* of a physical quantity; we will make clear in the following that the use of the adjective *true* may be misleading, for example, when a quantity could have more than one *true value*. The *GUM* deems that the use of this adjective is redundant and therefore it suggests not using it. Nevertheless, in some circumstances the adjective *true* can help to clarify the concepts. Moreover, in the measurement of fundamental physical constants the use of the term *true value* may be appropriate.

The complex of uncontrolled phenomena that in a specific measurement influence its result is sometimes loosely referred to as the measurement error. However, the word *error* has a precise meaning in the theory of measurement. Indeed the error is defined as the difference between the (true) value of the quantity under test and the specific value resulting in the measurement underway. Indicating with μ the true value and with m the result of a particular measurement, the definition of measurement error is given as

$$e = \mu - m \tag{3.1}$$

The *measurement error*, defined by the previous relation, in analogy with the true value of a physical quantity, is therefore in principle *unknown*.

We can ideally divide the measurement error in two parts, traditionally referred to, respectively, as the random error and the systematic error.

Random Error. *The random error is the error component that changes in an unpredictable way in repeated measurements.*

Random errors are caused by variables, referred to as influence quantities[3] that are not the measurand but that can change the result of the measurement. If stochastic fluctuations affect these quantities, random errors are observed when repeating a measurement under otherwise identical conditions. This kind of errors cannot be corrected but it is possible to reduce their impact by increasing the number of observations when *their average value is expected to be null.*

Systematic Error. *The systematic error is the component of the measurement error that in a series of repeated observations either remains constant or varies in a predictable way.*

Systematic errors can be reduced by significant amount when their origin is known and a procedure to calculate the correction to apply to the measurement value has been worked out.[4] However, similar to the case of random errors, they cannot be canceled.

Example. When we connect a voltmeter to an electrical circuit, the finite internal impedance of the instrument affects the potential difference between the two contact points A and B. The correction of this effect is possible when we know both the circuit impedance between A and B and the voltmeter impedance. However, both these two values have uncertainties that will propagate to the evaluation of the potential difference obtained after correcting for the systematic effect.

---

[3]For example, the temperature of a ruler is an influence quantity in the measurement of a length because of thermal expansion of materials.

[4]Common examples of systematic errors are the *offset* and the *calibration* of a measuring instrument. In the first case, if x_v is the true value, the instruments yield a value $y = x_0 + x_v$, while in the second case it yields $y = \alpha x_v$ with $\alpha \neq 1$.

3.3 Uncertainty

The uncertainty of the result of a measurement reflects the lack of exact knowledge of the value of the measurand. The result of a measurement, even after correcting for known systematic effects, remains only an estimate of the value of the measurand because of the uncertainty that arises from random effects and from the imperfect correction of systematic effects.

The definition of uncertainty recommended by the working group that compiled the *GUM* is:

a parameter, associated with the result of a measurement that characterizes the dispersion of the values that could reasonably be attributed to the measurand

The *GUM* refers to a set of causes that are responsible for generating uncertainty in the measurements, which we report below:

(a) *Incomplete definition of the measurand.* For example, the definition of gravity acceleration at the sea level is incomplete since it does not take into account that this acceleration also changes with the latitude. Therefore, there are an infinite number of values satisfying this (incomplete) definition.
(b) *Imperfect realization of the definition of the measurand.* The practical implementation of any physical situation is always subject to imperfections that prevent a strict adherence to the ideal prescriptions. For example, the study of the free fall of a heavy body in vacuum requires some knowledge of the residual pressure along the body trajectory.
(c) *Non representative sampling.* The sample measured may not represent the defined measurand. Suppose, for example, that a manufacturer should measure the mean value of the resistance of resistors of a given nominal value. If the resistors under test originate from only one of the different production machines, then the sample under test may not be representative of the quantity that he wants to measure.
(d) *Inadequate knowledge of the effects of environmental conditions on the measurement or imperfect measurement of environmental conditions.* For example, nuclear physics experiments performed in underground laboratories to reduce environmental radioactivity may be affected by the presence of radioactive elements in ground waters whose behavior is often unpredictable.
(e) *Incorrect reading of analog instruments.* The typical example here is the *parallax effect* when reading the position of an index on a graduated ruler. This effect can be reduced using a mirror placed below the indicator and reading the ruler only when the index and its mirror image are superposed.
(f) *Finite resolution or discrimination threshold.* For example, the least significant digit of a digital instrument puts a lower limit to its uncertainty.
(g) *Inexact values of measurement standards and reference materials.*
(h) *Inexact values of constants and other parameters obtained from external sources and used in the data reduction algorithm.* The uncertainty on the resistance value of a copper bar, of length l and cross section area A, deduced from the

second Ohm's law $R = \rho l/A$, in addition to the contribution of uncertainties in the geometric parameters l and A, is affected by the uncertainty of copper resistivity ρ that is usually obtained from specialized handbooks. In particular the *Handbook of Chemistry and Physics* (1990 edition) quotes $\rho = 1.678\,\mu\Omega\,\text{cm}$ (at a temperature $T = 20\,°\text{C}$) without an explicit quantification of its uncertainty. With the information obtained from this source, we can only assume that the value of copper resistivity is included in the interval between 1.6775 and 1.6785 $\mu\Omega\,\text{cm}$ with a flat probability distribution. This uncertainty contributes to the overall uncertainty on the value of R.

(i) *Approximations and assumptions incorporated in the measurement method and procedure.*

(j) *Variations in repeated observations of the measurand under apparently identical conditions, as already mentioned in previous items.*

These sources of uncertainty are not necessarily independent and some or all of items from (a) to (i) can contribute to the variations in repeated observations (j).

It is worth at this point to remark once again that in the theory of measurement the term "*error*" should not be confused with the term "*uncertainty*". The two terms are not synonymous and indicate completely different concepts.

Classification of measurement uncertainties. Broadly speaking, the uncertainty of a measurement consists of different contributions that are conventionally grouped in two categories, according to the method adopted for their evaluation. Some of these contributions can be computed from the statistical distribution of results obtained in repeated observations and can be quantified by the sample estimate of the standard deviation of the underlying probability distribution function. The remaining contributions, those that do not lead to any variation in repeated observations with the same equipment, must also be quantified through a standard deviation. In this case, the relevant probability distributions need to be inferred on the basis of previous experience and/or of any other relevant available information, such as for example the technical specifications and user manual provided by the manufacturer of a measuring instrument.

For these reasons, the following classification was introduced in the *GUM* and is nowadays widely adopted:

- *TYPE A UNCERTAINTIES*, whose magnitude is an estimate derived from the statistical analysis of experimental data.
- *TYPE B UNCERTAINTIES*, whose evaluation cannot rely on the availability of a representative sample of the relevant probability distribution function.

It should be remarked that this classification refers only to the method used to evaluate the different uncertainty contributions but does not have any bearing on the probabilistic meaning of the measurement result. In other words, if the results of a particular measurement follow a normal distribution with variance σ^2, the method used to estimate σ^2, different for type A or B, does not change the nature and the content of the probabilistic information that can be deduced from the results obtained.

The two categories introduced above do not have a simple relation with the distinction, widely used before the *GUM*, between random and systematic errors. In particular, the reader should avoid the temptation to identify type B uncertainty contributions as systematic components of uncertainty.

3.3.1 Type A Evaluation of Standard Uncertainty

We can apply the methods for type A evaluation of the uncertainty when we can rely on an adequate number of independent measurements of the same quantity under (apparently) similar conditions. In case the measurement process has sufficient resolution, we will detect a dispersion of measured values. If n observations $\{x_i\}$, $i = 1, \ldots n$ are available for the quantity X, probability theory indicates [4] that the *best estimate* of the quantity X is provided by the *arithmetic average*, $\bar{x}$, of the individual results:

$$\bar{x} = \frac{1}{n}\sum_{i=1}^{n} x_i$$

We can evaluate the measurement uncertainty associated to the estimate $\bar{x}$ with the following steps. The *best estimate of the variance* $s^2(x)$ of the probability distribution describing the measurement process is obtained from the sample $\{x_i\}$ and is given by

$$s^2(x) = \frac{1}{n-1}\sum_{i=1}^{n}(x_i - \bar{x})^2$$

The square root of the variance is the standard deviation. The *best estimate of the variance of the simple average* is:

$$s^2(\bar{x}) = \frac{s^2(x)}{n}$$

The standard uncertainty $u(\bar{x})$ associated to the estimate $\bar{x}$ is the *experimental standard deviation of the average*

$$u(\bar{x}) = s(\bar{x})$$

3.3.2 Type B Evaluation of Standard Uncertainty

Since the statistical methods applied are well established, the evaluation of type A uncertainties does not offer specific difficulties. On the contrary, we must base the evaluation of type B uncertainties on an accurate analysis of the experimental methods adopted. This type of uncertainties cannot be deduced using repeated observations but the experimenter needs to collect all the available information about the variability of the measurand. Following the *GUM*, this information must be obtained using the "*scientific judgement*" of people involved in the measurement. This concept includes the use of such sources of information as:

- data previously obtained in similar conditions
- theoretical or experimental knowledge of instruments behavior
- technical specifications supplied by instrument manufactures
- calibration data
- uncertainties associated to reference data obtained from published literature.

It is appropriate that, as for type A, also the evaluation of the uncertainty of type B consists of an estimate of the *standard* deviation. This choice permits to evaluate in a consistent way the *combined uncertainty* of quantities obtained in indirect measurements, see next section.

As a final observation, it is obvious that the quality of the evaluation of the uncertainty obtained in this way is a function of the degree of completeness of the information used and of the investigator's capability to use it critically.

The examples that we will propose in the following sections will help the reader to best understand the methods to use in the evaluation of type B uncertainties.

3.4 Combined Uncertainty

In most circumstances, the value of a measurand is not obtained directly but is rather determined by N other measured quantities $X_1, X_2, \ldots, X_N$, through a functional relationship f:

$$Y = f(X_1, X_2, \ldots, X_N) \tag{3.2}$$

that, in metrology, is referred to as the *mathematical model* of the measurement. The expression (3.2) is a relationship between the stochastic variables X_i and Y, and the *Theory of Probability* allows obtaining the expressions linking the average values and the variances of the variables X_i (identified as input variables) to the average value and the variance of the variable Y (the output variable).

Let's denote with η the average, or expected, value of Y

$$E[Y] = \eta = E[f(X_1, X_2, \ldots, X_N)]$$

where $E[\ldots]$ represents the expected value operator. Furthermore, denoting with ξ_i the expected values of the input variables X_i and using a Taylor series expansion of f in the neighborhood of the ξ_i, we have to the first order

$$E[Y] = \eta = E[f(\xi_i + \Delta_i)] = E[f(\xi_i) + \sum_i \frac{\partial f}{\partial \xi_i}\Delta_i + \cdots] \tag{3.3}$$

where $\Delta_i = X_i - \xi_i$ while the symbol $\partial f / \partial \xi_i$ represents the partial derivative of f with respect to X_i computed in ξ_i, in this context referred to as the *sensitivity coefficient* of the output variable Y with respect to the input quantity X_i. Taking into account that the operator *expected value* $E[\ldots]$ is linear and that $E[\Delta_i] = 0$, we get:

$$E[Y] = \eta = f(\xi_i) \tag{3.4}$$

This relation means that, to the first order, the average value of the output variable Y can be obtained by computing the function f using the average values of its input quantities X_i. It should be noted that the expression (3.4) has been obtained by means of a first order approximation of a Taylor series expansion; therefore, it may become inadequate to evaluate η in case the standard deviations of the input quantity X_i are large enough to show the effect of possible non-linearity of the function f.

Moving to the evaluation of the variance σ_Y^2 of Y, we have by definition $\sigma_Y^2 = E[(Y - \eta)^2]$. Using again the first order Taylor expansion as given above in the expression (3.3), we can write that

$$\sigma_Y^2 = E[(Y-\eta)^2] = E[(f(X_i) - f(\xi_i))^2] = E\left[\sum_{i,j} \frac{\partial f}{\partial \xi_i}\frac{\partial f}{\partial \xi_j}\Delta_i \Delta_j\right]$$
$$= \sum_{i,j} \frac{\partial f}{\partial \xi_i}\frac{\partial f}{\partial \xi_j} E\left[\Delta_i \Delta_j\right]$$

The quantity $E\left[\Delta_i \Delta_j\right] = E[(X_i - \xi_i)(X_j - \xi_j)]$ is referred to as the *covariance matrix*:

$$\text{cov}(X_i, X_j) = E[(X_i - \xi_i)(X_j - \xi_j)]$$

It is easy to recognize that the diagonal elements $\text{cov}(X_i, X_i)$ of this covariance matrix are the variances of the variables X_i. Its off diagonal elements $\text{cov}(X_i, X_j)$ quantify the *degree of mutual dependence*, or *correlation*, between the input variables X_i and X_j. In conclusion, the expression of the variance of Y becomes

$$\sigma_Y^2 = \sum_{i,j} \frac{\partial f}{\partial \xi_i} \cdot \frac{\partial f}{\partial \xi_j} \text{cov}(X_i, X_j) \tag{3.5}$$

As already commented above about the expected value, the variance σ_Y^2 too is an approximated value obtained from a first order expansion of f.

Propagation of uncertainties. We can extend the results obtained above for the average value and the variance of the variable Y (see Eqs. (3.4) and (3.5)) to the *sample estimates* of both the average value and the variance (i.e., the square value of standard uncertainty). We therefore can write the relations equivalent to Eqs. (3.4) and (3.5) yielding the *estimate* of y obtained from the measured values x_i:

$$y = f(x_i)$$

and of the *variance estimate* of the output quantity y:

$$u_c^2(y) = \sum_{i,j}^{N} \frac{\partial f}{\partial x_i} \cdot \frac{\partial f}{\partial x_j} u(x_i, x_j) \tag{3.6}$$

where $u(x_i, x_j) = u(x_j, x_i)$ is the estimate of the covariance matrix. The relation (3.6) above is referred to as the *law of uncertainty propagation*. The effects of possible correlations among the input quantities becomes more evident when rewriting Eq. (3.6) as

$$u_c^2(y) = \sum_{i=1}^{N} \left(\frac{\partial f}{\partial x_i}\right)^2 u^2(x_i) + 2\sum_{i=1}^{N-1} \sum_{j=i+1}^{N} \frac{\partial f}{\partial x_i} \cdot \frac{\partial f}{\partial x_j} u(x_i, x_j) \tag{3.7}$$

where we have used $u^2(x_i) = u(x_i, x_i)$.

It is worth at this point calling the reader attention to the fact that the correlation represented by the covariance matrix does not refer to the physical variables X_i but rather to the values x_i resulting from the measurements. To elucidate this remark, let us suppose of measuring the effective value of the input V_i and output V_{out} voltage of an amplifier to measure its gain. If the major uncertainty contribution comes from the voltmeter calibration, the results of the measurements are positively correlated when performed with the same instrument and the correlation coefficient would increase if the two values approach each other. On the contrary, when the two signals are measured with completely different instruments, manufactured, for example, by different producers, it is safe to assume that the two values are not correlated and the off diagonal elements of the covariance matrix can be taken equal to zero.

In absence of correlations, the combined uncertainty on the value of Y simplifies to

$$u_c^2(y) = \sum_{i=1}^{N} \left(\frac{\partial f}{\partial x_i}\right)^2 u^2(x_i) \tag{3.8}$$

In this case, the variance of the output variable is simply given by the sum of the variances of the input variables weighted by the square value of the pertaining sensitivity coefficient.

It is useful to quote correlations in a dimensionless fashion. This is achieved by manipulating the covariance matrix to obtain the so-called correlation matrix whose elements ρ_{ij} are referred to as the correlation coefficients between the two variables x_i and x_j. They are given as

$$\rho_{ij} = \frac{u(x_i, x_j)}{u(x_i)\, u(x_j)} \tag{3.9}$$

(more precisely this relation is an estimate of the correlation coefficient). From its definition it follows that $-1 < \rho < 1$.

We must remark that when two variables are independent from each other, their correlation coefficient ρ is null but on the contrary, a vanishing correlation coefficient does not imply that the two variables are independent. In other words the condition $\rho_{ij} = 0$ is necessary but not sufficient to guarantee the independence of two variables. Indeed, it can be shown that the correlation coefficient only accounts for a linear dependence among two variables.[5] Therefore, this is usually adequate when dealing with small variations in the framework of a *Taylor* expansion as done in the preceding discussion.

3.4.1 Combined Uncertainty for Monomial Functions

It is often the case that the functional relationship represented by Eq. (3.2) consists of a monomial expression of the kind

$$Y = cX_1^{p_1} X_2^{p_2} \dots X_N^{p_N} \tag{3.10}$$

with c being a constant. When we know the exponents in this equation without uncertainty, the expression for the variance estimate of y can be simplified as it is possible to show easily that, in absence of correlations,[6] the following relation holds:

$$\left[\frac{u_c(y)}{y}\right]^2 = \sum_i^N \left[p_i \frac{u(x_i)}{x_i}\right]^2 \tag{3.11}$$

[5] Suppose, for example, that the stochastic variable x has a symmetric probability distribution function so that $E[x] = 0$. The variable $y = x^2$ is obviously *totally dependent* upon x, nevertheless since $E[y] = \sigma_x^2$ we get $cov(x, y) = E[(x - 0)(y - \sigma_x^2)] = E[x^3] - \sigma_x^2 E[x] = 0$, exploiting once again the symmetry of the x distribution. This shows that, although the two variables are maximally correlated, their correlation coefficient is null.

[6] The case with finite correlations is dealt with in Problem 16 at the end of this chapter.

The Eq. (3.11) shows that the relative variance $\left[\frac{u_c(y)}{y}\right]^2$ of y is equal to the sum of the similar quantities of the x_i weighted with the square value of the pertaining exponent. Equation (3.11) yields directly the relative variance that is often a more direct way to quantify the accuracy of a measurement. The value of the y variance is simply obtained by the product of the right hand side of Eq. (3.11) with y^2. The result obtained above with Eq. (3.11) also shows that, when we combine different measurements, the input variables with the higher relative uncertainty are responsible for the largest contribution to the uncertainty of the derived quantity. Therefore, when we plan such a measurement, we should possibly avoid combining data having widely different relative uncertainty. For example, to measure a resistance by exploiting the Ohm's law, it is pointless to plan for a current relative uncertainty below one part over a thousand in case we know that we cannot achieve a voltage relative uncertainty better than one part over a hundred.

3.5 Expanded Uncertainty and Coverage Factor

Although the (combined) *standard* uncertainty u_c can be used to express the uncertainty of a measurement result, it does not say anything about its probability distribution. In some commercial, industrial, or regulatory applications, or when health and safety are concerned, it is often necessary to give a measure of uncertainty that defines an *interval*[7] containing the *true value* of a measurand with a probability p[8] near to 100 %. Such an interval takes the name of *expanded uncertainty* and it is customary to use the symbol U to denote it. The expanded uncertainty is obtained multiplying the standard uncertainty by a constant k: $U = ku_c$, and k takes the name of *coverage factor*. The value of the coverage factor k corresponding to a given probability level p depending upon the probability distribution function of the measurand.

Typical values of the coverage factor fall in the interval $(2 \div 3)$. When it is reasonable to assume that the estimate of a measurand follows a Gaussian distribution, a coverage factor $k = 2$ yields an interval corresponding to a $\simeq$95 % of probability, while $k = 3$ correspond to an interval with probability exceeding 99 %. In case of a uniform probability distribution, a coverage factor $k = 1.7$ corresponds to $p \simeq 100\,\%$.

The *combined expanded uncertainty* can be obtained using Eq. (3.6) after increasing all standard uncertainties $u(x_i)$ by the same coverage factor k. However, the interval obtained in this way *does not have the same probability content* of the corresponding intervals for the input variables unless all the x_i are described by a Gaussian

---

[7] Sometimes, this interval is referred to as the *confidence interval*. However, the *GUM* does not recommend its use since in statistics it has a precise meaning and, strictly speaking, it could only be used for type A uncertainties.

[8] This probability value is sometimes referred to as the *confidence level*. Here too, the considerations of the previous note apply.

distribution. It is not easy to associate probability content to the extended combined uncertainty as defined above and therefore, its use is not recommended.

3.6 Compatibility of Different Measured Values

A recurring problem in the professional life of an experimentalist is the comparison between two values of the same quantity. These values could have been obtained with different measurement setups or one of them could have been derived from other measured quantities, as it happens when trying to validate a physical law.

As an example of the first kind of situation, we can check the calibration of a voltmeter using it to measure a voltage whose value has already been obtained with an instrument whose calibration has been recently certified. For the second situation, assume for example that we need to validate the second Ohm's law: we will compare the resistance value of a copper bar as measured by an ohmmeter with the value resulting from the knowledge of copper resistivity and the measures of the geometrical parameters of the bar as obtained for example with a caliper.

In all these circumstances, we need to compare quantitatively the values of two different measurements. Let us call μ_1 the value of the first one and u_1 its uncertainty while μ_2 indicates the value of the second with u_2 its uncertainty. The question we need to answer is whether we can consider the two values compatibles with each other after taking into account their respective uncertainties.

When the values of our measurements are real numbers, the probability that they are exactly equal is obviously zero and therefore the absolute value of their difference $\delta = |\mu_1 - \mu_2|$ is always finite. The question we must answer then is "which value of δ is sufficiently large to lead us to disbelieve that the two measurements are both correct and pertain to the same quantity?"

A quantitative approach to answering this question consists in evaluating the probability of observing an absolute difference of measured values exceeding δ when the two quantities we are measuring are the same and the difference between the observed values is entirely due to the uncertainties of the two measurements. When this probability turns out to be too small, we have to conclude that it is unlikely that the difference observed is due to uncertainties and we will have to admit that either the two quantities are different, or at least one of the two measurements is not correct.

However, to evaluate properly this probability, we would need to have a full knowledge of the probability distribution functions describing the outcome of the two measurements. This information is seldom fully available and often we will have to rely only on the availability of standard uncertainties. In this case, we can only evaluate the estimate of the standard uncertainty of the difference δ as $u_\delta = \sqrt{u_1^2 + u_2^2}$.

If we can plausibly assume that uncertainties are due to a large number N of small and uncorrelated effects, the central limit theorem [4] can be invoked to assume, in first instance, that the distribution expected for the difference is a Gaussian[9] with standard deviation equal to u_δ.

In these circumstances, it is possible to compute the probability that the absolute value of the measurement's difference exceeds δ, or, in other words, that the observed difference δ is entirely due to measurement uncertainties. In formulas, we have

$$p = \frac{1}{\sqrt{2\pi}} \int_{-\infty}^{-\delta} \mathrm{e}^{-x^2/2u_\delta^2} dx + \frac{1}{\sqrt{2\pi}} \int_{\delta}^{+\infty} \mathrm{e}^{-x^2/2u_\delta^2} dx = 1 - \mathrm{erf}\left(\frac{\delta}{\sqrt{2}u_\delta}\right)$$

where $\mathrm{erf}(x) = 2/\sqrt{\pi} \int_0^x \mathrm{e}^{-t^2} dt$ is the well-known *error function*.

When the difference between the two measured values is small with respect to u_δ, we compute a probability of nearly 100 % and we will not have sufficient motivations to believe that they do not refer to the same quantity. In case the measured difference is equal to its standard uncertainty u_δ, we compute a 32 % value for the probability that such an observed difference is due entirely to uncertainties. This probability would fall below 5 % when the measured difference becomes larger than twice u_δ and, in this case, it is more appropriate to admit that the two measured quantities are different since we now can expect that further measurements would confirm our conclusion in more than 95 % of occurrences.

It is apparent from the previous discussion that, even in case the probability distributions would be available, we can never ascertain the compatibility between two measurements with absolute certainty. All we can do, by following the approach illustrated above, is to assign a probability to the hypothesis that they are *not* compatibles. Moreover, it is not easy to establish from first principles how large this probability should be to conclude that the two measured values are not compatible. It is very reasonable that this level should depend on the purpose that the two measurements should serve. For example, when we need to decide whether the readings of two thermometers are sufficiently in agreement to guarantee the uniformity of the temperature in our flat we can accept a risk of being wrong higher than the one we could afford when using the two thermometers in an experiment to validate the principle of energy conservation.

How to behave then in the laboratory practice when one has to judge the compatibility of two different measurements of the same quantity? Our practical suggestion is to calculate the ratio between the absolute value of their difference and its standard uncertainty obtained by propagating the uncertainty of the two measures in question. Values of this ratio lower than unity will not require a review of the measures while

---

[9]It is worth recalling that the Gaussian shape is an asymptotic limit. Moreover, the further we move away from the average value, the higher the value of N required to approach this limit. For this reason, with our approach, we cannot obtain reliable values for the integrated probability when its value becomes too small.

values greater than two indicate that at least one of the measurement procedures needs to be revised. For intermediate values, we will refer to the goodwill of the experimenter!

3.7 Evaluating Uncertainties

Summing up, the basic idea behind the methods described in the *GUM* consists in assuming that every measurand, that is the quantity to be evaluated, and all others variables that can affect its value should be dealt with as random variables and not as uniquely valued quantities. Therefore, we can only describe the measurand value with a distribution of probability density.

To deal correctly with the problem of evaluating the measurement uncertainty it is useful to adopt the procedure consisting of the following points:

1. make a mathematical model of the measurement system, see Eq. (3.2).
2. list all input variables subject to uncertainties.
3. calculate the standard uncertainty for each of them using type A analysis for those with repeated measurements and type B for the others.
4. calculate the sensitivity coefficients $\frac{\partial f}{\partial x_i}$.
5. compute the combined uncertainty taking into account all known correlations among the input variables.

Before presenting some practical examples of uncertainty evaluation, it is worthwhile that we bring to the reader attention the following important recommendation taken from the GUM:

Although the Guide provides a framework for assessing uncertainty, it cannot substitute for critical thinking, intellectual honesty and professional skill. The evaluation of uncertainty is neither a routine task nor a purely mathematical one; it depends on detailed knowledge of the nature of the measurand and of the measurement. The quality and utility of the uncertainty quoted for the result of a measurement therefore ultimately depend on the understanding, critical analysis and integrity of those who contribute to the assignment of its value.

3.8 Examples of Uncertainty Evaluation

When dealing with electrical measurements, the uncertainties to assign to the results are often those of type B. These are the cases when the value obtained in repeated measurements remains the same. Type A uncertainties, although always present in any measurement, in these cases are smaller than the sensitivity of the instruments in use and therefore cannot be detected. This kind of measurement are *precise* insofar as they always give the same result when repeated; nevertheless the main interest of the experimenter is how *accurate* the measurement is, that is how much it is different from the *true value*[10] of the measurand. In the following sections, we will assume that all the uncertainties discussed are of type B unless the contrary is explicitly stated.

3.8.1 Voltage Measurements with an Analog Voltmeter

The uncertainty associated to measurements performed with analog electrical instruments (typically multimeters) has been established in the recommendations of the International Electro technical Commission (IEC) that have adopted the concept of class of accuracy[11] to classify instruments. This class C is *the maximum possible value of the uncertainty at any point of the range of the measuring device expressed as the percentage of its full-scale deflection.* If FS is the value corresponding to the full-scale deflection and C is the accuracy class, then the maximum value of the uncertainty Δ_{max} that we can associate to a measurement obtained with this scale is:

$$\Delta_{\text{max}} = \frac{C}{100} \cdot FS \tag{3.12}$$

For example, the measurement performed with a voltmeter of *accuracy class* $C = 0.5$ and used with the full-scale corresponding to $FS = 250\,\text{V}$ yields a maximum uncertainty of $(0.5/100) \times 250 = 1.25\,\text{V}$ *at every point of its range*. This implies that the probability that the *true value* falls in the interval of width $2\Delta_{\text{max}}$ centered at the measured value is nearly one, i.e., it is (almost) certain. In absence of additional specifications for the instrument in use, we can only assume that the probability distribution of the true value is *uniform* over the interval $2\Delta_{\text{max}}$ and centered on the measured value. It is easy to show that the variance of this distribution is $\Delta^2_{\text{max}}/3$ that yields a standard uncertainty

$$u = \frac{\Delta_{\text{max}}}{\sqrt{3}} = \frac{1.25}{1.7320} V = \pm 0.7\,\text{V}$$

As a further exemplification, suppose we have to measure DC voltages with an instrument of class $C = 1.5$, using a range with full-scale deflection $FS = 50\,\text{V}$. We measure two different voltage levels and obtain $V_1 = 30\,\text{V}$ and $V_2 = 40\,\text{V}$. From the definition of accuracy class, we compute the maximum uncertainty to associate to the two measurement values: $\Delta_{\text{max}} = (C/100)FV = 1.5 \times 50/100 = 0.75\,\text{V}$; finally

---

[10]Here, the adjective *true* added to the noun *value* has to be intended as an intensifier; instead of *true value* it could be said: measurement taken with an instrument more accurate than the one in use or a reference measure (when available).

[11]The term *precision*, traditionally used for the analog instruments, has been more recently replaced by *accuracy* to be consistent with the recommendations issued by the working group and described in the *GUM*.

we estimate the relative standard uncertainty to associate to the two measurements, (with the term relative uncertainty we intend the ratio of the uncertainty to the measured value):

$$\frac{\Delta_{max}}{V_1} = \frac{100 \times 0.75}{30} = 2.50\,\%$$

$$\frac{\Delta_{max}}{V_2} = \frac{100 \times 0.75}{40} = 1.88\,\%.$$

Note that the relative uncertainty is higher in the first measurement because of the lower value; therefore when using instruments whose uncertainty is a fixed percentage of the full-scale deviation, it is advisable to set them up to work as near as possible to the full-scale deviation.

3.8.2 Voltage Measurements with a Digital Multimeter

The uncertainty associated to a measurement obtained with a digital multimeter is usually obtained consulting the user manual that must accompany every instrument. Most manufacturers of professional multimeters declare a value of the expanded uncertainty U such that the measurand is contained with a probability of 99 % in the interval of width of $2U$ centered at the measured value.[12] Assuming that the values obtained measuring the same voltage with different realizations of the same instrument are well described by a Gaussian distribution, we calculate that this extended uncertainty corresponds to a coverage factor $k = 2.6$; the standard uncertainty is therefore: $u = U/2.6$.

In the user manual of digital multimeters, the manufacturers provide instructions to compute (expanded) uncertainty for each range of the instrument. Many manuals report that the uncertainty is given by:

$$\pm(\text{percentage of the reading} + \text{number of } \textit{digit}) \tag{3.13}$$

where the word *digit*, sometimes exchanged with the word *count* of similar meaning, indicates the value of the less significant digit for the range in use. The number of digit represents the *resolution* of the instrument for that range. The first term in Eq. (3.13) represents the uncertainty of the instrument calibration while the second term is independent from the measured value and results from the combined effect of digital resolution and electrical noise in the instrument circuits. These two contributions are mutually independent and therefore they should be added quadratically when

[12]The interested reader can find many examples of user manuals and application notes on line with a simple query.

computing the total uncertainty of the measurement.[13] For example, suppose we are measuring a voltage of about 10 V working with a digital multimeter on a full-scale of 20.0000 V. From the user manual, we read that for this scale the uncertainty is given by $\pm(0.003\,\%\ \text{reading} + 2\,\text{digit})$; in this case the least significant digit is equivalent to 0.0001 V and we obtain: $U = [(0.003\,\% \times 10)^2 + (2 \times 0.0001)^2]^{1/2}\ V = 0.36\,\text{mV}$. Taking into account the coverage factor the *standard uncertainty* is given by

$$u = \frac{0.36}{2.6}\,\text{mV} = 0.14\,\text{mV}$$

3.8.3 Standard Resistor

The following example, taken from the *GUM*, shows how to use data from a calibration certificate. Suppose we have a standard resistor whose resistance R_s, nominally $10\,\Omega$, has been certified to be $10.000742 \pm 0.000129\Omega$ at a temperature $T_o = 23\,°\text{C}$. Let also assume that the value of R_s belongs to a normal distribution and that the uncertainty quoted above is an *expanded uncertainty* ($U = 129\,\mu\Omega$) corresponding to a coverage factor $k = 2.58$; this means that the value of R_s falls, with a probability of 99 %, in the interval of width $2U$ centered on the quoted value.[14] The *standard uncertainty* pertaining to the value of R_s is given by $u_{R_s} = 129\,\mu\Omega/2.58 = 50\,\mu\Omega$. We remark that the value of R_s reported by the calibration certificate is only valid when the temperature of the resistor is $T_o = 23\,°\text{C}$. Suppose now that the resistor is used at a temperature $T \neq T_o$. Let us assume that $T = (0.0 \pm 0.3)\,°\text{C}$, and that the value of the temperature coefficient of the resistor is $\alpha = (7.4 \pm 0.3)10^{-5}\,°\text{C}^{-1}$. Under these hypotheses the mathematical model describing the resistance of our resistor is

$$R_s(T) = R_s(T_o)[1 + \alpha(T - T_o)]$$

The sensitivity coefficients are:

$$\frac{\partial R_s(T)}{\partial T} = R_s(T_o)\alpha; \quad \frac{\partial R_s(T)}{\partial \alpha} = R_s(T_o)(T - T_o); \quad \frac{\partial R_s(T)}{\partial R_s(T_o)} = 1 + \alpha(T - T_o).$$

[13] Unfortunately, many user manuals implicitly suggest adding them linearly thereby leading to an overestimate of total uncertainty that can become important at the bottom of the range in use.

[14] Indeed we have:

$$\int_{-2.58}^{+2.58} 1/(\sqrt{2\pi})\mathrm{e}^{-t^2/2}dt = 0.990005.$$

The best estimate of the resistance value is:

$$R_s(T)=R_s(T_o)[1+\alpha(T-T_o)] = 10.000742 \times (1 - 0.000074 \times 23) = 9.983721\,\Omega$$

and finally the combined uncertainty to associate to $R_s(T)$ is:

$$u_c = \sqrt{\left[\frac{\partial R_s(T)}{\partial T}\right]^2 u_T^2 + \left[\frac{\partial R_s(T)}{\partial \alpha}\right]^2 u_\alpha^2 + \left[\frac{\partial R_s(T)}{\partial R_s(T_o)}\right]^2 u_{R_s}^2}$$

The uncertainties on the temperature T, the temperature coefficient α and the resistance value $R_s(T_o)$ are not correlated and therefore the correlations terms in the previous formula have been safely neglected. It follows that we need to evaluate just three contributions respectively due to:

(1) uncertainty of the temperature T

$$\begin{aligned}[R_s(T_o)\alpha]^2 u_T^2 &= \left(10.000742 \times 7.4 \times 10^{-5}\right)^2 0.3^2 = \left(2.22 \times 10^{-4}\right)^2 \\ &= 4.94 \times 10^{-8}\,\Omega^2\end{aligned}$$

(2) uncertainty of the temperature coefficient α

$$\begin{aligned}[R_s(T_o)(T-T_o)]^2 u_\alpha^2 &= (10.000742 \times (-23.0))^2 \left(0.3 \times 10^{-5}\right)^2 = (-6.90 \times 10^{-4})^2 \\ &= 4.76 \times 10^{-7}\,\Omega^2\end{aligned}$$

(3) uncertainty of the resistance value R_s at temperature T_0

$$\begin{aligned}[1+\alpha(T-T_o)]^2 u_{R_s}^2 &= \left(1 - 7.4 \times 10^{-5} \times 23\right)^2 (5 \times 10^{-5})^2 = (4.99 \times 10^{-5}) \\ &= 2.49 \times 10^{-9}\,\Omega^2\end{aligned}$$

Note that the most important contribution to the combined uncertainty comes from the temperature coefficient. Adding the three contributions, we get

$$u_c = \sqrt{4.94 \times 10^{-8} + 4.76 \times 10^{-7} + 2.49 \times 10^{-9}} = 0.7\,\text{m}\Omega$$

Finally, our result for the value of the resistor resistance when used at a temperature $T = 0\,°\text{C}$ is:

$$R_s(T = 0\,°\text{C}) = 9.9837 \pm 0.0007\,\Omega$$

3.8.4 Examples of Correlated Measurements

Example 1 (*Voltage measurements*) The measurement of the voltage in two points A and B of a given circuit, performed with the *same* digital voltmeter, yields the following results: $V_A = 3.512\,\text{V}$ and $V_B = 3.508\,\text{V}$. The same two values are always obtained in repeated measurements. In the user manual of the voltmeter, we read that the expanded uncertainty of the instrument, in the configuration in use, is given by 0.2 % of the value read on the digital display. We need to evaluate the potential drop between the two points A and B.

The best estimate for this voltage difference is easily obtained and is $\Delta V = V_A - V_B = 4\,\text{mV}$ to which we must associate the correct uncertainty value. Assume now that the quoted uncertainty (of type B), as discussed above in the Sect. 3.8.2, corresponds to a confidence level of 99 % and that the relevant probability distribution function is Gaussian. In these hypothesis the standard uncertainty of both measured voltages is $u(V) = 7/2.6 = 3\,\text{mV}$. If we naively, and wrongly, assume that the two uncertainties are not correlated we would get an uncertainty for ΔV given by $u(\Delta V) = \sqrt{2}u(V) = 4\,\text{mV}$, yielding $\Delta V = (4 \pm 4)\,\text{mV}$! This result is obviously wrong insofar, for example, it implies that the value $\Delta V = 0$ is plausible. Moreover, it would imply that it is possible to observe a voltage at point 2 higher than in point 1 with a probability of 32 %.

To proceed correctly, we first note that

(i) the measured values do not change when repeated, ruling out any random component to the uncertainty and

(ii) the proximity of the measured values (they differ only of 0.1 %), strongly suggests that systematic uncertainty due to the calibration procedure are equal for the two measurements since we used the same instrument. From these considerations it follows that the uncertainty of ΔV is due only to the quantization contributions, that is $u_d = 1/\sqrt{12}\,\text{mV} = 0.3\,\text{mV}$; this yields $u(\Delta V) = \sqrt{u_d^2 + u_d^2} = \sqrt{2u_d^2} = 0.4\,\text{mV}$.

Finally, the result of the measurement is:

$$\Delta V = (4.0 \pm 0.4)\,\text{mV}$$

Let us remark once again that the two measurements are strongly correlated since we used the same voltmeter for both of them. Should we have used two different instruments, this correlation would be lost and the two uncertainties should be composed in full quadratically.

Let us now consider again the previous example with a more formal approach. The uncertainty $u(V)$ to be associated to the measurements performed with the digital voltmeter mentioned above should be correctly expressed as $u^2(V) = (0.02V/2.6)^2 + (1\ \mathit{digit})^2$, where the first term represents the calibration contribution while the second is due to the quantization of the voltage value, the two terms being obviously independent from each other. The two measured values are different by a small amount, much smaller than the full-scale of the instrument, and therefore it is very reasonable to assume that the correlation coefficient ρ of the two calibration contributions to the uncertainties to attach to V_A and V_B, is nearly one. On the contrary, the correlation coefficient of the two quantization contributions (the second term in the expression given above) is null. In formulas, we have:

$$u(V_A) = \sqrt{\left(\frac{0.02V_A}{2.6}\right)^2 + \left(\frac{0.001}{\sqrt{12}}\right)^2}\ V; \quad u(V_B) = \sqrt{\left(\frac{0.02V_B}{2.6}\right)^2 + \left(\frac{0.001}{\sqrt{12}}\right)^2}\ V$$

At this point, separating calibration from quantization contribution, we can use the expression (3.7) for uncertainty propagation and the definition (3.9) for the correlation coefficient to obtain:

$$u(\Delta V)=\sqrt{\left(\frac{0.02V_A}{2.6}\right)^2+\left(\frac{0.02V_B}{2.6}\right)^2-2\rho\frac{0.02V_A}{2.6}\frac{0.02V_B}{2.6}+\left(\frac{0.001}{\sqrt{12}}\right)^2+\left(\frac{0.001}{\sqrt{12}}\right)^2} \simeq 0.4\,\mathrm{mV}$$

This example should warn the reader once again that a correct evaluation of the uncertainty to be associated with a measurement always requires a careful critical analysis before applying any formula.

Example 2 (*Interpolation*) In the previous example, taking into account correlations did lead to a lower uncertainty for the value of the output variable. This is not the case in the following example that discusses a method to estimate the value of an output variable by a linear interpolation between two measured quantities.

Assume that we measure the quantity y as a function of the variable[15] x. Let $(x_1, y_1 \pm u_1)$ and $(x_2, y_2 \pm u_2)$ be two such measurements and suppose that we need to find the value of x_o corresponding to a given value of y_o falling inside the interval (y_1, y_2) as shown in Fig. 3.1. In addition, let us assume that the interval (x_1, x_2) is small enough to approximate $y(x)$ with its first order expansion around x_o.

Before discussing correlation effects, let us briefly recall the definitions and the applications of the linear interpolation procedure. This method allows filling voids in a table of data pair x_i, y_i, in the hypothesis that the dependence of y upon x is linear. Figure 3.1 shows how to implement a linear interpolation between two data points (x_1, y_1) and (x_2, y_2) to find the value x_o corresponding to an *assigned* value of $y = y_o$.

Using a linear approximation, we can easily deduce:

$$\frac{y_o - y_1}{x_o - x_1} = \frac{y_2 - y_1}{x_2 - x_1} \quad \text{from which:} \quad x_o = \frac{x_2 - x_1}{y_2 - y_1}(y_o - y_1) + x_1 \tag{3.14}$$

[15] In a practical example, x may represent the frequency of a sinusoidal input signal that is attenuated by a factor $y(x)$ at the output of a filter circuit.

Fig. 3.1 Linear interpolation

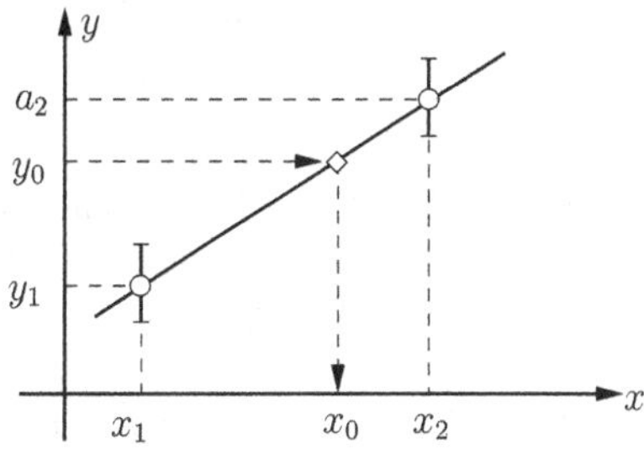

To compute the uncertainty associated to the value of x_o given above, we first calculate the sensitivity coefficients:

$$\frac{\partial x_o}{\partial y_1} = -\frac{(x_2 - x_1)(y_2 - y_o)}{(y_2 - y_1)^2}$$

$$\frac{\partial x_o}{\partial y_2} = \frac{(x_2 - x_1)(y_1 - y_o)}{(y_2 - y_1)^2}$$

Since we are *interpolating*, the point y_o falls within the interval defined by y_1 and y_2 and the sign of the two derivatives is the same.

Propagating the uncertainties defined above, we obtain for the uncertainty of x_o the following relation

$$u(x_o) = \sqrt{\left(\frac{\partial x_o}{\partial y_1}\right)^2 u_1^2 + \left(\frac{\partial x_o}{\partial y_2}\right)^2 u_2^2 + 2\rho \left(\frac{\partial x_o}{\partial y_1}\right)\left(\frac{\partial x_o}{\partial y_2}\right) u_1 u_2} \tag{3.15}$$

where ρ is the correlation coefficient between uncertainties u_1 and u_2. Inserting the expressions for the sensitivity coefficients, we obtain:

$$u(x_o) = \frac{(x_2 - x_1)}{(y_2 - y_1)^2}\sqrt{(y_2 - y_o)^2 u_1^2 + (y_o - y_1)^2 u_2^2 + 2\rho(y_o - y_1)(y_2 - y_o)u_1 u_2} \tag{3.16}$$

If we define a dimensionless variable θ through the relation $y_o = y_1 + (y_2 - y_1)\theta$, its value falling in the range (0, 1), the previous expression becomes:

$$u(x_o) = \frac{(x_2 - x_1)}{(y_2 - y_1)}\sqrt{(1 - \theta)^2 u_1^2 + \theta^2 u_2^2 + 2\rho\theta(1 - \theta)u_1 u_2} \tag{3.17}$$

When the uncertainties u_1 and u_2 are mutually independent (this is the case for example when y_1 and y_2 are counts of random events such as radioactive decays, or when the two measurements have been performed with completely different setups), we can assume $\rho = 0$ and the uncertainty of x_o becomes:

$$u(x_o) = \frac{(x_2 - x_1)}{(y_2 - y_1)}\sqrt{(1-\theta)^2 u_1^2 + \theta^2 u_2^2} \tag{3.18}$$

On the contrary, in case the uncertainties u_1 and u_2 are strongly correlated, as it happens for example when y_1 and y_2 are only slightly different and result from voltage measurements obtained with the same instrument, then $\rho \simeq 1$ and Eq. (3.17) becomes:

$$\begin{aligned} u(x_o) &= \frac{(x_2 - x_1)}{(y_2 - y_1)}\sqrt{(1-\theta)^2 u_1^2 + \theta^2 u_2^2 + 2\theta(1-\theta)u_1 u_2} \\ &= \frac{(x_2 - x_1)}{(y_2 - y_1)}\left[(1-\theta)u_1 + \theta u_2\right] \end{aligned} \tag{3.19}$$

It is very easy to verify that the value resulting from the formula (3.19), valid for correlated uncertainty, is always larger than the one obtained from the expression (3.18) that applies in absence of correlations. The reader is encouraged to find a graphical explanation for this finding.

Finally, let us note that, in case of *extrapolation* the value of y_o falls outside the interval (y_1, y_2) and θ is not included in the interval $(0, 1)$. Therefore, our previous result is not valid for the extrapolation method.

Problems

Problem 1 A voltage divider has been built with two resistors whose resistance values are known with the same relative uncertainty (for example $\varepsilon_R = 1\,\%$). Assuming that uncertainties are not correlated, it is required to calculate the uncertainty on its partition ratio $A = R_2/(R_1 + R_2)$. [A. $u_A/A = (1 - A)\sqrt{\varepsilon_{R_1}^2 + \varepsilon_{R_2}^2}$.]

Problem 2 With two resistors of equal resistance value and the same relative uncertainty (for example $\varepsilon_R = 1\,\%$) we need to build a voltage divider to reduce by a factor 2 the voltage $V \pm u_V$ provided by a known source. Assuming that uncertainties are not correlated, compute the uncertainty on the partition ratio and on the output voltage V_A. [A. $u_A/A = \varepsilon_R/\sqrt{2}$, $u_{V_A}/V_A = \sqrt{\varepsilon_R^2/2 + (u_V/V)^2}$.]

Problem 3 Consider a set of 100 resistors, each with resistance equal to 10 Ω, and a resistor of resistance value 990 Ω, all values having the same relative uncertainty ε_R. It is necessary to build a voltage divider to reduce signal amplitude by about a factor 100. Which one of the two following solution should we adopt to obtain the lowest uncertainty on the partition ratio:

(a) the 990 Ω resistor in series with a 10 Ω resistor whose terminals supply the output voltage
(b) the series of all 10 Ω resistors using the last one for the output voltage?
[A. solution b unless all uncertainties of the 10 Ω resistors are completely correlated, in which case the two solutions are equivalent.]

Problem 4 A simple balance is made of a spring whose elongation is measured on a linear scale. It is calibrated using five standard weights as shown in the following table. Each of the table entries provides a determination of the calibration coefficient. Evaluate the uncertainty of each of these determinations when the relative uncertainty to associate with the reading on the linear scale and that of the standard weights is 1 %. Use the data obtained to assess if the response of the spring can be assumed linear in the given range of elongation.

Weight (gr)	100	300	700	900	1400
Elongation (mm)	10.89	32.76	72.81	91.19	147.65

Problem 5 Repeat the previous problem with the measurements obtained with a spring of better quality and make sure that its behavior can be assumed linear. In this case, provide the best estimate of the calibration coefficient valid for the range explored. [A. 9.95 ± 0.07 mm/gr.)

Weight (gr)	100	300	700	900	1400
Elongation (mm)	10.00	30.21	70.56	90.50	140.39

Problem 6 Two positive currents of intensity, respectively, 199 and 301 mA and a negative current of 496 mA converge in a node of an electrical circuit. Assess whether the Kirchoff's law of currents is being violated bearing in mind that the currents were measured with an amperometer that provides values with a standard relative uncertainty of 1.5 %. [A. Data are compatible with the quoted law.]

Problem 7 A resistor whose nominal value is 15 Ω is measured by the same experimenter with 10 different ohmmeters all providing results with a relative standard uncertainty of 1 % of the reading, all results being listed in the table. Calculate the best estimate of the value of the resistance with its uncertainty. [A. 15.0 Ω with standard uncertainty of 0.03 %.]

Measurement #	1	2	3	4	5	6	7	8	9	10
Resistance (Ω)	14,90	15,01	14,98	14,87	15,04	14,96	15,12	15,07	15,01	15,10

Problem 8 The same resistance of the previous problem is measured by different experimenters using the same ohmmeter providing results with a relative standard uncertainty of 1.5 % of the reading, obtaining the results listed in the table. Calculate the best estimate of the value of the resistance with its uncertainty. [A. 15.0 Ω with standard uncertainty of 1.5 %]

Measurement #	1	2	3	4	5	6	7	8	9	10
Resistance (Ω)	15.01	14.99	14.99	15.01	15.02	14.99	15.00	15.00	14.99	15.02

Problem 9 To check the linearity of an ammeter the following measurements are carried out. The instrument is placed in series with the parallel of 10 variable resistors, see the figure, which are preliminarily inserted individually in the circuit and adjusted to obtain a reading for 1.50 mA on the instrument with a relative standard accuracy equal to 1.5 %. Finally, the 10 resistances are fed simultaneously. How much should the measure differ from the expected value of 15 mA to rule out that the instrument is linear? [R. 0.45 mA.]

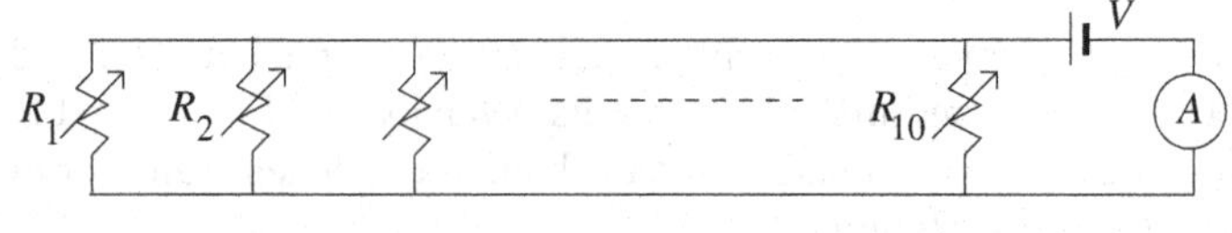

Problem 9

Problem 10 A voltage of 4.7 ± 0.2 V is measured across a resistor when a current of 0.32 ± 0.02 mA flows through it. Determine the value of its resistance with the associated uncertainty. [A. R = 14.7 kΩ with relative uncertainty of 7.5 %.]

Problem 11 The voltage measurement of a battery of nominal value 4.5 V is performed with a digital voltmeter that provides results with a relative uncertainty of one part in a thousand. The measurement must be repeated because of the presence of radio interference and the values in the table are obtained. Give the best estimate of the voltage and of its standard uncertainty. [A. (4.53 ± 0.01) V.]

#	*Voltage (V)*	#	*Voltage (V)*	#	*Voltage (V)*	#	*Voltage (V)*
1	4.53	8	4.58	15	4.61	22	4.57
2	4.45	9	4.43	16	4.47	23	4.57
3	4.45	10	4.51	17	4.57	24	4.50
4	4.57	11	4.61	18	4.53	25	4.50
5	4.56	12	4.55	19	4.59	26	4.49
6	4.42	13	4.42	20	4.61	27	4.55
7	4.55	14	4.52	21	4.50	28	4.57

Problem 12 Given the circuit in the figure, the current flowing in the resistors R_1 and R_2 is measured by inserting in sequence the same ammeter whose impedance can be assumed negligible. The uncertainty of this instrument is 1.5 % and is due to the accuracy of the calibration procedure. The measured current values are $I_{R_1} = 54\,\mu\text{A}$ and $I_{R_2} = 56\,\mu\text{A}$. Evaluates the best estimate of the current flowing in the resistor R_3 and its uncertainty. [A. $I_{R_3} = 110\,\mu\text{A}$ with relative uncertainty of 1.5 %.]

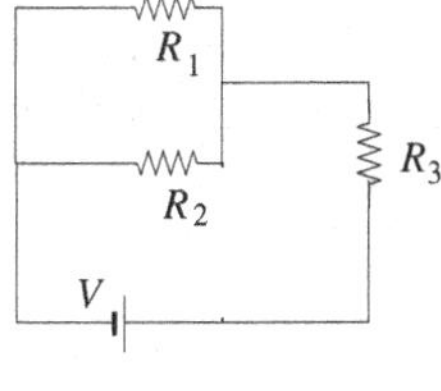

Problem 12

Problem 13 In the circuit shown in the figure, the current flowing in the resistors R_1 and R_2 is measured by inserting in sequence the same ammeter whose impedance can be assumed negligible. The uncertainty of this instrument is 1.5 % and is due to the accuracy of the calibration procedure. The measured current values are $I_{R_1} = 498\,\mu\text{A}$ and $I_{R_2} = 482\,\mu\text{A}$. Knowing that the resistance of R_2 is 998 Ω with an uncertainty of 1.5 % of its value, compute the best estimate of the value of the resistance R_3 and its uncertainty. [A. $R_3 = 30.1\,\text{k}\Omega$ with relative uncertainty 1.5 %.]

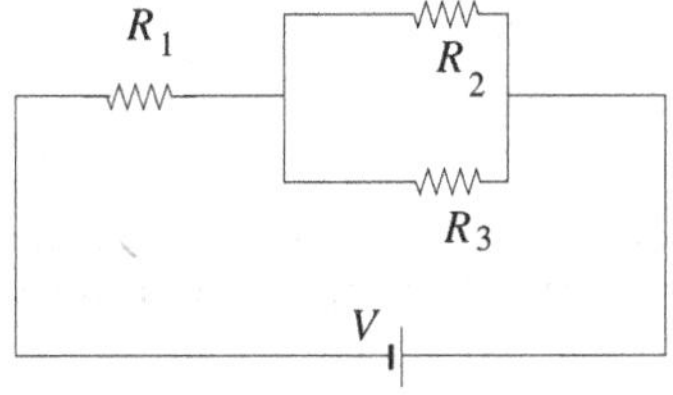

Problem 13

Problem 14 In the circuit shown in the figure, the current flowing in the resistors R_1 and R_2 is measured with two different instruments both with a relative uncertainty of one percent. The measured current values are $I_{R_1} = 498\,\mu\text{A}$ and $I_{R_2} = 482\,\mu\text{A}$. Knowing that the resistance of R_2 is 998 Ω with an uncertainty of 1.5 % of its value, compute the best estimate of the value of the resistance R_3 and its uncertainty. [A. $R_3 = 30.1\,\text{k}\Omega$ with relative uncertainty higher than 100 %.]

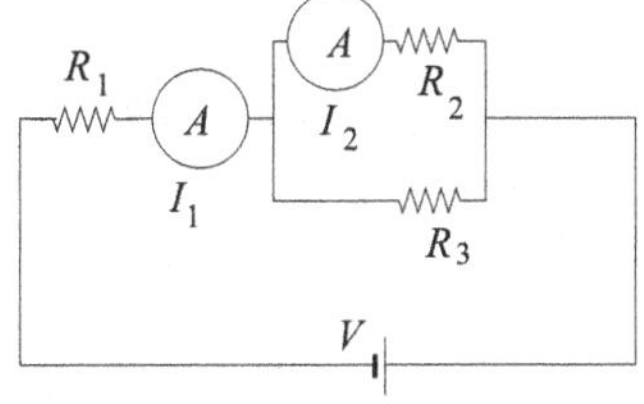

Problem 14

Problem 15 Two resistors in series (see Figure) form a resistive voltage divider. The value of the two resistances was measured with the same ohmmeter and the values obtained were $R_1 = 911\,\text{k}\Omega$ and $R_2 = 1030\,\text{k}\Omega$ both with a standard relative uncertainty of the $3\,\%$. However, the two measurements were carried out without recording the temperature of the resistors. The only information available is that this temperature was certainly in the interval between $T_1 = 15\,°\text{C}$ and $T_2 = 25\,°\text{C}$. Assuming to operate at a temperature intermediate between T_1 and T_2 and knowing that the material of the two resistors has a temperature coefficient $\alpha = 5 \times 10^{-3}\,°\text{C}^{-1}$, evaluate the uncertainty to assign to the resistances during operation and to the partition ratio A of the voltage divider. [A. $u(R_1) = 30\,\text{k}\Omega,\ u(R_2) = 34\,\text{k}\Omega,\ A = 0.5331,\ u(A) = 5 \times 10^{-3}$.]

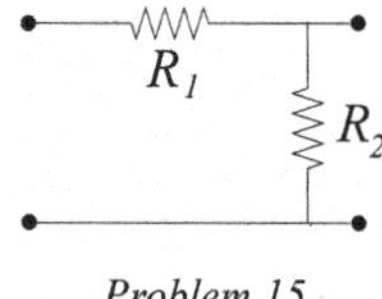

Problem 15

Problem 16 Compute the expression of the combined uncertainty for monomial functions taking into account correlations among independent variables. [A. $[u_c(y)/y]^2 = \sum_{i=1}^{N} [p_i u(x_i)/x_i]^2 + 2 \sum_{i=1}^{N-1} \sum_{j=i+1}^{N} \rho_{ij} p_i p_j (u_i/x_i)(u_j/x_j)$.]

References

1. BIPM, IEC, IFCC, ILAC, ISO, IUPAC, IUPAP and OIML, Evaluation of measurement data - Guide to expression of uncertainty in measurement. JCGM 100:2008 (2008)
2. Les Kirkup, Eur. J. Phys. **23**, 483 (2002)
3. JCGM- Joint Committee for Guide in Metrology. Technical report (2008)
4. J.R. Taylor, An Introduction to Error Analysis. University Science Books - Sausalito California (1996)
5. P.B.D. Keith Robinson, Data Reduction and Error Analysis for the Sciences. McGraw-Hill Education, New York (2002)

Chapter 4
Direct Current Electrical Measurements

4.1 Introduction

The physical quantities of interest in continuous current (DC) circuits are the voltage of nodes and the current flowing through branches. In addition, for the circuit analysis, the knowledge of the resistance value of its components is necessary. The instruments used to measure current intensity are the *ammeters*; those to measure voltage are the *voltmeters* while the *ohmmeters* provide a direct measurement of component resistance. Instruments capable of all these three functions, and many others needed for circuits with time-dependent currents, are known as *multimeters* and are available in all laboratories. These devices can be realized in many different ways and, broadly speaking, are categorized as analog or digital instruments. The first kind provide the value of the physical quantity through an index that can move continuously on a graduated scale, the second kind provide the result of measurements as a number appearing on a digital display.

In turn, analog instruments can be subdivided into electromechanical instruments, wherein all the energy required to move the index (or any other movable unit) is supplied by the circuit under test, and electronic ones, wherein a large part of the needed energy is given by an internal energy source. Digital instruments always fall into this second category.

This chapter is devoted to explain the principles underpinning the operation of instruments (ammeters, voltmeters, and ohmmeters) used in the study of DC circuits. We first recall in the next section some general properties of measuring instruments. Then in Sect. 4.3, we discuss tools and methods for the measurement of DC electrical currents. There we describe in full detail the functioning of the moving coil ammeter with the purpose of giving the reader an example of the depth of analysis required for a proper design of a measuring instrument. In Sect. 4.4, we illustrate the use of the analog voltmeter, with special attention to the perturbation caused on the measured voltage by the insertion of the instrument in the circuit under test. The section continues with a description of the scheme of principle of a digital voltmeter

R. Bartiromo and M. De Vincenzi, *Electrical Measurements in the Laboratory Practice*, Undergraduate Lecture Notes in Physics,
DOI 10.1007/978-3-319-31102-9_4

and focusses on the evaluation of uncertainties arising from the output quantization. Finally, Sect. 4.5 is devoted to resistance measurements and we describe there both the implementation of voltmeter–ammeter method and the properties of the analog ohmmeter.

4.2 Properties of Measuring Instruments

Measuring instruments[1] consist of devices that transform the quantity to be measured into another quantity that is easily measurable, such as for example the position of an index that can move on a graduated ruler. The instruments used in electrical measurements are, in general, tools that need a calibration. This means that their reading scale must be defined by comparing their measurement results with absolute quantities or with results of instruments already calibrated. The calibration procedure should be repeated with a frequency defined by the manufacturer, so that the instrument maintains the accuracy stated in the specifications.

The properties of interest in a measuring instrument can be numerous and dependent on its particular use. However, some features are of a general nature, and valid for all instruments. Among them, we have

- **Sensitivity**. Sensitivity is defined as the ratio between the variation of the response and the variation of the solicitation. If G is the quantity under test (*the measurand*), a voltage for example, and R is the response of the instrument, for example the angular deviation of the index in an analog voltmeter, the sensitivity S of the instrument is

$$S = \frac{dR}{dG} \tag{4.1}$$

  Generally speaking, the sensitivity in not constant across the whole working range but can depend on the value of G. We will see an example of this kind of behavior when discussing the analog ohmmeter in the last section of this chapter.
- **Accuracy**. The accuracy of an instrument consists in its ability to provide for the physical quantity under test a value as close as possible to its *true value*.[2]
- **Precision**. The precision is the capability of an instrument to reproduce the same result when used in the same experimental conditions. The precision is a qualitative concept. In the most recent publications of the Institute for Standardization (ISO), the term precision is replaced by the terms "repeatability" and "reproducibility". More information is given in the original literature [1] on this subject.

---

[1] Here instrument also stands for a complex measuring system.

[2] Here the adjective *true* added to the noun *value* has to be intended as an intensifier; instead of *true value* it could be said: measurement taken with an instrument more accurate than the one in use or a reference measure (when available).

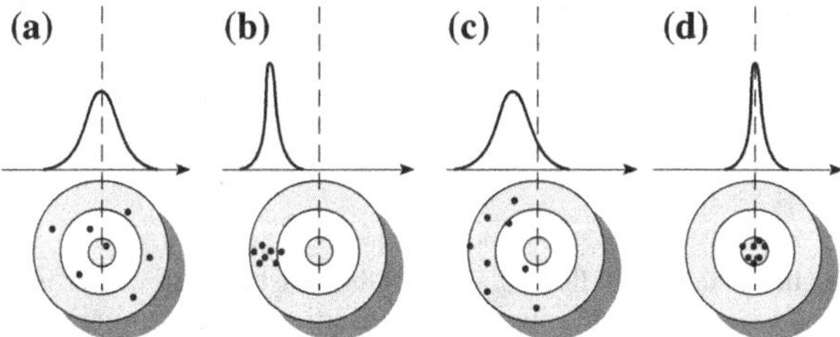

Fig. 4.1 Analogy between various distributions of shots on target and properties of accuracy and precision of measurements. *A* measures accurate but not precise, *B* measures precise but not accurate, *C* measures nor accurate nor precise, and *D* precise and accurate measurements. On *top* of each target, we show the parent distribution of the horizontal position of the hits

- **Working range**. The working range is the interval of measurand values for which the instrument is able to perform the measurement. The maximum value is called *full-scale* while the minimum is called *threshold.*
- **Promptness**. The promptness is linked to the response time of the instrument to the input quantity. The lower this time, the higher the promptness of the instrument. The promptness remains a qualitative concept unless it refers to a mathematical model describing the instrumental response as a function of time. For example, the characteristic time of a mercury thermometer is a quantitative definition of its promptness and makes it possible the comparison of different instruments based on the same physical principle.

A note on accuracy and precision. In the terminology used in metrology, *accuracy* and *precision* describe two properties of a measurement completely separate and independent: an instrument can be accurate but not precise, and vice versa. To illustrate the difference between precision and accuracy is helpful to use an analogy with the sport of shooting by interpreting the center of the target as the "true value" of a measurand and the different trials as the measurements. The Fig. 4.1 shows four different distributions of shots (A, B, C, and D) on the target that can be commented as follows: (A) averaging the positions of the different shots we come very near to the center, on the other hand the individual shots are rather dispersed. We can conclude that in this case, the measures are accurate but not precise, (B) the measures are precise but not accurate, (C) the measures are neither accurate nor precise, and (D) the measures are accurate and precise.

In conclusions, the accuracy is related to the proximity of the measurement to the true value while the precision is linked to the dispersion of the measurements. From this discussion, it is obvious that a "good" instrument must be both accurate and precise.

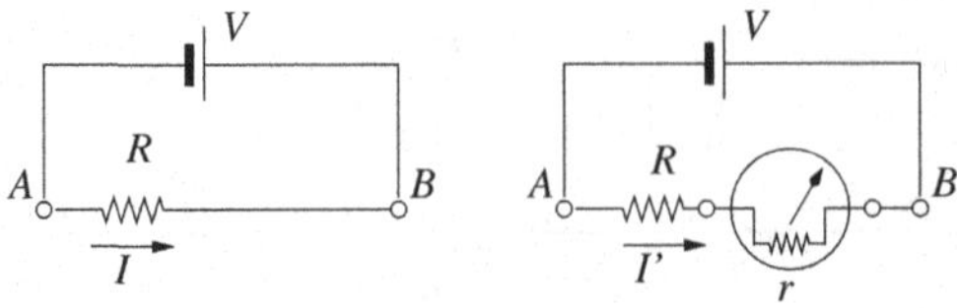

Fig. 4.2 Measurement of the current flowing through the resistance R by inserting the ammeter in series; r is the internal resistance of the ammeter. The insertion of the instrument perturbs the value of the quantity to be measured

4.3 Electrical Current Measurements

The instrument used to measure an electrical current is the *ammeter*, which allows reading, directly on a graduated ruler or through a digital display, the intensity of the current flowing through it. To produce accurate results, the ammeter needs a careful calibration usually performed by the manufacturer using precise and stable current sources of known intensity.

How to use an ammeter. When we need to measure the current flowing through a given conductor (or a series of conductors), we must open the original circuit and insert the ammeter as shown in Fig. 4.2. In other words, the ammeter must be inserted in series to the conductor through which flows the current we want to measure. The insertion of the ammeter *modifies the original circuit* because of its internal resistance r and modifies the current under measurement. Suppose we have to measure the intensity of the electrical current flowing through the resistance R in the circuit of Fig. 4.2. If V is the voltage difference between the two points A and B, then the current through R has an intensity $I = V/R$. Once the ammeter, with its internal resistance r has been inserted, the current intensity becomes:

$$I' = \frac{V}{R+r} = \frac{V}{R}\left(\frac{1}{1+\frac{r}{R}}\right)$$

The term in brackets in the previous expression quantifies the *systematic effect* due to the insertion of the ammeter in the circuit. Obviously, only when $r = 0$ the current flowing in the perturbed circuit is equal to the one we are trying to measure. In practice, however, we can neglect the effect of the presence of the instrument in the circuit only when $r \ll R$. Using of Thévenin's theorem, we can reduce any kind of linear circuit to the scheme of Fig. 4.2. In this case, R is equal to the resistance of the circuit as seen from the two nodes connected by the branch through which flows the current we are measuring. These observations help us to understand that a merit of an ammeter is to have an internal resistance as small as possible.

Shunt Resistance. Like all measuring instruments, ammeters have a *full-scale* limit, i.e., a maximum current intensity that can be measured properly. With a simple trick, we can increase the value of the full-scale limit of an ammeter adding a resistance

Fig. 4.3 Shunt resistance

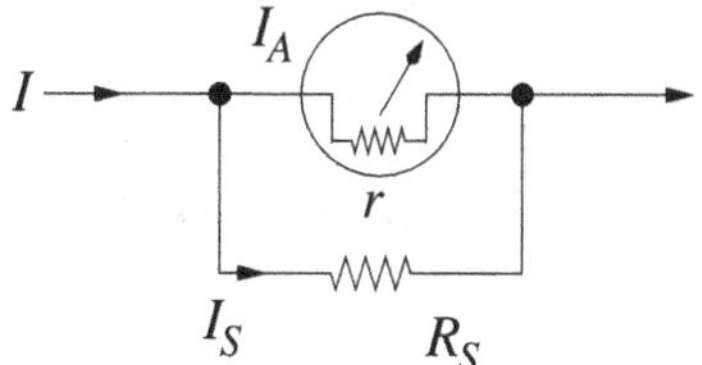

R_s in parallel to it, as shown in the Fig. 4.3. In this circumstance, the resistance R_s is referred to as the *shunt resistance*. Let us denote with I the current we want to measure and with r the internal resistance of the ammeter; then using the two Kirchhoff's laws, we can write

$$I = I_A + I_s \qquad \text{and} \qquad R_s I_s = r I_A$$

from which we can obtain

$$I_A = \frac{R_s}{R_s + r} I = \frac{1}{1 + \dfrac{r}{R_s}} I$$

This relation gives the current I_A flowing through the ammeter as a function of the shunt resistance. Now it is easy to see that to increase by a factor k the full-scale of an ammeter, we need a shunt resistance $R_s = r/(k-1)$. In conclusion,

to increase by a factor k the full-scale of an ammeter it is sufficient to connect it in parallel to a shunt resistance equal to a fraction $1/(k-1)$ *of its internal resistance* r.

As already mentioned above, there exist various types of devices capable of measuring a current intensity; we will discuss in the following two of them, namely:

- *the moving coil ammeter*, an analog electromechanical instrument exploiting the mechanical interaction between a current passing through a coil and the magnetic field of a permanent magnet.
- *the digital ammeter*, a digital instrument that measures the voltage across a known resistance. Application of the first Ohm's law yields the current intensity.

The use of analog devices is becoming less common as these instruments are supplanted by digital devices, which are more versatile and economical. Nevertheless, we will give some details of the physics of the analog ammeter to show the deepness of analysis necessary for the proper design of a measuring device.

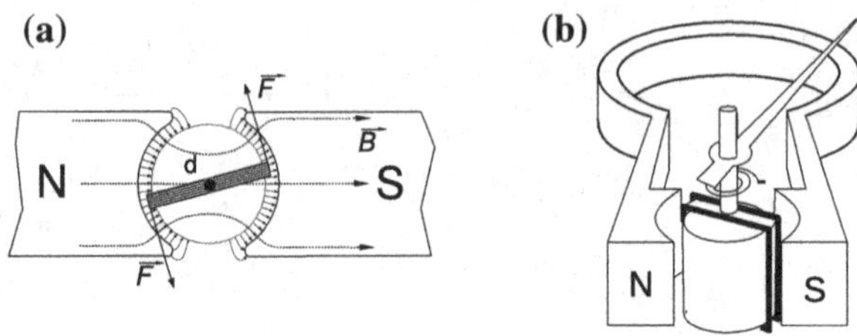

Fig. 4.4 Schematization of the moving coil ammeter: **a** section of the air gap of the permanent magnet with the soft iron core around which the coil can rotate; note that in this gap the B field is radial and of constant amplitude. Therefore, the torque on the coil does not depend on θ. In **b**, we show a perspective view of the instrument

4.3.1 Moving Coil Ammeter

This instrument, also known as D'Arsonval's ammeter from the name of its inventor,[3] consists of a coil immersed in a radial magnetic field generated by a permanent magnet of a suitable shape and a soft iron cylinder placed in the air gap of this magnet. In the space between the magnet and the cylinder, the magnetic induction field **B** is radial in direction (relative to the center of the cylinder) and constant in module. The coil can rotate around the axis of the soft iron cylinder, perpendicular to the direction of the magnetic field, see Fig. 4.4a. The elastic force generated by a spiral spring maintains the coil in a stable rest position, as shown in Fig. 4.4b. The coil supports an index to measure its angular position θ on a linear scale.

When an electric current of intensity I flows in the coil (consisting of n turns of area $S = hd$), it generates a mechanical torque, directed along the rotation axis and *not dependent on* θ, whose module is given by:

$$\mathcal{M} = n\,hd\,B\,I = \Phi^* I \tag{4.2}$$

where we used[4] $\Phi^* = nhdB$. The previous equation shows that *the mechanical torque generated by the current flowing through the coil is directly proportional to its intensity*. This torque causes a rotation of the coil until a new equilibrium position is reached where the elastic torque due to the spring balances it. Since the elastic torque of spiral spring is given by $-C\theta$, where θ is the rotation angle and C a constant, the new equilibrium position is given by:

$$\theta_0 = \frac{\Phi^*}{C} I \tag{4.3}$$

[3]This instrument is also known under the name of D'Arsonval's galvanometer. The name galvanometer, until the early years of last century, was used to indicate the most sensitive instruments for measuring electric currents.

[4]Φ^* is the flux linked with the n turns of the coil when immersed in a uniform field of magnetic induction B perpendicular to its surface.

This proves that the angular position of the moving coil at steady state is proportional to the current flowing through it. Therefore, *the value of the current I can be obtained by the index position after calibrating the scale directly in units of electrical current intensity*.

Equation of motion of the moving coil. For the proper design of the instrument, the mere knowledge of the steady state is not sufficient. It is also very important to know how the equilibrium is attained. Considering the moving coil as a rigid body, its equation of motion is

$$\mathscr{I}\ddot{\theta} = \mathscr{M}_{ext}$$

where $\mathscr{I}$ is the inertia moment of the moving coil with respect to its rotation axis, θ is the rotation angle from the rest position, and $\mathscr{M}_{ext}$ is the module of the torque due to the forces acting on the coil. $\mathscr{M}_{ext}$ is the sum of three terms:

- the torque, due to the interaction of the coil with the magnet, whose module is given by Eq. (4.2) and will be referred to as $\Phi^* I$,
- the torque of the elastic force of the spiral spring that, as we said before, is given by $-C\theta$ and tends to return the moving coil in the rest position, and
- the torque of a viscous force $-\beta\dot{\theta}$, proportional to the angular rotation speed $\dot{\theta}$. This force is caused only for a small fraction by the air friction but is rather mainly due to the effect of magnetic induction (Foucault's current).

Let us consider in more details the origin of this component of the viscous force. We choose to measure the angular position θ of the moving coil starting from the position where its plane is parallel to the direction of the magnetic induction field **B** in the soft iron core, see Fig. 4.4a. The magnetic flux ϕ linked to the coil is a function of θ and, recalling that the **B** field is divergence-less, we can use the Gauss theorem and show that ϕ is equal to the magnetic flux linked to the lateral surface of the half cylinder subtended by the coil in the air gap. In this gap, **B** is purely radial and with simple geometrical arguments, we can show that $\phi = nhdB\theta = \Phi^*\theta$.

This result implies that, during the motion of the coil, the magnetic flux linked to it changes and gives origin to an electromotive force $f = -d\phi/dt = -\Phi^* d\theta/dt = -\Phi^*\dot{\theta}$. Using the superposition theorem, we can see that, in these conditions, the current flowing through the coil is $I + f/R$ where R is the resistance of the circuit under test as seen by the ammeter and I is the current intensity we want to measure. Putting this new current intensity in the expression (4.2) for the mechanical torque, we obtain $\mathscr{M} = \Phi^* I - \Phi^{*2}\dot{\theta}/R$, which leads us to the expression for the electromagnetic friction coefficient $\beta = \Phi^{*2}/R$. We remark that this value depends not only on the instrument parameters but also on the resistance of the circuit under test.

In conclusion, the equation describing the motion of the coil is

$$\mathscr{I}\ddot{\theta} + \beta\dot{\theta} + C\theta = \Phi^* I \tag{4.4}$$

In the next section, we will illustrate in detail the solutions of this equation. However, it is immediate to deduce again its steady-state solution when the current I does not depend on time, neglecting all time derivatives we obtain:

$$\theta_0 = \frac{\Phi^*}{C} I \tag{4.5}$$

Ammeter sensitivity. Equation (4.5) allows obtaining the ammeter sensitivity, as defined in Sect. 4.2; we identify the response of the instrument with the index angular position θ and we get the ammeter sensitivity value as

$$S_{amp} = \frac{\partial \theta}{\partial I} = \frac{\Phi^*}{C} \tag{4.6}$$

4.3.2 *Considerations on the Motion of the Moving Coil*

Coming back to Eq. (4.4), with a constant current we can assume without loss of generality that the instrument is connected to the circuit at the time $t = 0$. Consequently, the current in the ammeter coil is given by $I = 0$ when $t < 0$ and $I =$ constant for $t > 0$.[5] The general integral of Eq. (4.4) is:

$$\theta(t) = A_1 e^{m_1 t} + A_2 e^{m_2 t} + \frac{\Phi^* I}{C} \tag{4.7}$$

where the constants A_1 and A_2 depend on the initial conditions and will be computed in the following. m_1 and m_2 are the solutions of the characteristic equation $m^2 + \frac{\beta}{\mathcal{I}} m + \frac{C}{\mathcal{I}} = 0$ and are given by:

$$m_{1,2} = -\frac{\beta}{2\mathcal{I}} \pm \sqrt{\frac{\beta^2}{4\mathcal{I}^2} - \frac{C}{\mathcal{I}}} \tag{4.8}$$

The dynamical properties of the motion described by Eq. (4.7) depend upon the sign of its discriminant $\Delta = (\beta/2\mathcal{I})^2 - C/\mathcal{I}$. The following cases are possible:

- $\Delta < 0$. *The motion is a damped oscillation.* In this case, the solutions of the characteristic equation m_1 and m_2 are two *complex conjugate numbers* yielding a solution representing a damped sinusoidal motion, see Fig. 4.5 (plot *a*).
- $\Delta > 0$. *The motion is aperiodic.* Indeed, in this case m_1 and m_2 are *real and negative* numbers and the motion is approximatively exponential, see Fig. 4.5 (plot *c*). The time needed for $\theta(t)$ to reach its equilibrium position θ_0 will depend upon the value of $\beta^2 - 4\mathcal{I}C$; the higher its value, the longer the coil will take to reach its equilibrium position.

[5]The fact that the current changes from zero to a finite value in zero time is an approximation. In fact, the inductance of the coil of the ammeter does not allow sharp variations of the current that flows through it. If L_B is the inductance of the coil and R is the value of the resistance "seen" by the ammeter, then $\tau = L_B/R$ is the order of magnitude of the time needed for the current to reach the stationary value. This result will be deduced in Sect. 9.7.

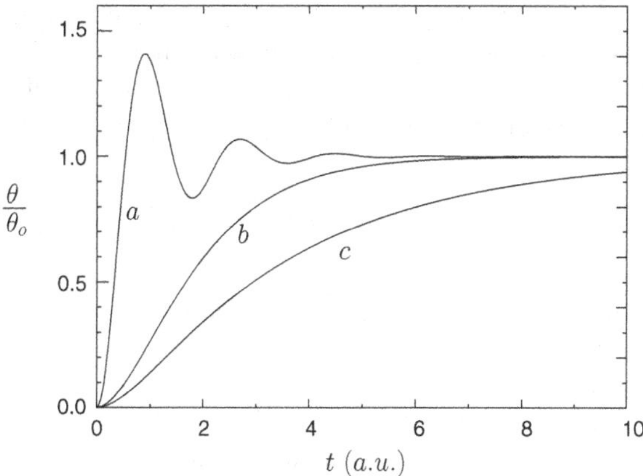

Fig. 4.5 Motion of the moving coil: **a** damped oscillation, **b** critically damped and **c** aperiodic

- $\Delta = 0$. *The motion is critically damped.* In this case $\beta = \sqrt{4\mathscr{I}C}$ and the two roots of the associate equation are *real negative and coincident*. The resulting motion is intermediate between a damped oscillation and an aperiodic motion, see Fig. 4.5 (plot b). For practical reasons, as we shall see, this is the regime of operation chosen by manufacturers for moving coil ammeters.

Solution of the differential equation of motion. Detailed calculations. In this section, we elaborate the general solution (4.7) of the differential equation (4.4) to obtain the details of the different kinds of motion for the different values of the discriminant Δ.

$\boldsymbol{\Delta} < 0$: **Damped oscillation.** To ease the notation, we use in the expression (4.7) the following positions:

$$a = \frac{\beta}{2\mathscr{I}}, \quad b = \sqrt{\frac{C}{\mathscr{I}} - \frac{\beta^2}{4\mathscr{I}^2}} \tag{4.9}$$

and we obtain for the equation of motion:

$$\theta(t) = \mathrm{e}^{-at}\left[A_1\mathrm{e}^{jbt} + A_2\mathrm{e}^{-jbt}\right] + \theta_0 \tag{4.10}$$

Since θ is a real number, the two constants A_1 and A_2 must be complex conjugate numbers: $A_1 = A' + jA''$, $A_2 = A' - jA''$ (where A' and A'' are real numbers). With a simple mathematical manipulation[6] we get the following expression for the function $\theta(t)$:

$$\theta(t) = A\mathrm{e}^{-at}\sin(bt + \alpha) + \theta_0 \tag{4.11}$$

[6]Besides the Euler's formula, we need the following trigonometric relation:

$$A_1 \cos x - A_2 \sin x = A \sin(x + \phi) \quad \text{with } A = \sqrt{A_1^2 + A_2^2},\ \phi = -\arctan(A_2/A_1).$$

To obtain the values of α and θ_0, we have to impose the initial conditions. We assume that, at the time $t = 0$ when we insert the ammeter in the circuit, the moving coil is at rest in the position corresponding to the zero of the scale so that $\theta(0) = 0$ and $\dot{\theta}(0) = 0$. From expression (4.11), we get

$$0 = A\sin\alpha + \theta_0$$

and

$$0 = -Aa\sin\alpha + Ab\cos\alpha$$

that yield

$$\alpha = \arctan\frac{b}{a}; \quad \text{and} \quad A = -\theta_0\sqrt{\frac{a^2+b^2}{b^2}}$$

and finally:

$$\theta(t) = \theta_0\left[1 - \sqrt{\frac{a^2+b^2}{b^2}}\mathrm{e}^{-at}\sin(bt+\alpha)\right] \tag{4.12}$$

The motion represented by this equation consists of an oscillation around the equilibrium position θ_0 with a (pseudo) period $T = 2\pi/b$ and an amplitude decreasing exponentially with a characteristic time $1/a$ (see Fig. 4.5a). Because of this oscillation, the index excursion can be disruptive; therefore, manufacturers tend to avoid operation in these conditions.

$\boldsymbol{\Delta} > 0$: **Aperiodic motion**. When $\Delta > 0$, it is easy to see that the two solutions m_1 and m_2 yielded by relation (4.8) are real and negative. Consequently, the motion of the coil is represented by the sum of two exponential functions and $\theta(t)$ approaches the equilibrium position θ_0 with a time-constant longer the higher the value of Δ. When this time-constant increases, the speed with which the coil approaches the equilibrium position decreases and the residual mechanical friction can alter the final position of the coil thereby affecting the result of the measurement. For this reason, the manufacturers tend to avoid operation in these conditions.

$\boldsymbol{\Delta} = 0$: **Critically damped motion**. When $\Delta = 0$, the solution of differential equation (4.4) becomes

$$\theta(t) = \theta_0 + \mathrm{e}^{-at}(A_1 + A_2 t) \tag{4.13}$$

and with the usual initial conditions: ($\theta(0) = 0, \dot{\theta}(0) = 0$), we get

$$A_1 = -\theta_0; \qquad A_2 = -a\theta_0$$

and finally:

$$\theta = \theta_0\left[1 - (1+at)\exp(-at)\right] \tag{4.14}$$

When we disconnect the ammeter from the circuit ($\theta = \theta_0,\ \dot{\theta} = 0$), the return of moving coil to the initial position (the zero of the scale) is described by

$$\theta = \theta_0 (1 + at) \exp(-at) \tag{4.15}$$

The constant a, whose expression is given by position (4.9) as $a = \beta/2\mathscr{I} = \sqrt{C/\mathscr{I}}$ (the last equality deriving from the condition $\Delta = 0$), has the dimensions of the inverse of a time[7] and is a measure of the *promptness* of the ammeter. In the following, we give a quantitative estimate of a.

We assume typical values for the instrument parameters and namely:

- Number of turns in the coil $n = 100$.
- Magnetic induction field in the permanent magnet $B = 0.2\,\text{T}$.
- Geometrical dimensions of the coil $d = 2 \times 10^{-2}\,\text{m}$, $h = 1.5 \times 10^{-2}\,\text{m}$.
- Intensity of the current flowing through the coil $I = 40\,\mu\text{A}$

We begin by computing the inertia moment, $\mathscr{I}_b$, of a coil made with a copper wire (density $\rho_{Cu} = 8.96\,\text{g/cm}^3$) of diameter $2r = 0.1\,\text{mm}$, neglecting the mass of its support frame:

$$\frac{1}{2}\rho_{Cu} n \pi r^2 d^2 \left(h + \frac{d}{3}\right) \simeq 3.1 \times 10^{-8}\,\text{kg m}^2$$

On top of this, we need to consider the important contribution of the index, used to read the angular position of the coil, whose length is about 10 cm and whose mass we assume of the order of 0.1 g:

$$\mathscr{I} = (0.1 \times 10^{-3})\frac{(0.1)^2}{3} = 2.7 \times 10^{-7}\,\text{kg m}^2$$

For the electromagnetic torque acting on the coil, we get:

$$nB(hd)I = 100 \times 0.2 \times 2 \times 10^{-2} \times 1.5 \times 10^{-2} \times 40 \times 10^{-6} = 2.4 \times 10^{-7}\,\text{N m}$$

When conditions for *critically damped motion* are satisfied, from the relation (4.9), we obtain that $a = \sqrt{C/\mathscr{I}}$. The spring constant C can be obtained assuming that for an electrical current in the coil of 40 μA the index rotates of an angle equal to $\pi/2$:

$$a = \sqrt{\frac{C}{\mathscr{I}}} = \sqrt{\frac{nB(hd)I}{\mathscr{I}\pi/2}} = 0.75\,\text{s}^{-1}$$

The inverse of a is equal to 1.3 s and, as said in the previous section, represents the order of magnitude of the time needed for the instrument to respond to the input (promptness of the ammeter). With this promptness, the ammeter will take

[7]It would be inappropriate to refer to the quantity $1/a$ as the characteristic time of the moving coil ammeter in conditions of critical damping since the Eqs. (4.14) and (4.15) do not have an exponential form because of the presence of a term linear in t.

a few seconds to reach its steady state. On one hand this is a limitation, since the instrument cannot detect a current changing with a frequency greater than a; but on the other hand, it can turn out to be an advantage in case high frequency *electrical noise* interferes with the measure but is not detected by the ammeter because of insufficient promptness.

4.4 Measurement of Voltage Difference: Analog Voltmeters

The instruments used to measure a voltage difference are called *voltmeters*. An analog voltmeter allows obtaining the value of a voltage difference by means of the measurement of a current intensity and the application of the first Ohm's law. In other words, *the analog voltmeter is nothing else but an ammeter used in an appropriate fashion*. Suppose we need to measure the voltage drop between the points A and B of a circuit, as schematically shown in Fig. 4.6. To this purpose, we connect between the two points an ammeter in series with a suitable resistance R'. With reference to Fig. 4.6, we can write:

$$V_A - V_B = (R' + r)I_A = r_V I_A$$

where r is the ammeter internal resistance and I_A the intensity of the current flowing through it. Once we calibrate appropriately the scale of the ammeter in *volt*, we can read directly the voltage difference between the points A and B. The instrument consisting of the ammeter with in series the resistance R' is an analog *voltmeter*. The quantity $r_V = R' + r$ is the *internal resistance* of this voltmeter. It is an important parameter to evaluate the perturbation of the circuit under test due to the connection with the voltmeter.

The perturbation introduced by the voltmeter. For a linear circuit, we can evaluate the effects due to the connection with the voltmeter by making use of Thévenin's theorem. It states that any linear network is equivalent to the circuit of Fig. 4.7, where $V_0 = V_A - V_B$, the open circuit voltage, coincides with the voltage difference we need to measure. The insertion of the voltmeter amounts to connecting nodes A and B with the voltmeter internal resistance r_V. Consequently, the current flowing in r_V is given by $I_V = V_0/(R_{eq} + r_V)$. The voltage drop across r_V, i.e., the voltage difference measured by the voltmeter, is given by:

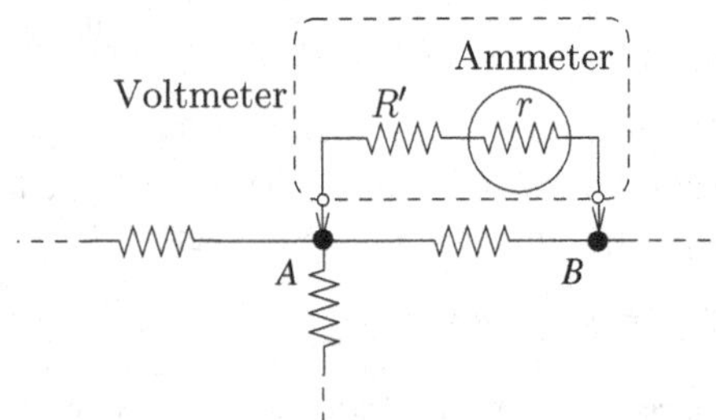

Fig. 4.6 Measure of the voltage with a voltmeter. The voltmeter is shown schematically by the components contained in the *dashed path*, namely the ammeter A in series with the resistance R'

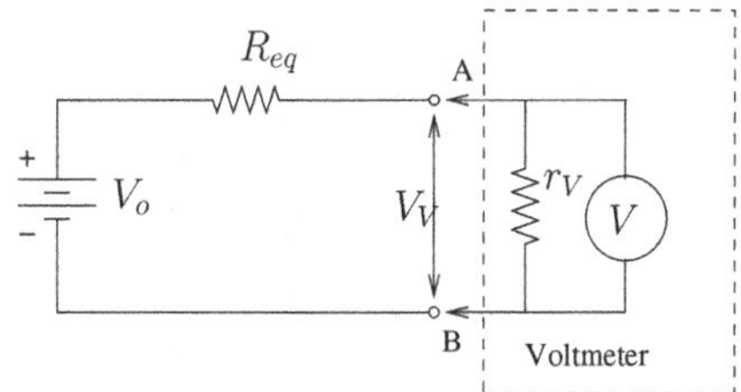

Fig. 4.7 Voltmeter connected to the Thévenin's equivalent of the circuit under test. This scheme allows computing the perturbation to the circuit introduced by the voltmeter internal resistance r_V

$$V_V = I_V r_V = \frac{r_V}{R_{eq} + r_V} V_0 \tag{4.16}$$

This result shows that the connection of a voltmeter to a circuit causes a perturbation of the voltage difference to be measured whose magnitude is quantified by relation (4.16). Note that the higher the internal resistance of the instrument, the lower the perturbation it causes. An internal resistance as high as possible is therefore a factor of merit of a voltmeters. To correct the *systematic effect* introduced by the voltmeter, we need to know the value of both r_V and R_{eq} and use them in Eq. (4.16) to obtain the unperturbed voltage difference V_0 as:

$$V_0 = \frac{R_{eq} + r_V}{r_V} V_V = \left(1 + \frac{R_{eq}}{r_V}\right) V_V$$

This shows that the value of the systematic effect is determined by the ratio of the equivalent circuit resistance to the voltmeter internal resistance.

Voltmeter "sensitivity". It is customary for manufacturers of analog voltmeters to quote the value of the internal resistance by means of a parameter, improperly called "sensitivity" of the voltmeter. This parameter, quoted in units of Ω/V, yields the internal resistance when multiplied by the value of the full-scale in use. For example, if a voltmeter with sensitivity equal to 20 000 Ω/V is used with a full-scale value of 5 V, its internal resistance is equal to $20000 \times 5 = 100\,\text{k}\Omega$. Usually, the sensitivity is printed on the scale of the voltmeter as in Fig. 4.14 that shows the scale of an analog multimeter: in the corner at the bottom left two values of the "sensitivity" are printed: 20 000 Ω/V for DC operation and 4000 Ω/V for measurements with sinusoidal currents (AC operation).

4.4.1 Digital Voltmeters

The development of integrated electronics has led to the gradual replacement of analog instrumentation, mostly based on electromechanical sensors, with fully electronic instruments that replace the continuous graduated ruler with a digital device providing directly a reading of the measured quantity for discrete numerical values. With reference to electrical measurements, the pillar of digital instrumentation is

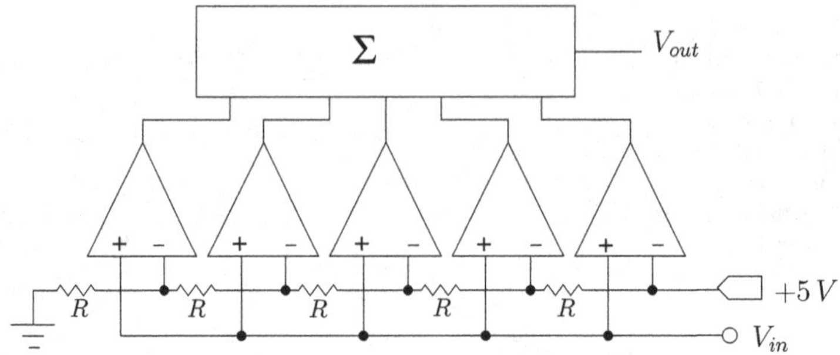

Fig. 4.8 Block diagram of an ADC (FLASH). In the diagram, the comparators are represented with *triangles* and the circuit that adds inputs is indicated with Σ

Table 4.1 Analog to digital conversion

Input voltage (V)	Output voltage (V)	Average value of V_{in} (V)
$0 < V_{in} < 1$	$V_u = 0$	$\langle V_{in} \rangle = 0.5$
$1 < V_{in} \leq 2$	$V_u = 1$	$\langle V_{in} \rangle = 1.5$
$2 < V_{in} \leq 3$	$V_u = 2$	$\langle V_{in} \rangle = 2.5$
$3 < V_{in} \leq 4$	$V_u = 3$	$\langle V_{in} \rangle = 3.5$
$4 < V_{in} \leq 5$	$V_u = 4$	$\langle V_{in} \rangle = 4.5$
$5 < V_{in}$	$V_u = 5$	$\langle V_{in} \rangle = ??$

the voltage comparator. It is a nonlinear circuit[8] with two inputs (marked with $+$ and $-$) and one output whose behavior is the following: as long as the voltage V_+ of the positive input is lower than the voltage V_- of the negative input, the output voltage is equal to a low value, nominally zero. On the contrary, the output voltage rises to a positive and constant value, usually a few volts, when V_+ becomes higher than V_-. We now introduce a simple scheme of principle to illustrate the characteristics of a voltmeter capable to read only a discrete number of values. As shown in Fig. 4.8, we assume to have five voltage comparators with an upper output level of 1 V. We connect in parallel their positive inputs to the voltage to be measured while the negative inputs are separately connected to the sequential outputs of a voltage divider evenly spaced 1 V apart from each other. The comparators' outputs are used as inputs to a 5-way circuit able to count the number of positive inputs and show it on a display.[9]

The behavior of this device for positive unipolar inputs is described in the Table 4.1 We can better understand the information obtained when measuring an unknown

[8]In this section, we limit ourselves to the functional description of the electronic devices mentioned, leaving aside any discussion on their practical implementation that would require concepts and notions learned later in the course of study.

[9]Alternatively, we could consider an analog version of this circuit obtaining an output voltage proportional to the sum of the input voltages (see Problem 16 at the end of this chapter for a practical embodiment of such a circuit). An instrument whose accuracy is just enough to distinguish the number of positive outputs may then read this voltage.

voltage value with such an instrument using the following observations. Assume to obtain a given output value, for example the value 2. Any input signal of amplitude comprised between 2 and 3 can generate such an output *with equal probability*. For this kind of probability distribution, we can compute an expected value of 2.5 V with a standard deviation of $1/\sqrt{12}$ V. Therefore, there is an *offset* between the instrument reading and the expected value of the input signal that in this simple example amounts to 0.5 V for all measured values. This is obviously a systematic uncertainty that in principle we can easily correct. Moreover, in practice, this offset is often, but not always, sufficiently small to have a negligible effect on the measurement result. We can conclude that the principal consequence of the output quantization is the random statistical uncertainty due to the output level separation. To generalize this result, denoting with Δ this separation, the *offset* is equal to $\Delta/2$ while for the standard uncertainty we get $\Delta/\sqrt{12}$.

However, a closer look at Fig. 4.8 shows that there are two other potentially important sources of uncertainty. First, since the resistances of the divider are never equal, the separation between the output levels cannot be strictly constant. Consequently, the offset now acquires a random component and the correction to the expected output depends in an unknown manner upon the instrument reading. Fortunately, in a device of good quality this fluctuation can be maintained at a small value compared to the average width Δ of the channels.

Second, and far more important, the uncertainty on the value of the divider supply voltage V affects directly the value of all the levels. In other words, if we know the value of V with 1 % relative uncertainty, all output levels will have the same value of relative uncertainty. It is important to note that this component of uncertainty is a feature of the instrument in use and that all measurements taken with that particular instrument will be affected by a correlated component of uncertainty. In contrast, the uncertainties due to the channel amplitude Δ turn out, by their nature, to be independent from each other.

The previous discussion has been summarized graphically in Fig. 4.9, which shows on the abscissa the value of the input voltage and on the ordinate the value read by the measuring device. The continuous line represents the ideal case when all output levels have the same width and their lower edges are perfectly aligned on the first bisectrix (dotted line). The fluctuation of the partition ratio causes the variation of the amplitude of the channel with deviations from the first bisectrix. Taking account the uncertainty on the divider supply voltage, the average deviation of the lower edges of the channels from the first bisectrix is not zero (dashed straight line in the Fig. 4.9).

The scheme of principle hitherto used, though useful to illustrate the salient features of a digital voltmeter, cannot generate the large number of levels needed for an accurate measurement. Such a device can be obtained only by changing from a parallel to a serial architecture, that is, by adopting a scheme in which all levels of output are generated by the same comparator. To this purpose, we must make full use of the resources of digital electronics.

Many digital devices are based on the ability of binary logic circuits to count voltage pulses of standard form and to show the result as a decimal number on a display. To convert a voltage level in a number proportional to its amplitude, one can

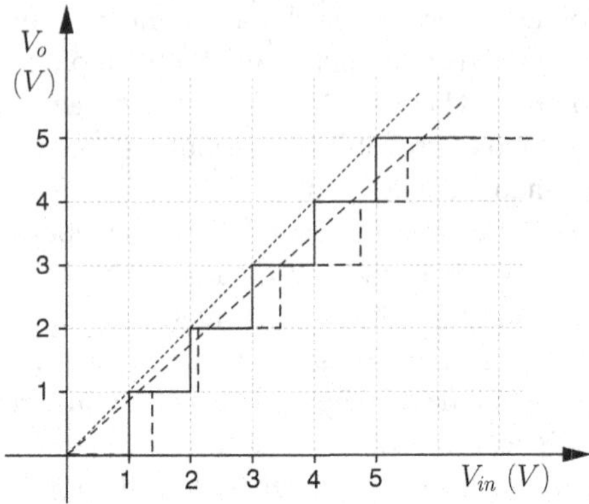

Fig. 4.9 Transfer function (digital) of an ADC, ideal (*continuous line*), and real (*dashed line*)

use the device whose layout is shown in Fig. 4.10. It consists of a binary counter, which can be periodically reset, counting the pulses generated by a precision clock via a two inputs logic circuit, referred to as AND, which allows the passage of pulses only when both its inputs are at a high voltage level.[10] While the first input of the AND circuit is connected to the clock, the second input is connected to the output of a voltage comparator similar to the one described above. The positive input of this comparator connects to the source of the voltage to be measured while the negative input connects to the output of a digital to analog converter (DAC). This last device takes as an input the digital output of the binary counter and returns at the output a voltage level proportional to the number represented.[11] As long as the output of the DAC is lower than the voltage to measure, the output of the comparator is high thereby enabling the passage through the AND circuit of pulses from the clock to the binary counter that increments its digital output. As soon as this level is sufficiently high to make the DAC output higher than the voltage under test, the counter stops and the display shows the result of the measurement until the counter is reset.

A detailed analysis of this kind of voltmeter shows that the uncertainty of its measurements can be characterized in a fashion very similar to the scheme of principle previously discussed. This is obvious for the random component caused by the finite-level separation. The output of the DAC circuit is affected by the uncertainty of the conversion constant, similar to that caused in our scheme of principle by the divider voltage supply, and by a fluctuation of the elementary voltage step similar to that caused by differences in the resistor value in that divider.

The device we just discussed belongs to the category of analog-to-digital converters, often referred to with the acronym ADC. They are classified according to the number n of bits used for the digital conversion and to the value of the full-scale volt-

[10]In digital electronics, we always need to use two voltage levels, a "low" one corresponding to "0" and a "high" one corresponding to "1". A particular and widespread standard of digital electronics, called TTL, admits voltage values in the range $0 \div 0.8\,\text{V}$ as "0" and voltage values in the range $2.0 \div 5\,\text{V}$ as "1".

[11]Problem 17 at the end of this chapter illustrates a scheme of principle to obtain a digital to analog conversion.

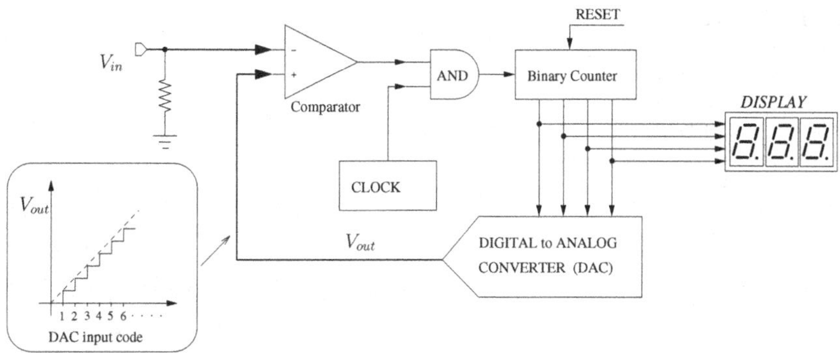

Fig. 4.10 Building a digital voltmeter with a binary counter and a digital to analog converter. The *thin lines* transmit digital signals while analog signals travel on *thick lines*. The *insert* shows the output voltage of the DAC as a function of the decimal encoding of the binary number at its input

age V_{fs}, that is the input voltage causing the resetting of the counter (overflow). The channel spacing Δ will therefore be equal to $V_{fs}/2^n$ and corresponds to the variation of one bit of the digitized value. For this reason, Δ is often referred to as the value of the least significant bit (LSB). It is evident that, for the same full-scale, instruments with a higher number of bits yield measurements that are more accurate.

Before concluding this section, we emphasize that in an ADC converter, the signal to be measured is sent to the input of a comparator consisting of an integrated operational amplifier with high input impedance, approximatively of $1 \div 10\,\text{M}\Omega$. For this reason often it is possible to neglect the perturbation of the measured value induced by the internal resistance of a digital voltmeter.

4.4.2 Electrostatic Voltmeter

The electrostatic voltmeter, also said *electrometer*, is an electromechanical device used to measure voltage by exploiting the electrostatic force exchanged by two electrically charged conductors. Although the electrometer, similar to all the rest of electromechanical instrumentation, is becoming obsolete and is being displaced by digital devices, nevertheless it has the considerable advantage of not absorbing current when measuring a DC voltage.[12]

This kind of instrument is still used when measuring high voltages with low currents such as those produced by Van de Graaff generators.

---

[12]When used with a sinusoidal signal of angular frequency ω, the electrometer will have an impedance equal to $1/\omega C$, with C capacitance of the electrodes. The current absorption in this case is not null and depends on the frequency of the signal to be measured, as it will be explained in the following of this book.

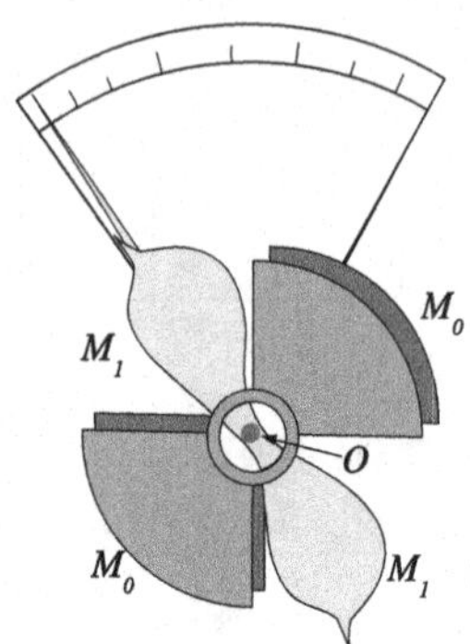

Fig. 4.11 Scheme of the Thomson voltmeter. M_0 are two fixed sectors, M_1 is the mobile sector that can rotate around a pivot. M_1 is shaped so as to make the instrument response linear for large angular deviations

Now we describe in more detail the principles of operation of an electrometer, as depicted in Fig. 4.11. A mobile sector M_1 rotates around the pivot O and penetrates the gap between two fixed sectors M_0. When sectors are shaped in such a way to makes their mutual capacity C proportional to the rotation angle, we will have $C(\theta) = k\theta$. In the absence of voltage difference between the sectors, a spiral return spring maintains the mobile sector outside the gap between the two fixed sectors. When a voltage is applied, the electrostatic force between negative and positive charges generates a mechanical torque whose module can be obtained by the derivative of the electrostatic energy $\mathscr{E}$ with respect to the rotation angle θ:

$$\mathscr{M} = -\left.\frac{\partial \mathscr{E}}{\partial \theta}\right|_{V=cost} = \frac{1}{2}V^2\frac{dC(\theta)}{d\theta} = \frac{1}{2}V^2 k$$

(note that we must take into account that the potential V is constant when taking the derivative of $\mathscr{E}$). Denoting by $\mathscr{M}_r = -h\theta$ the torque due to the return spring, the new angular position of the moving sector at steady state is obtained when $\mathscr{M} + \mathscr{M}_r = 0$ and corresponds to:

$$\theta = \frac{kV^2}{2h}$$

Note that the response of such an instrument is not linear but *quadratic*. It is possible to obviate this problem, at least in the large deviation region, with a more elaborate shaping of sectors, as in the Thomson electrometer shown in Fig. 4.11. It is easy to show that if we manage to obtain a logarithmic dependence of C on the rotation angle, a linear response of the instrument can be recovered.

4.5 Resistance Measurements

Subject of this section are the methods used to measure the resistance of resistors or, in general, of electric components having a finite resistance value.

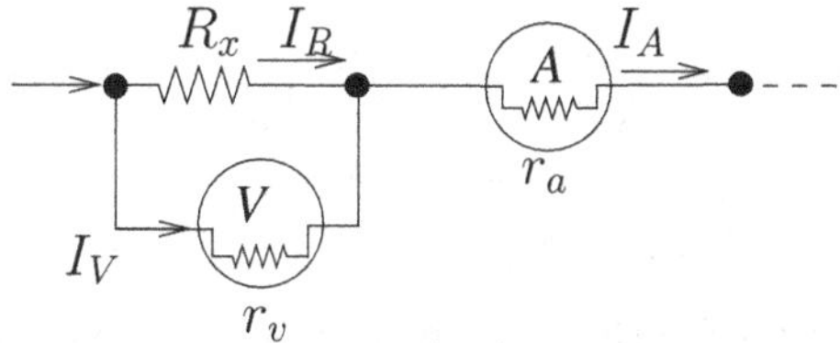

Fig. 4.12 Resistance measurement with voltmeter–ammeter method

4.5.1 *Voltmeter–Ammeter Method*

The voltmeter–ammeter method requires the simultaneous measurement of the voltage drop ΔV across the resistor and of the intensity I_R of the electrical current flowing through it. The resistance value R_x is deduced in principle by using the first Ohm's law:

$$R_x = \frac{\Delta V}{I_R}$$

In Fig. 4.12, we show a possible setup for the measurement of R_x. The voltmeter is connected in parallel to the resistor and yields the correct[13] value of the voltage difference ΔV across its terminals, while the current I_A, as measured by the ammeter, consists of the sum of the current I_R flowing through R_x and of the current I_v flowing through the voltmeter. The presence of this latter current is a neat example of a systematic effect that needs to be corrected. Using the two Kirchhoff's laws, we can evaluate I_R from the measurement of I_A and the knowledge of the voltmeter internal resistance r_v: $I_A = I_R + I_v$ and $I_R R_x = I_v r_v$ that together yield $I_v = I_R R_x / r_v$.

The intensity of the current flowing in the resistor under test, in terms of the current intensity I_A measured by the ammeter is given by:

$$I_R = I_A \frac{r_v}{r_v + R_x} \tag{4.17}$$

This allows us to write for R_x the following equation:

$$R_x = \frac{\Delta V}{I_R} = \frac{\Delta V}{I_A}\left(1 + \frac{R_x}{r_v}\right)$$

Solving for R_x we finally get

$$R_x = \frac{\Delta V}{I_A}\left(\frac{1}{1 - \dfrac{\Delta V}{r_v I_A}}\right) \tag{4.18}$$

[13] Here the qualifier *correct* means that this kind of setup is immune from *systematic effects* in the voltage measurement.

where we have isolated in brackets the correction to apply to the ratio of the measured quantities ΔV and I_A. When $r_v \gg \Delta V/I_A$, i.e., when $r_v \gg R_x$, the correction decreases and eventually becomes negligible. Note that, in the setup of Fig. 4.12, the internal resistance of the ammeter does not have any effect on the value of the resistance resulting from the measurement.

Example. The measurement of an unknown resistance is performed with the setup of Fig. 4.12. The voltage drop ΔV across its terminals is obtained by an analog voltmeter, with sensitivity equal to 20 000 Ω/V, used with a full-scale of 5 V. Therefore the internal resistance of this voltmeter is $r_v = 20\,000 \times 5 = 100\,\text{k}\Omega$. The values provided by the voltmeter and the ammeter are, respectively, $\Delta V = 3.40\,\text{V}$ and $I_A = 48.5\,\mu\text{A}$. Their ratio yields $\Delta V/I_A = 70.1\,\text{k}\Omega$. This value is very different from the correct value of R_x, as obtained from Eq. (4.18)

$$R_x = \frac{\Delta V}{I_A}\left(\frac{1}{1-\dfrac{\Delta V}{r_v I_A}}\right) = 70.1 \times 10^3 \left(\frac{1}{1-\dfrac{70.1\times 10^3}{10^5}}\right) = 234\,\text{k}\Omega$$

In this case, the correction for the systematic effect amounts to a factor 3.34, very large because the internal resistance of the voltmeter is lower than the resistance to be measured.

Consider now a different example with the same setup. Now the same voltmeter measures again $\Delta V = 3.40\,\text{V}$ while, on the contrary, the ammeter yields a current intensity $I_A = 1.58\,\text{mA}$, much higher than before. In this case, $\Delta V/I_A = 2.15\,\text{k}\Omega$ and Eq. (4.18) gives

$$R_x = \frac{\Delta V}{I_A}\left(\frac{1}{1-\dfrac{\Delta V}{r_v I_A}}\right) = 2.15 \times 10^3 \left(\frac{1}{1-\dfrac{2.15\times 10^3}{10^5}}\right) = 2.20\,\text{k}\Omega$$

The systematic correction is now only worth 2 % of the measured value because the voltmeter resistance is five times higher than R_x.

When the voltmeter internal resistance is comparable to R_x, it may be convenient to change the measurement setup by connecting the voltmeter in parallel to the series of R_x and the ammeter. However, in this case too, we need to correct for a systematic effect and the reader is requested to discuss it in Problem 12 at the end of this chapter.

4.5.2 *Ohmmeter*

The moving coil ammeter can be used within a device, the *ohmmeter*, which allows measuring the value of an unknown resistance. The layout of principle of the moving

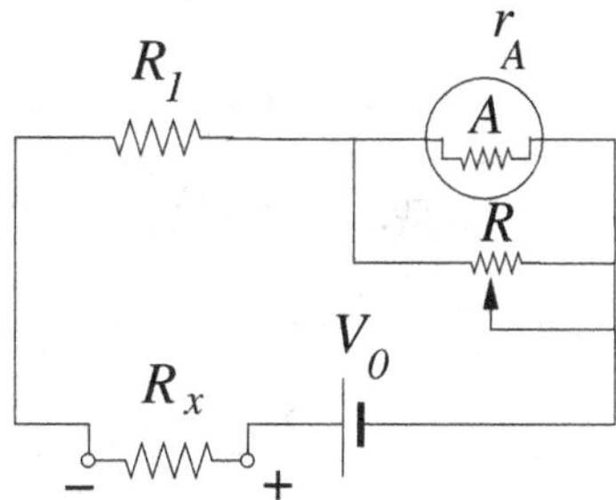

Fig. 4.13 Layout of principle of the analog ohmmeter

coil ohmmeter is shown in Fig. 4.13. The ammeter A is shunted with a variable resistor R and is connected in series with a voltage generator V_0, a current limiting resistor R_1 and the resistance under test R_x. It is easy to show[14] that the intensity i_x of the electrical current flowing through the moving coil is given by

$$i_x = \frac{V_0 R}{(R_x + R_1)(R + r_A) + R r_A} \tag{4.19}$$

where r_A is the internal resistance of the ammeter A. The evaluation of R_x is simplified by the variable resistor R. Its value is adjusted in such a way that the current in the ammeter is equal to its full-scale value i_{FS} when R_x is short-circuited. In other words, the value of R is such that the right-hand side of Eq. (4.19) yields the value i_{FS} when $R_x = 0$. Using this result and a little algebra, we can recast that equation as it follows:

$$\frac{i_x}{i_{FS}} = \frac{R'}{R_x + R'} \tag{4.20}$$

where R' is the series of R_1 with the parallel of R and r_A, see Fig. 4.13. This shows that the current intensity i_x measured by the ammeter and the value R_x of the unknown resistance are linked by a *hyperbolic* relationship. Consequently, the scale of the ohmmeter is not linear but reproduces the behavior described by Eq. (4.20). An example of such a scale is shown in the upper part of the Fig. 4.14.

Ohmmeter sensitivity. Recalling the definition (4.1), the sensitivity of the ohmmeter should be given by $d\theta/dR_x$, where θ is the angular rotation of the moving coil corresponding to the resistance R_x. It is customary, however, to use, instead of θ, the variable $f = \theta/\theta_{\text{Max}}$, i.e., the fraction of the maximum possible rotation that corresponds to the full-scale value of the ammeter.[15] Therefore, we define the *sensitivity of the ohmmeter* with the expression

[14]It is sufficient to note that the current supplied by the generator is $\frac{V_0}{R_x+R_1+r_A||R}$ and to use the formula for the current divider to evaluate the fraction flowing through r_A.

[15]From the previous discussion, it is obvious that $\theta = \theta_{\text{Max}}$ is obtained when $R_x = 0$.

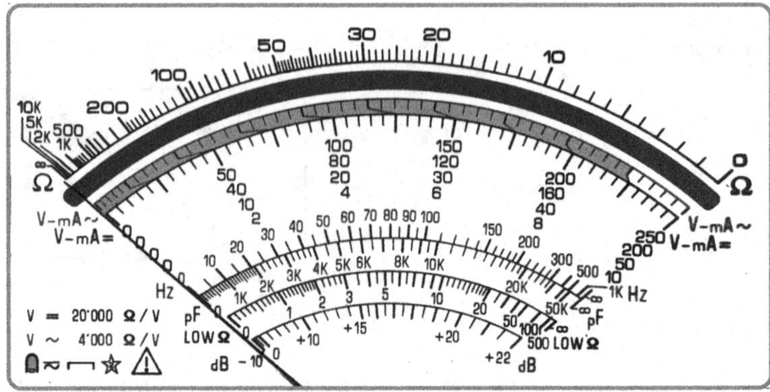

Fig. 4.14 Typical scale of an analog multimeter. The *upper ruler* pertains to the instrument used as Ohmmeter. Two values of voltmeter sensitivity are reported at the *lower left corner* 20 000 and 4 000 Ω/V, respectively, for continuous and alternating current operation

$$S_{oh} = \left|\frac{df}{dR_x}\right| = \frac{1}{\theta_{\text{Max}}}\left|\frac{d\theta}{dR_x}\right| \tag{4.21}$$

Two factors contribute to the ohmmeter sensitivity, one depending upon the parameters of the moving coil, the other linked to the characteristics of the ohmmeter circuit. Indeed, we can rewrite Eq. (4.21) as:

$$S_{oh} = \frac{1}{\theta_{\text{Max}}}\left|\frac{d\theta}{dR_x}\right| = \frac{1}{\theta_{\text{Max}}}\left|\frac{d\theta}{di_x}\cdot\frac{di_x}{dR_x}\right| = \frac{1}{\theta_{\text{Max}}}\frac{\Phi^*}{C}\left|\frac{di_x}{dR_x}\right| = \frac{1}{i_{FS}}\left|\frac{di_x}{dR_x}\right| \tag{4.22}$$

where we have used the ammeter sensitivity as given by relation (4.6). Computing di_x/dR_x from Eq. (4.20), we obtain the expression of the analog ohmmeter sensitivity as

$$S_{oh} = \frac{R'}{(R_x + R')^2} \tag{4.23}$$

Using this result, the reader can easily verify that the sensitivity of the ohmmeter has a maximum for $R_x = 0$ and decreases monotonically until it vanishes null for $R_x = \infty$.

Uncertainty of the ohmmeter. We conclude this section discussing the relative (standard) uncertainty associated with the measurements obtained with the analog ohmmeter. We make use of the relation (4.20) to express R_x as a function of $f = i_x/i_{FS}$:

$$R_x = \frac{1-f}{f}R'$$

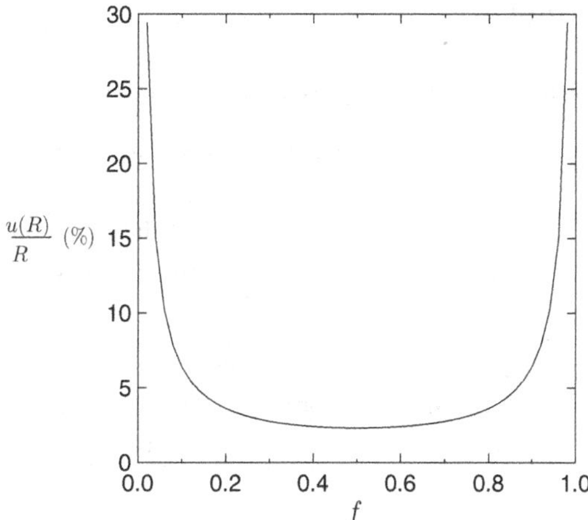

Fig. 4.15 The relative uncertainty of the measurement of a class 1 analog ohmmeter as a function of the ratio $f = i_x/i_{FS}$

Denoting with $u(i_x)$ the standard uncertainty associated to the current value i_x and with $u(R_x)$ the standard uncertainty associated to the measurement of the resistance, the relative uncertainty resulting for R_x is:

$$\frac{u(R_x)}{R_x} = \frac{1}{f(1-f)} \cdot \frac{u(i_x)}{i_{FS}} \tag{4.24}$$

where we have assumed negligible uncertainty for i_{FS}. Recalling now that the ammeter uncertainty is given by its class C through the relation $u(i_x) = C \times 10^{-2}/\sqrt{3}\, i_{FS}$, we finally obtain:

$$\frac{u(R_x)}{R_x} = 10^{-2}\frac{C}{\sqrt{3}} \cdot \frac{1}{f(1-f)} \tag{4.25}$$

The plot of the relative uncertainty of results obtained with an analog ohmmeter of class 1 is shown in the Fig. 4.15 as a function of the ratio $f = i_x/i_{FS}$. Note that this uncertainty has a minimum when the measurements are obtained around the middle of the scale while it becomes very high approaching both the lower and the higher end of the scale.

Digital ohmmeters. Nowadays, the resistance of components used in the laboratory is measured almost exclusively by means of digital *Ohmmeters*, which are more reliable and accurate than analog devices. The digital ohmmeters exploit the voltmeter–ammeter method to obtain the resistance of components connected to their terminals and show the numerical value of the result on a digital display.

Problems

Problem 1 A class 2 analog multimeter performs the following measurements:

(a) a current of 35 mA on a full-scale of 50 mA
(b) a voltage of 7.5 V on a full-scale of 10 V
(c) a resistance of 8.3 kΩ at 40 % of the full-scale.
Calculate the standard uncertainty to associate with each of these measurements. [A. 0.58 mA, 0.12 V, 339 Ω.]

Problem 2 Find the value of the shunt resistor R_s needed to measure a current of 4 A with an ammeter with full-scale range equal to 5 mA and internal resistance $r_a = 5\,\Omega$. Compute the internal resistance of the shunted ammeter. [A. $R_s = 6.26 \times 10^{-3}\,\Omega$, $r'_a = 6.25 \times 10^{-3}\,\Omega$.]

Problem 3 With reference to the previous problem, compute the uncertainty u_I to associate to the current value due to uncertainty u_{R_s} associated to the value of the shunt resistance R_s. [A. $\frac{u_I}{I} = \frac{r_a}{R_s+r_a}\frac{u_{R_s}}{R_s}$.]

Problem 4 Transform an ammeter with a range of 50 μA and internal resistance 12 Ω in a voltmeter with a range of 100 V. Find the value of the resistance R_s to connect in series to the ammeter and the internal resistance of the voltmeter thus produced. [A. $R_s = 2 \times 10^6\,\Omega = 2\,\mathrm{M}\Omega$.]

Problem 5 With reference to the previous problem, compute the uncertainty u_V to associate to the voltage value due to uncertainty u_{R_s} associated to the value of the series resistance R_s. [A. $\frac{u_V}{V} = \frac{u_{R_s}}{R_s}$.]

Problem 6 In the circuit shown in the figure, an ammeter with internal resistance equal to 10 Ω measures a current of 15.4 mA. Using Thévenin's theorem, correct this current value for the perturbation induced by the measuring instrument. [A. 16.9 mA.]

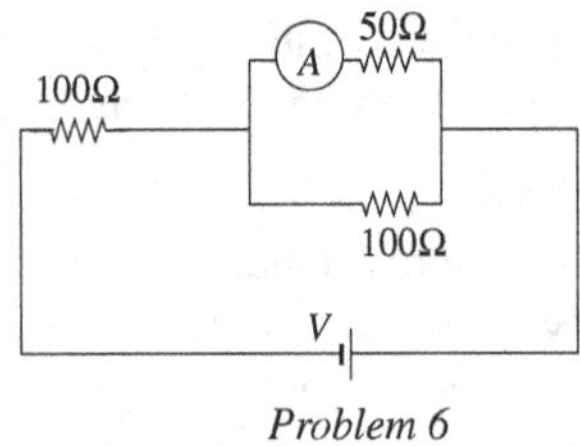

Problem 6

Problem 7 In the circuit shown in the figure, a voltmeter with sensitivity 20 000 Ω/V measures a voltage of 7.75 V on a full-scale range of 10 V. Using Norton's theorem, compute the value of the voltage undisturbed by the instrument. [A. 8.72 V.]

Problem 8 With a voltmeter of known internal resistance r_v and with an auxiliary resistance of known value R, it is possible in principle to recover the parameters of the Norton's equivalent circuit of any unknown linear electrical network as seen by two terminals. The necessary experimental procedure consists in measuring first the voltage V_1 at the output terminals and then to measure its value V_2 when modified by the insertion of the resistance R in parallel to the voltmeter. Find the expression of the Norton's current I_{eq} and the equivalent resistance R_{eq}. [A. $I_{eq} = \frac{1}{R}\frac{V_2 V_1}{V_1 - V_2}$, $R_{eq} = \frac{R}{\frac{V_2}{V_1 - V_2} - \frac{R}{r_v}}$.]

Problem 9 With an ammeter of known internal resistance r_a and with an auxiliary resistance of known value R, it is possible in principle to recover the parameters of the Norton's equivalent circuit of any unknown linear electrical network as seen by two terminals. The necessary experimental procedure consists in measuring first the current I_1 after connecting the ammeter at the output terminals and then to measure its value I_2 when modified by the insertion of the resistance R in series to the ammeter. Find the expression of the Norton's current I_{eq} and the equivalent resistance R_{eq}. [A. $I_{eq} = \frac{I_1 R}{R - \frac{r_a (I_1 - I_2)}{I_2}}$, $R_{eq} = \frac{R I_2}{(I_1 - I_2)} - r_a$.]

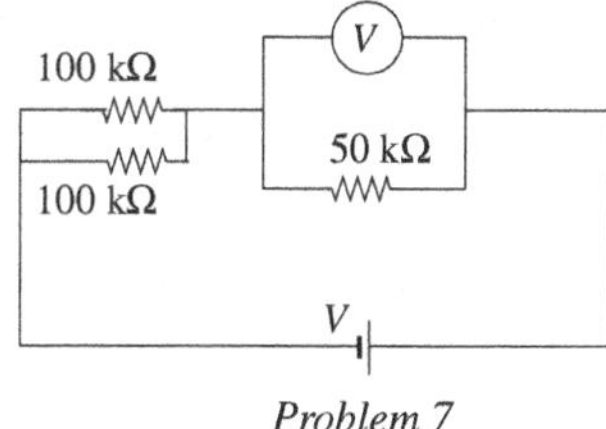

Problem 7

Problem 10 Obtain the parameters of Thévenin's equivalent circuit of an unknown linear network by measuring the short-circuit current I with an ammeter of known internal resistance r_a and the open circuit voltage V with a voltmeter of known internal resistance r_v. Find the expression of Thévenin's voltage V_{eq} and the equivalent resistance R_{eq}. [A. $V_{eq} = V\frac{1 + \frac{r_a}{r_v}}{1 - \frac{V}{I}\frac{1}{r_v}}$, $R_{eq} = \frac{\frac{V}{I} - r_a}{1 - \frac{V}{I}\frac{1}{r_v}}$.]

Problem 11 Obtain the parameters of Norton's equivalent circuit of an unknown linear network by measuring the short-circuit current I with an ammeter of known internal resistance r_a and the open circuit voltage V with a voltmeter of known internal resistance r_v. Find the expression of Norton's current I_{eq} and the equivalent resistance R_{eq}. [A. $I_{eq} = I\frac{\frac{V}{I}(r_v - r_a)}{r_v(\frac{V}{I} - r_v)}$, $R_{eq} = \frac{\frac{V}{I} - r_a}{1 - \frac{V}{I}\frac{1}{r_v})}$.]

Problem 12 Compute the expression of the measured value of the unknown resistance R_x when the voltmeter–ammeter method is implemented with the setup shown in figure. [A. $R_x = V/I_a - r_a = V/I_a(1 - I_a r_a/V)$.]

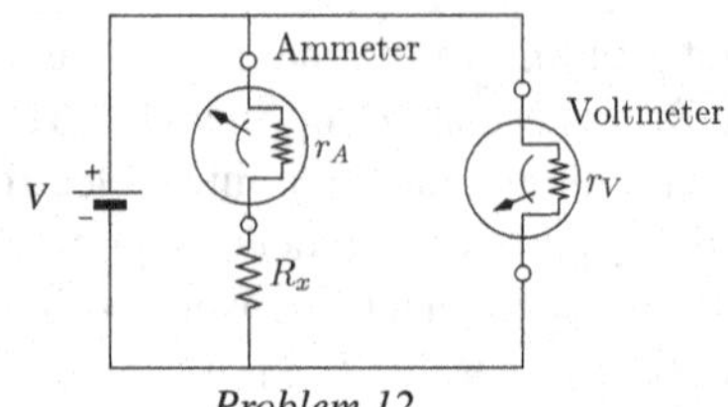

Problem 12

Problem 13 The Wheatstone bridge shown in the figure is used for the measurement of the unknown resistance R_x. Recalling that the bridge is balanced when the current in the ammeter A is null, compute the relative uncertainty associated to the value of R_x due to the standard uncertainty u_A of the measurement performed by the ammeter. Evaluate this uncertainty for an analog instrument of class 2 used on a full-scale of $50\,\mu$A and the following values of remaining parameters: $V = 10\,$V, $R_1 = R_2 = R_3 = 1\,\text{k}\Omega$. [A. $u_R/R = (u_A/V)(R_2R_3/R_1 + R_1 + R_2 + R_3) = 0.02\,\%$.]

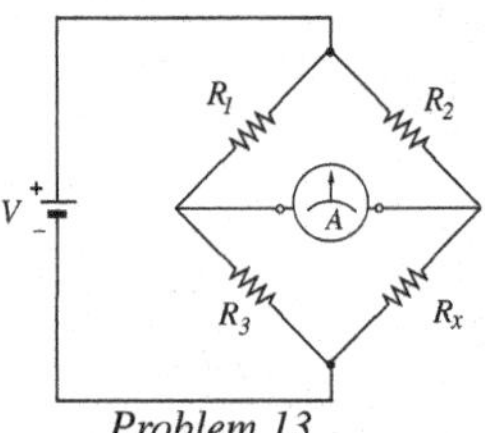

Problem 13

Problem 14 A 9 V battery with negligible internal resistance powers a chain of $50\,\text{k}\Omega$ resistors. The measurement of the potential difference between the nine connections and the earthed negative pole of the battery is obtained with an analog voltmeter of sensitivity equal to 20 000 Ω/V used on a full-scale of 10 V. Use Thévenin's theorem to compute the expression for the measured voltage as a function of the order number of the connection. [A. $V_n = V/[(N - n)R/R_V + N/n]$.]

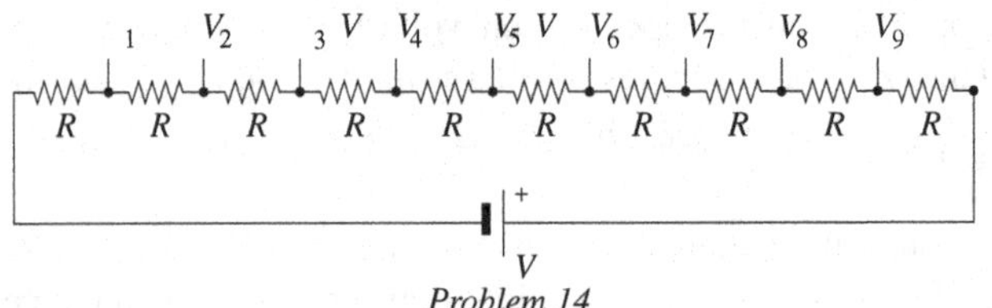

Problem 14

Problem 15 An analog voltmeter, whose internal resistance is r_v, is used to measure the tension ΔV between two points A and B of a circuit whose characteristics are unknown. To get the unperturbed value of ΔV, we perform two measurements: (1) we connect the voltmeter to the points A and B and measure the value ΔV_1; (2) we insert in series with the voltmeter a known resistance r^* and obtain the value ΔV_2. Compute the value of ΔV from the results of these two measurements and determine the value of r^* that minimizes the relative uncertainty to associate to the value obtained. [A. $\Delta V = \Delta V_1 \Delta V_2 r^* / (\Delta V_1 - \Delta V_2) r_v$; $r^* = r_v/2$.]

Problem 16 Show that the voltage V_u across the resistance R_0 of the circuit shown in the figure is proportional to the sum of the voltages of its N inputs. [A. $V_u = \sum_i V_i/(R/R_0 + N)$.]

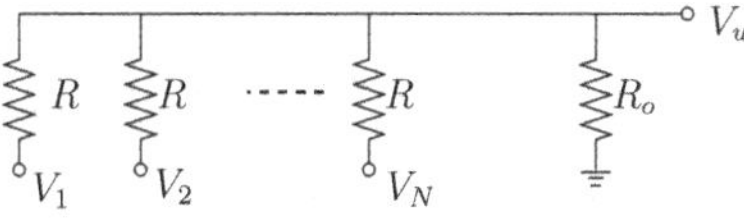

Problem 16

Problem 17 In the circuit in the figure, known as the "$R-2R$" ladder, the switches, indicated with the numbers from 0 to 3, can be connected to ground or to voltage V. The number of switches, in this case 4 (that is the length of the ladder) may increase according to the need. Show that the output voltage is: $V_u = \sum_{n=0}^{3} b_n V/2^{4-n}$, where n is the number that identifies the switch and b_n is 1 when the switch is connected to the generator or 0 otherwise.
Hint: replace with the Thévenin's equivalent the two leftmost resistors of value $2R$ *continuing with the same technique for the remainder of the circuit. Note also that the resistance viewed from the output terminals is always R, regardless of the length of the ladder.*

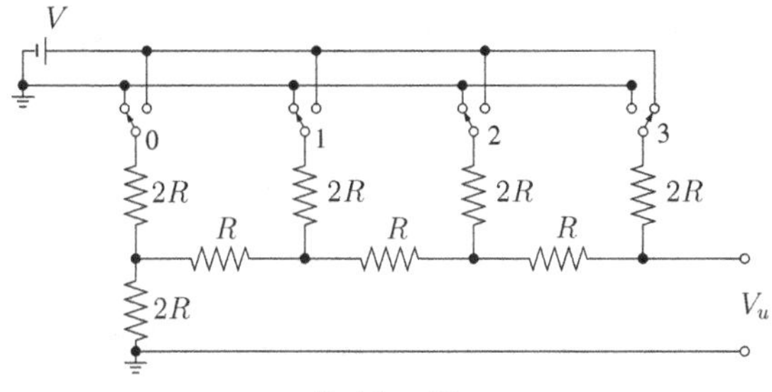

Problem 17

Problem 18 The circuit in the figure is called "$R - 2R$ inverted ladder", and is used, similarly to the one described in the previous problem, for the conversion of digital signals to analog signals. Differently from the previous problem, in this circuit the generator V delivers a constant current regardless of the position of the switches. Show that the current measured by the (supposed ideal) ammeter is $I = \sum_{n=0}^{3} b_n (V/R) 2^{n-4}$ where n is the number of the switch as shown in the figure and b_n is 1 when the switch is connected to the ammeter or 0 otherwise.

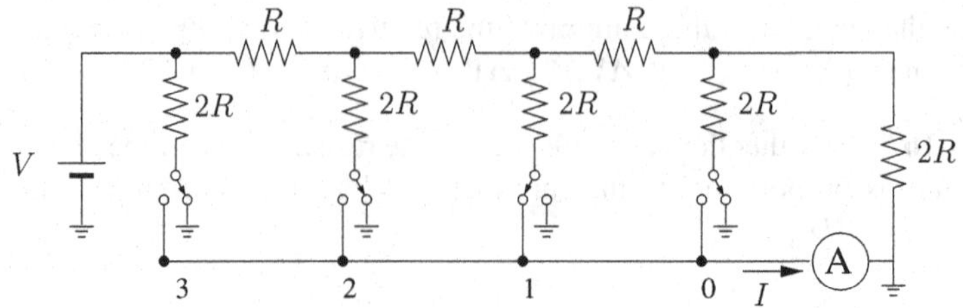

Problem 18 Circuit R − 2R inverted ladder " with 4 inputs (or bit).

Hint: apply the reciprocity theorem exchanging the ammeter with generator V and assume that only one switch is connected to the generator. Applying the rules of the current divider (see Problem 8 *of Chap. 1) and the principle of superposition one quickly comes to the solution.*

Reference

1. BIPM, IEC, IFCC, ILAC, ISO, IUPAC, IUPAP and OIML, International vocabulary of metrology - Basic and general concepts and associated terms (VIM). JCGM 200:2012 (2012)

Chapter 5
Alternating Current Circuits

5.1 Introduction

In previous chapters, we exhaustively studied methods and techniques to deal with continuous current in electrical circuits. However, in most applications, electrical circuits are exploited with time-dependent voltages and currents. In these circumstances, we cannot work with the simple algebraic relations we used so far to model circuit behavior but we need to deal with equations that are differential with respect to the time variable t.

Electric signals with time dependence can be divided into two categories[1]: transient signals, having finite time duration, and stationary signals that are periodic in time. In the first case, we have no other option than solving directly the differential equation describing the circuit behavior working with the independent variable in the time domain. We will deal with this subject in Chap. 9. It turns out that in the second case a special role is played by sinusoidal waveforms and circuit behavior can be more conveniently described by specifying the signal frequency.

We start this chapter in the next section by giving definitions useful to characterize periodic signals. Then, in Sect. 5.3, after a reminder of the behavior of the most important electric components with time-dependent voltage and current, we use Kirchhoff's laws to write the differential equation describing the current behavior in a simple circuit with a time-dependent voltage generator. This equation is solved in Sect. 5.4 for a sinusoidal voltage waveform and we come to the conclusion that this waveform does not change when it propagates in a linear circuit. In Sect. 5.5, we show that sinusoidal waveforms can be usefully associated with complex functions to reduce differential equations to algebraic equations. In Sect. 5.6, we illustrate the symbolic method that allows recovering most methods and theorems used for DC currents for the solution of circuits under sinusoidal excitation. The formulas needed

[1]There exists one more category, random signals, not considered in this book devoted only to deterministic signals.

R. Bartiromo and M. De Vincenzi, *Electrical Measurements in the Laboratory Practice*, Undergraduate Lecture Notes in Physics,
DOI 10.1007/978-3-319-31102-9_5

for the handling of electric power are given in Sect. 5.7 and finally, in the last section, we write the most important theorems for linear network, previously discussed in Chap. 2, in a formulation more appropriate to circuits with periodic currents.

5.2 Periodic and Alternating Signals

In dealing with time-dependent currents, periodic waveforms play an important role. It is therefore appropriate to give definitions of parameters relevant for their characterization. In the following, we will indicate with $s(t)$ a generic time-dependent signal that can be either a voltage or a current.

Periodic signal. A time-dependent signal $s(t)$ is periodic when it repeats the same pattern after a given time interval T, the *period* of the signal. This implies that the following relation must be satisfied:

$$s(t) = s(t+T) \quad \text{for every } t \in (-\infty, \infty)$$

The *frequency* ν of a periodic signal is the number of periods contained in a time unit. It is equal to the inverse of the period: $\nu = 1/T$. The frequency is measured in *hertz* in the SI system, its symbol is Hz; 1 Hz corresponds to one cycle per second $1\,\text{Hz} = 1\,\text{s}^{-1}$. In electronics, where very fast signals are often encountered, we find a widespread use of multiples of the hertz, such as the kilohertz ($1 \cdot \text{kHz} = 1000\,\text{Hz}$), the megahertz ($1 \cdot \text{MHz} = 10^6\,\text{Hz}$), the gigahertz ($1 \cdot \text{GHz} = 10^9\,\text{Hz}$), and the terahertz ($1 \cdot \text{THz} = 10^{12}\,\text{Hz}$).

Alternating signal. A periodic signal is qualified as alternating when its average value is null:

$$\begin{cases} s(t) = s(t+T) & \text{for every } t \in (-\infty, \infty) \\ \dfrac{1}{T}\displaystyle\int_t^{t+T} s(t')\,dt' = 0 & \end{cases} \tag{5.1}$$

Alternating signals are often indicated with the acronym AC, standing for alternating current. Signals symmetric with respect to the time axis are obviously alternating.

All periodic signals can be made alternating by adding a constant level equal and opposite to its average value. In Chap. 7, we will show how it is possible to obtain an alternating signal from a generic periodic signal with a very simple device.

Signal effective amplitude. Many electrical phenomena depend upon the square value of voltage or current. For a few examples, the energy dissipated by Joule effect in a resistor is proportional to the square of the applied voltage or of the current flowing through it; the force exchanged by two conductors with the same current is proportional to the square of its intensity; and the force exchanged by the two plates of a capacitor is proportional to the square of their voltage difference. In all these examples, the net amount of the induced effects depends on a time average of the driving term: in the first case, the temperature increase is determined by the average performed by the resistor thermal capacity while, in the two remaining examples, the conductors displacement is an average performed by their mechanical inertia.

These considerations show that useful information can be gained by measuring the *root mean square value*, often referred to as the *effective value*, of an alternating quantity as defined by

$$s_{rms} = \sqrt{\frac{1}{T}\int_t^{t+T} s^2(t')\,dt'} \tag{5.2}$$

where we use the suffix *rms* as an acronym for *root mean square*. This acronym is widely adopted to indicate the effective value of alternating quantities and can be found on the display of most measuring instruments.

The effective value is the usual choice when quoting the amplitude of an alternating voltage or current. For example, when in Europe we quote the value of 220 V for the voltage available from the commercial electric power distribution grid, we refer to its effective value. This voltage has a sinusoidal waveform described by the relation $V(t) = V_o \sin\, 2\pi t/T$, with a period $T = 20\,\text{ms}$ and a peak amplitude $V_o = 311\,\text{V}$. Using its definition (5.2), we obtain

$$V_{rms} = \sqrt{\frac{1}{T}\int_t^{t+T} V_o^2 \sin^2(2\pi t'/T)\,dt'} = V_o/\sqrt{2} = 220\,\text{V}$$

5.3 Alternating Current Circuits and Their Solution

Electric and magnetic fields are always associated to the presence of electrical currents. They are the main responsibility for the different properties of AC circuits with respect to their DC counterparts. Indeed, a time-dependent magnetic field can induce an electromotive force in conductors that adds to the effect of the circuit generators, while a time-dependent electric field can change the electric charge on conductor surfaces giving origin to displacement currents in the circuits. Furthermore, radiative phenomena may become very important at high signal frequency causing a loss of power as electromagnetic radiation in the open space.

A full description of a circuit consisting of conductors and dielectric insulators in the presence of alternating currents can only be obtained by solving Maxwell equations in the appropriate geometry and with the relevant boundary conditions. This is a formidable task in most circumstances requiring considerable efforts. However, it is possible to come up with a more manageable approach that makes use of simpler equations when the following conditions are satisfied:

1. The physical dimensions of circuits and their elements are "small" with respect to the wavelength of electromagnetic signals propagating in the circuit, see Chap. 1.
2. Each circuit element can be satisfactorily described by a single electrical property (resistance, capacitance, inductance ...).

3. Radiative phenomena can be neglected implying that power lost through electromagnetic waves is negligible with respect to the power circulating in the circuit.

Circuits where the first of the previous conditions is satisfied are referred to as *lumped parameters* circuits and can be solved with the methods illustrated in the rest of this chapter; on the other hand, circuits with dimensions larger than the signal wavelength, as for example the transmission lines, are referred to as circuits with *distributed parameters* and will be dealt with only in the last chapter of this book. In the rest of this chapter we will show that with the three approximations listed above, we can construct a theory of electrical circuits that is well verified by experimental measurements and has innumerable and important practical applications.

5.3.1 Components of Alternating Current Circuits

Electrical components used in circuits have been already described in detail in the first chapter of this book. Here we just give a reminder of their most important parameters (R, C, and L) and of the physical relations defining them.
Electrical resistance is defined by the *Ohm's law*, which states that the voltage difference $v(t)$ across terminals of a conductor is proportional to the current $i(t)$ flowing through it:

$$v(t) = Ri(t) \tag{5.3}$$

Ohm's law is not included in the fundamental equations of electrodynamics, as formulated by Maxwell. It is *phenomenological* in nature and describes the properties of electrical conduction in the matter. There are materials that do not comply with the Ohm's law and almost all materials can show significant deviations from it at very high or very low applied voltages and currents.
Electrical capacitance is defined by the law stating that, in *conditions of complete electrostatic induction*, the electrical charge on the plates of a capacitor is proportional to their voltage difference $v(t)$. Expressing the electrical charge as the integral over time of the current $i(t)$, and denoting with C the capacitance, we have

$$v(t) = \frac{1}{C} \int i(t)\, dt \tag{5.4}$$

Note that for continuous currents a capacitance is equivalent to an open circuit.
Inductance is defined by the law of Faraday–Newman–Lenz describing electromagnetic induction. This law states that the electromotive force (*e.m.f.*) induced in a circuit with self-induction coefficient L is given by

$$v(t) = L\frac{di(t)}{dt} \tag{5.5}$$

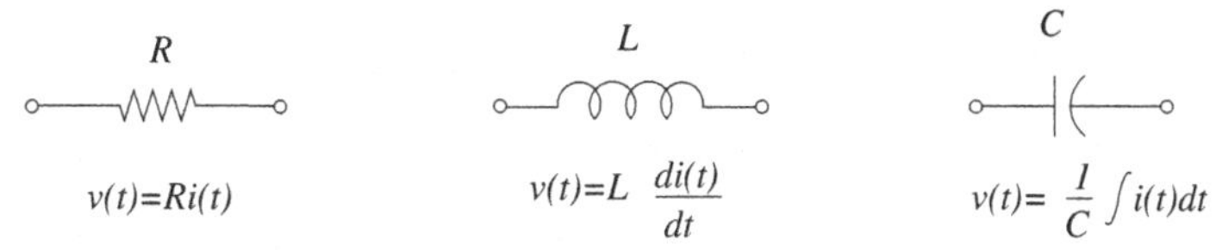

R L C

$v(t)=Ri(t)$ $v(t)=L\frac{di(t)}{dt}$ $v(t)=\frac{1}{C}\int i(t)dt$

Fig. 5.1 AC circuit components and their symbols

Note that for continuous currents an inductance is equivalent to a short circuit: $v(t) = 0$ with $i(t) =$ const. In Fig. 5.1 we summarize the properties of these three components together with the electrical symbols used to represent them in circuit diagrams.

There is one more component, the *voltage transformer*, exploiting the phenomenon of mutual magnetic induction as described by the law of Faraday–Newman–Lenz. This component is more complex than those described above and we will study it with mathematical methods appropriate for the solution of alternating current circuits in next chapter.

5.3.2 The Solution of Alternating Current Circuits

As in the case of continuous currents, the two Kirchhoff's laws are the basic tools for the solution of AC circuits. However, these laws need to be suitably reformulated for circuits where currents and voltages are time dependent.

First, consider Kirchhoff law of currents, based on the principle of electric charge conservation. The presence in the circuit of components, such as for example a capacitor, wherein electric charge can accumulate, implies that the charge density ρ, on each of its plates, becomes time dependent. In this case, it is necessary to include the *displacement current*[2] in the charge continuity equation $\mathbf{J_s} = \partial \mathbf{D}/\partial t$. The continuity equation becomes

$$\nabla(\mathbf{J} + \mathbf{J_s}) = 0 \tag{5.6}$$

Therefore, with the inclusion of $\mathbf{J_s}$, *the Kirchhoff's law of currents holds also in the presence of alternating currents.* Since the displacement current only exists within components and the internal structure of components is neglected in the lumped parameters approximation, we can disregard displacement currents in our approach to the solution of AC circuits. *Therefore, we can use first Kirchhoff's law in the formulation given for DC currents.*

The Kirchhoff's law of voltages is based on the principle of energy conservation. In the case of DC currents, we exploit the conservative nature of the electrostatic

[2]The interested reader can find a detailed discussion of this subject in manuals of electromagnetism, as for example [1].

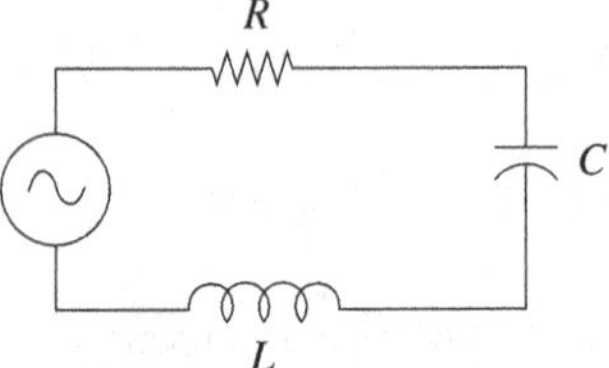

Fig. 5.2 The RLC series circuit

field and obtain that in a closed loop the sum of voltage drops is equal to the sum of all *e.m.f.* $\mathscr{E}_k$ present in the loop: $\sum \mathscr{E}_k = \sum R_k I_k$. However, we know that in the presence of time-dependent electromagnetic field there is an energy transfer from the magnetic to the electric field and vice versa. The third of Maxwell's equations

$$\nabla \times \mathbf{E} = -\frac{\partial B}{\partial t}$$

shows that the electric field is no more conservative. However, *by taking into account the e.m.f. due to variable magnetic field*, as computed by the previous equation, in inductances, where magnetic induction is important, *the Kirchhoff's law of voltages maintains the same formulation of the case of DC currents.*

The circuit shown in Fig. 5.2 is a useful starting point to illustrate the problems posed by the task of solving[3] a circuit with time-dependent voltages and currents. This circuit consists of a unique loop with a generator of alternating voltage $v(t)$, a resistor with resistance R, an inductor of inductance L, and a capacitor of capacitance C all connected in series. Applying the Kirchhoff's law of voltages to this loop we obtain

$$v(t) = v_R(t) + v_L(t) + v_C(t)$$

Now we use Eqs. (5.3), (5.4), and (5.5) linking voltage and current in each of the components in the circuit and we obtain the following integro-differential equation for the current $i(t)$ flowing in the loop:

$$L\frac{di(t)}{dt} + Ri(t) + \frac{1}{C}\int_{-\infty}^{t} i(t')\,dt' = v(t) \tag{5.7}$$

Upon further differentiation with respect to time, we obtain a second-order differential equation with constant coefficients:

$$L\frac{d^2i(t)}{dt^2} + R\frac{di(t)}{dt} + \frac{1}{C}i(t) = \frac{dv(t)}{dt} \tag{5.8}$$

[3]We recall that we have solved a circuit once we know the current in each of its branches and the voltage of each of its nodes.

At this point, the following remarks are appropriate:

1. contrarily to the case of continuous current, we do not obtain an algebraic equation but we need to solve a differential equation, a more difficult task
2. the differential equation describing the circuit is linear, as expected since all three components in the circuit are linear
3. the solution of Eq. (5.8) consists of two contributions: the first is a particular solution that depends upon the waveform of the generator voltage $v(t)$; the second is the general integral of the associated homogeneous equation obtained by setting $v(t) = 0$.

We can easily verify that this last contribution is a *transient*. This can be seen considering the characteristic equation

$$Lm^2 + Rm + \frac{1}{C} = 0 \quad \text{whose solutions are:} \quad m_{1,2} = \frac{1}{2L}\left(-R \pm \sqrt{R^2 - 4\frac{L}{C}}\right)$$

Therefore, the general integral of Eq. (5.8) is $s(t) = Ae^{m_1 t} + Be^{m_2 t}$ with A, B integration constants. It is easy to see that *the real part of solutions* $m_{1,2}$ *is always negative*. Indeed R, L, and C are all positive and the quantity under the square root, $\Delta = R^2 - 4L/C$, is always lower than R^2. Therefore, $s(t)$ decreases exponentially for increasing elapsed time t and we can neglect it in the steady state.

In the following section, we will solve Eq. (5.8) for a voltage waveform of particular importance: the sinusoidal signal.

5.4 Sinusoidal Signals

Sinusoidal waveforms play a very special role among alternating signals and often they are considered, even if improperly, the alternating signals by definition. The sinusoidal signal has the following mathematical representation:

$$s(t) = s_p \cos(\omega t + \psi) = s_p \cos(2\pi \nu t + \psi) = s_p \cos\left(2\pi \frac{t}{T} + \psi\right) \tag{5.9}$$

In this equation s_p is the *amplitude* of the signal (sometimes more precisely referred to as *peak amplitude*), ω is the *angular frequency*, ν is the frequency, T is the period, and ψ is the *phase angle* of the sinusoid. We recall that the sine and cosine functions have the same shape and that it is possible to transform the first in the second with a simple translation of the time axis since $\cos(t) = \sin(t + \pi/2)$. Our choice of Eq. (5.9), strictly speaking a cosine function, to represent a sinusoidal signal is conventional and will be justified in the following.

We will see shortly that the sinusoidal signal maintains its waveform when passing through a linear circuit. For this reason, electric power distribution grids supply a sinusoidal voltage to their users. Moreover, we recall that, using the tools of *Fourier analysis*, many mathematical functions can be expressed in terms of a sum or an integral of sinusoidal functions. Therefore, the study of linear circuits under sinusoidal excitation with assigned frequency allows obtaining, in principle, their response to a signal with a generic waveform using the superposition theorem and Fourier decomposition.[4]

5.4.1 Differentiation and Integration of the Sinusoidal Signal

Given the sinusoidal signal (5.9), its time derivative is

$$\frac{d}{dt}s(t) = s_p \frac{d}{dt}\cos(\omega t + \psi) = -s_p \omega \sin(\omega t + \psi) = s_p \omega \cos(\omega t + \psi + \pi/2) \tag{5.10}$$

The time derivative of a sinusoidal signal remains sinusoidal with the amplitude multiplied by the angular frequency and the phase increased by $\pi/2$, i.e., the differentiated signal has a phase advance of $\pi/2$.

Integrating over time the sinusoidal signal (5.9) we obtain

$$\int s(t)\, dt = \int s_p \cos(\omega t + \psi)\, dt = \frac{s_p}{\omega} \sin(\omega t + \psi) = \frac{s_p}{\omega} \cos(\omega t + \psi - \pi/2) \tag{5.11}$$

The time integral of a sinusoidal signal remains sinusoidal with the amplitude divided by the angular frequency ω and the phase reduced by $\pi/2$, i.e., the integrated signal has a phase delay of $\pi/2$.

The importance of sinusoidal signals in circuits with alternating currents led to the development of specific and efficient methods for the solution of these circuits. Before addressing these methods, particularly the symbolic method, we solve the circuit of Fig. 5.2 by giving the solution of the differential equation (5.8) with a sinusoidal waveform for the generator output voltage: $v(t) = v_p \cos(\omega t + \psi_v)$.

As noted in the previous section, we are only interested to find a particular solution of Eq. (5.8). Therefore, we look for a sinusoidal solution that we express as

$$i(t) = i_p \cos(\omega t + \psi_i) = i_p \cos(\omega t + \psi_v - \phi)$$

where $\phi = \psi_v - \psi_i$ represent the phase difference between voltage and current. To simplify notation, we set to zero the phase ψ_v of the voltage signal, without

---

[4]For this reason we give in Appendix C.1.3 a summary of the most important formulas of the Fourier analysis.

any loss of generality, by changing the origin of the time axis. Using Eqs. (5.10) and (5.11), respectively, for the derivative and the integral of a sinusoidal signal, Eq. (5.7) becomes

$$-\omega L i_p \sin(\omega t - \phi) + R i_p \cos(\omega t - \phi) + \frac{1}{\omega C} i_p \sin(\omega t - \phi) = v_p \cos \omega t$$

from which we obtain

$$i_p \left[\left(\frac{1}{\omega C} - \omega L \right) \sin(\omega t - \phi) + R i_p \cos(\omega t - \phi) \right] = v_p \cos \omega t \tag{5.12}$$

We use simple trigonometry to separate the coefficients of $\cos \omega t$ and $\sin \omega t$ and we get

$$\begin{aligned} & i_p \left[-\left(\omega L - \frac{1}{\omega C} \right) \cos \phi + R \sin \phi \right] \sin \omega t \\ & + i_p \left[\left(\omega L - \frac{1}{\omega C} \right) \sin \phi + R \cos \phi \right] \cos \omega t = v_p \cos \omega t \end{aligned}$$

Since this relation must be satisfied for any value of the time t, the coefficient of the function $\sin \omega t$ must be null while the coefficients of the function $\cos \omega t$ on the left and on the right side must be equal to each other. Therefore, we obtain

$$\begin{aligned} & i_p \left[\left(\frac{1}{\omega C} - \omega L \right) \cos \phi + R \sin \phi \right] = 0 \\ & i_p \left[\left(\omega L - \frac{1}{\omega C} \right) \sin \phi + R \cos \phi \right] = v_p \end{aligned}$$

Dividing the first of these equations by $\cos \phi$, we can obtain the value of the unknown phase ϕ:

$$\tan \phi = \frac{\omega L - \dfrac{1}{\omega C}}{R}$$

In the second equation, we can express sine and cosine as a function of the tangent to obtain the value of i_p. The solution of our problem is therefore fully specified by

$$i_p = \frac{v_p}{\sqrt{R^2 + \left(\dfrac{1}{\omega C} - \omega L \right)^2}} \qquad \phi = \arctan \frac{\omega L - \dfrac{1}{\omega C}}{R} \tag{5.13}$$

The results obtained above for the RLC series circuit shown in Fig. 5.2 allow us to make the following remarks that can be shown to be valid for *all linear circuits under sinusoidal excitation*:

- when the excitation $v(t)$ is a sinusoidal function with angular frequency ω, the stationary solution for the circuit current is a sinusoidal function with the same angular frequency
- the method adopted to obtain the circuit solution is rather cumbersome requiring the repeated use of trigonometric relations. It can become much more complicated for circuits slightly more complex than the simple RLC series circuit.

In the next section, we will show how we can transform these complicated differential equations in simpler algebraic equations by adopting a representation of the quantities involved (voltages, currents, impedances, magnetic fluxes) with two-dimensional vectors or with complex numbers.

5.5 Sinusoidal Function Representation by Phasors and Complex Numbers

It is easy to see that a sinusoidal waveform, for example a voltage,

$$v(t) = v_p \cos(\omega t + \phi) \tag{5.14}$$

can be represented by a vector in a plane. To this purpose, consider a vector $\mathbf{V}$ with length v_p applied in the origin O of a system of orthogonal axes in a plane. Assume this vector lies at angle ϕ with the abscissa axis (see Fig. 5.3). Imagine now that, starting at time $t = 0$, this vector rotates counterclockwise with a constant angular speed ω. At time t, the angle the vector $\mathbf{V}$ makes with the abscissa axis is equal to $\omega t + \phi$ and we can verify easily that its projection on the axis yields the instantaneous value of $v(t)$. In this way, we have constructed a correspondence between the rotating

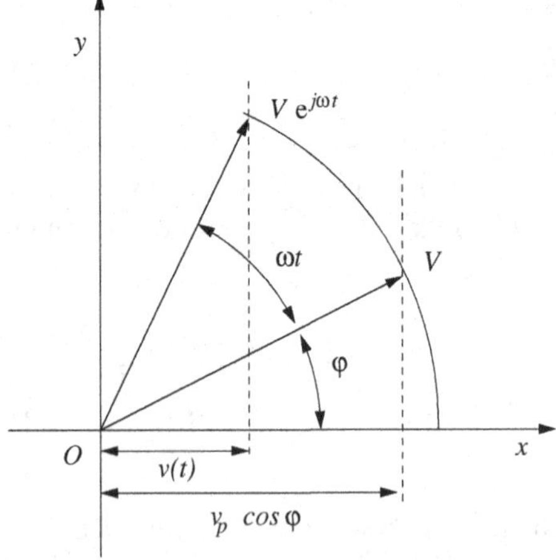

Fig. 5.3 Representation of a sinusoidal quantity with a vector (phasor)

vector **V** and the sinusoidal waveform. This waveform, once the angular frequency ω is given, is defined only by two parameters: its amplitude (v_p) and its phase (ϕ), the same two quantities used to define the vector **V** at time $t = 0$. This vector in the analysis of AC circuits is referred to as a *phasor* and is represented with the following notation:

$$V\angle\phi$$

where V is the magnitude of **V** and ϕ the angle the vector makes with the x-axis.

The phasor $V\angle\phi$ represents the sinusoidal waveform $v(t)$ defined by expression (5.14). This correspondence can be used to construct a *geometrical* solution of circuits under sinusoidal excitation. In this book, however, we will not pursue this method but prefer to adopt an algebraic method exploiting a representation of sinusoidal waveforms in terms of complex numbers.

Indeed the correspondence between sinusoidal waveforms and *phasors* is suggestive of a similar correspondence of these waveforms with complex numbers. It is well known that these numbers can be represented as a point in a two-dimensional plane, the Argand's complex plane, and therefore, it is possible to construct a correspondence between *phasors* and complex numbers. Furthermore, inspection of Euler's formula ($e^{j\alpha} = \cos\alpha + j\sin\alpha$, where j is the imaginary unit) suggests the idea to associate the waveform defined by relation (5.14) to the real part of the complex quantity $v_p e^{j(\omega t+\phi)}$.

Exploiting the notation with complex numbers, we can greatly simplify the analysis of circuits in the sinusoidal regime. This can be seen considering again the differential equation (5.7) with the following position:

$$v_p\cos(\omega t + \psi_v) \rightarrow v_p e^{j(\omega t+\psi_v)} = v_p e^{j\psi_v} e^{j\omega t} = V e^{j\omega t} \tag{5.15}$$

Here the complex quantity $V = v_p e^{j\psi_v}$ can be seen as a different way to indicate the phasor $V\angle\psi_v$ and is referred to as the *complex representation of the voltage*. Having introduced an imaginary part in their right-hand side of Eq. (5.8), we must now look for a solution, the current, with a similar complex expression:

$$i_p e^{j(\omega t+\psi_i)} = i_p e^{j\psi_i} \text{and}^{j\omega t} = I e^{j\omega t}$$

where $I = i_p e^{j\psi_i}$ is the *complex representation of the current*. Recalling that *the derivative and the integral of an exponential function remain exponential*

$$\frac{d}{dt}e^{at} = ae^{at}, \quad \int e^{at}\,dt = \frac{1}{a}e^{at} \tag{5.16}$$

in Eq. (5.7) we can simplify the time-dependent term $e^{j\omega t}$ and obtain an algebraic equation for the complex amplitude of the current:

$$I\left[R + j\omega L + \frac{1}{j\omega C}\right] = V \tag{5.17}$$

Solving this equation for I, we obtain

$$I = \frac{V}{R + j\omega L + \dfrac{1}{j\omega C}} \tag{5.18}$$

The module of I gives the peak amplitude i_p of the solution

$$|I| = \frac{|V|}{\left| R + j\omega L + \dfrac{1}{j\omega C} \right|} = \frac{v_p}{\sqrt{R^2 + \left(\omega L - \dfrac{1}{\omega C}\right)^2}} \tag{5.19}$$

while the phase difference ϕ between V and I is obtained as

$$\phi = \psi_v - \psi_i = \psi_v - \left(\psi_v - \arctan\frac{\mathfrak{I}[I]}{\mathfrak{R}[I]}\right) = \arctan\left(\frac{\omega L - \dfrac{1}{\omega C}}{R}\right) \tag{5.20}$$

where the functions $\mathfrak{R}[\ldots]$ and $\mathfrak{I}[\ldots]$ indicate, respectively, the real and the imaginary part of their argument. We have obtained again the solution of the previous section, see Eq. (5.13), but the procedure adopted is faster and simpler.

This procedure can be applied to a generic class of equations including (5.7) as a particular case. To this purpose, consider a physics problem requiring the solution of an equation, or a system of equations, of the kind

$$\mathscr{L}[x(t)] = f(t) \tag{5.21}$$

where $\mathscr{L}[\ldots]$ represents a generic linear integro-differential operator with constant coefficients. Using $f_1(t)$ and $f_2(t)$ to represent two different excitation functions and denoting with $x_1(t)$ and $x_2(t)$ the corresponding solutions of Eq. (5.21), because of the linearity of $\mathscr{L}$ we can easily show that the complex function $x^*(t) = x_1(t) + jx_2(t)$ is also a solution of the same equation when $f(t)$ is complex and equal to $f_1(t) + jf_2(t)$.

Similarly, it is easy to show that, if $y^*(t) = y_1(t) + jy_2(t)$ is the solution of Eq. (5.21) with excitation $f(t) = f_1(t) + jf_2(t)$, the real part $y_1(t)$ of $y^*(t)$ is its solution when $f(t) = f_1(t)$. The same is true for the imaginary part $y_2(t)$ when $f(t) = f_2(t)$. Therefore, a solution of this equation in the realm of complex numbers is equivalent to two real solutions for the underlying physics problem.

These considerations become very useful when the function $f(t)$ is sinusoidal $f(t) = f_0 \cos(\omega t + \phi)$. In this case, the use of complex exponentials allows transforming the original Eq. (5.21) in an algebraic equation in the realm of complex numbers. In these circumstances, it is useful to make the following position:

$$f^*(t) = f_0 \cos(\omega t + \phi) + j f_0 \sin(\omega t + \phi) = f_0 \exp[j(\omega t + \phi)]$$
$$= f_0 \exp(j\phi) \exp(j\omega t) = f_0^* \exp(j\omega t)$$

where $f_0^* = f_0 \exp(j\phi)$. Looking for a solution of the kind $x^*(t) = x_0^* \exp(j\omega t)$, we can use relations (5.16) to express derivatives and integrals and reduce the problem to the solution of an algebraic equation, as done above. In this way, we can get the expression of $x_0^* = x_0 \exp(j\phi_x)$ yielding amplitude and phase of the physical solution.

5.5.1 The Symbolic Method

We have now all the elements necessary to introduce the symbolic method,[5] a technique that greatly simplifies the analysis and the solution of linear circuits under sinusoidal excitation.

Consider again relation (5.17) obtained in the previous section using the complex representation of sinusoidal quantities. We remark that this relation has a structure very similar to the Ohm's law provided we regard:

- the complex representation of the current I as a current
- the complex representation of the voltage V as a voltage
- the quantity enclosed in the square brackets as an impedance composed by the series of three elements: the impedance of the resistance R, the impedance of the inductance $j\omega L$, and the impedance of the capacitance $1/j\omega C$.

This observation is not limited to the RLC series circuit but is generally valid for all linear circuits with sinusoidal currents. It is suggestive of an approach to the solution of AC circuits widely adopted and known as the *symbolic method*. In short, the solution of a generic linear circuit under sinusoidal excitation with the symbolic method includes the following steps, see Fig. 5.4:

1. Replace sinusoidal generators of voltage or current with their complex representation
2. Replace components in the circuit with their complex impedance using the following table:

   Resistance $\rightarrow Z_R = R$
   Inductance $\rightarrow Z_L = j\omega L$
   Capacitance $\rightarrow Z_C = \frac{1}{j\omega C}$

[5]The introduction of the symbolic method is largely due to the mathematician and engineer Charles P. Steinmetz, an advocate of the practical use of alternating current, who formalized the mathematics necessary for the study of circuits with sinusoidal currents [2].

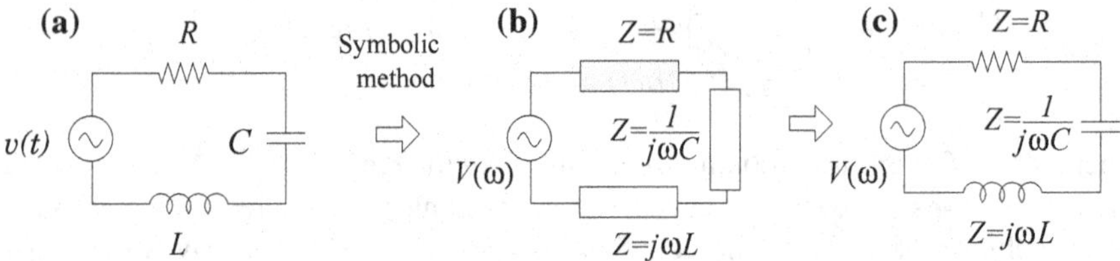

Fig. 5.4 Schematic of the symbolic method application; **a** diagram of the real circuit: voltages and currents are time dependent; **b** diagram of the equivalent circuit as transformed by the symbolic method: voltages, currents, and impedances are complex quantities depending on the angular frequency ω, not on time; **c** practical diagram where component symbols are the ones used in the real circuit

3. Use the two Kirchhoff's laws for currents and voltages *since they are also valid for the complex representations*[6]
4. Solve the circuit with the same procedures adopted for DC circuits, recalling that impedances in series are additive while for the parallel connection the equivalent impedance is given by the inverse of the sum of the inverse of each impedance.

In summary, the definition of complex impedance allows writing the relationship between complex representation of voltage and current in a way formally identical to the Ohm's law. In other words, the complex AC voltage must replace the DC voltage, the complex AC current the DC current, and the complex impedance the resistance. In formulas,

$$V = ZI \tag{5.22}$$

[6]We show here that this statement is true for Kirchhoff's law of currents. For sinusoidal currents this law yields at a node: $\sum_k i_k(t) = \sum_k i_{mk} \cos(\omega t + \phi_k) = 0$, for all time values t. Using simple trigonometry we obtain

$$\sum_k i_{mk} \cos(\omega t + \phi_k) = \sum_k i_{mk} (\cos \omega t \cos \phi_k - \sin \omega t \sin \phi_k) = 0$$

Since this relation must hold for all t values, the coefficients of both $\cos \omega t$ and $\sin \omega t$ must be null:

$$\sum_k i_{mk} \cos \phi_k = 0 \quad \text{and} \quad \sum_k i_{mk} \sin \phi_k = 0$$

We now add the second equation, multiplied by the imaginary unit j, to the first. We get

$$\sum_k i_{mk} (\cos \phi_k + j \sin \phi_k) = \sum i_{mk} e^{\phi_k} = \sum_k I_k = 0$$

This shows that *the complex representations of sinusoidal currents converging in a node fulfill Kirchhoff's law for currents.* With a similar procedure it is possible to show that Kirchhoff's law for voltages remains valid for the complex representation of sinusoidal voltages.

Equation (5.22) is referred to as the *generalized Ohm's law*. Its use allows extending to sinusoidal signals all solution methods used for DC circuits (method of nodes and method of loops).

The most important feature of the symbolic method is that it allows solving AC circuits, described by differential equations with time t as the independent variable, by means of algebraic equations, much simpler to solve, where the time plays no role and the frequency of the sinusoidal waveform is present as a parameter.

Impedance Composition. As illustrated above, the rules to apply in the composition of impedances are *similar to those used to compose resistances in DC circuits.*

Resistance and reactance. In the circuit *RLC series* of Fig. 5.2 the complex impedance "seen" by the generator can be expressed as

$$Z = R + j\left(\omega L - \frac{1}{\omega C}\right)$$

This expression for the impedance is not only true in the particular case under consideration but is valid in general. Whatever the connections of the individual components, we can always express the complex impedance as a sum of two terms: a real term, possibly dependent on ω, which we refer to as the *resistance*, and an imaginary term that we name *reactance*. Then, in general, we can write for a generic impedance:

$$Z(\omega) = R(\omega) + jX(\omega) = |Z|e^{j\phi}$$

where

$$|Z(\omega)| = \sqrt{R(\omega)^2 + X(\omega)^2} \qquad \phi_Z(\omega) = \arctan\frac{X(\omega)}{R(\omega)}$$

When we apply a voltage difference of complex amplitude V to a component with complex impedance $Z = R + jX$, the module[7] of the current flowing in the component is

$$|I| = \frac{|V|}{|Z|} = \frac{|V|}{\sqrt{R^2 + X^2}} \tag{5.23}$$

and the phase difference ϕ_I between current and voltage is

$$\phi_I = -\arctan\frac{X}{R}$$

Example. As a simple application of generalized Ohm's law, let us compute the current flowing through a resistance $R = 50.0\,\Omega$ in series with a capacitance $C = 20.0\,\mu\text{F}$ connected to a sinusoidal voltage generator with output effective value $V_{rms} = 7.10\,\text{V}$ at a frequency of 50.0 Hz.

---

[7]As shown in Sect. 5.5 the module of a complex quantity is equal to the peak value of the real sinusoidal quantity represented. Therefore, Eq. (5.23) yields the peak amplitude of the current. However, since this equation is linear, replacing $|V|$ with V_{rms} yields I_{rms}.

We can express the output of the generator as $v(t) = V_o \cos \omega t$. From its effective value, we obtain $V_o = \sqrt{2}V_{rms} = 10.0\,\text{V}$. By definition the angular frequency is $\omega = 2\pi\nu = 314\,\text{s}^{-1}$.

The complex impedance seen by the generator is $Z = R + jX$, with $R = 50.0\,\Omega$ and $X = -1/\omega C = -159\,\Omega$. Finally,

$$|I| = I_o = \frac{|V|}{|Z|} = \frac{V_o}{\sqrt{R^2 + X^2}} = \frac{10.0}{\sqrt{2.5 \times 10^3 + 25.3 \times 10^3}} = \frac{10.0}{167} = 59.9\,\text{mA}$$

The current effective value is $I_{rms} = 59.9/\sqrt{2} = 42.4\,\text{mA}$ and the current waveform has a phase difference of an angle $\phi_I = -\arctan(X/R) = 1.26\,\text{rad}$ with respect to the voltage waveform.

To see how a frequency dependence of the real part of the impedance comes about, consider the same two components connected in parallel. In this case we have

$$\frac{1}{Z} = \frac{1}{R} + j\omega C \quad \text{yielding} \quad Z = \frac{R}{1 + (\omega RC)^2} - j\frac{\omega R^2 C}{1 + (\omega RC)^2}$$

The real part of this impedance is equal to R in DC conditions and becomes progressively lower as the generator frequency increases.

5.6 Electric Power in AC Circuits

Consider a generic circuit element with a voltage drop $v(t)$ at its terminals and a current $i(t)$ flowing through it. Using the definition of electric power given for DC circuits, we can calculate the *instantaneous power* transferred to the component as $p(t) = v(t)\,i(t)$. When the component is used under sinusoidal excitation, we can write $v(t) = V_o \cos \omega t$ and $i(t) = I_o \cos(\omega t + \phi)$, where ϕ is the phase difference between current and voltage in the component we are considering. The expression for the instantaneous power $p(t)$ *transferred to the component* becomes

$$p(t) = v(t)i(t) = V_o \cos \omega t\, I_o \cos(\omega t + \phi) \tag{5.24}$$

It is easy to verify that when the phase shift is not null, this power can become negative, as shown in Fig. 5.5. When this happens, energy is transferred *by the component to the rest of the circuit*. The average power P, transferred to the component, is obtained by averaging expression (5.24) over a period $T = 1/\nu = 2\pi/\omega$

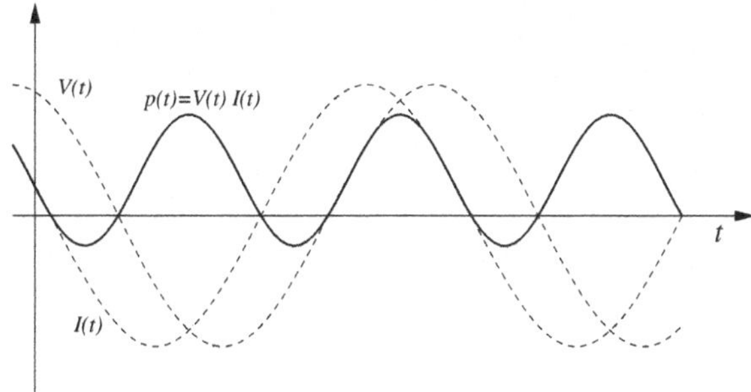

Fig. 5.5 The continuous line represents the time evolution of the instantaneous power corresponding to the voltage and to the current shown by the *dotted lines*. When it becomes negative (*below the horizontal line*), power is transferred from the component to the rest of the circuit. In this figure, the phase shift between voltage and current (the latter being in phase advance with respect to the former) is 1 radian. Note that the instantaneous power has a frequency twice that of the voltage (or current)

$$\begin{aligned} P &= \frac{1}{T}\int_0^T p(t)\,dt = \frac{1}{T}\int_0^T v(t)i(t)\,dt = \frac{V_oI_o}{T}\int_0^T \cos\omega t\cos(\omega t+\phi)\,dt \\ &= \frac{V_oI_o}{T}\left[T\frac{\cos\phi}{2} + \frac{1}{2}\cos\phi\int_0^T \cos 2\omega t\,dt - \sin\phi\int_0^T \sin\omega t\cos\omega t\,dt\right] \\ &= \frac{V_oI_o}{2}\cos\phi = V_{rms}I_{rms}\cos\phi \end{aligned} \tag{5.25}$$

This result shows that the average power transferred to a component under sinusoidal excitation is the product of three factors: the effective voltage, the effective current, and the cosine of the phase shift between current and voltage. The factor $\cos\phi$ is referred to as the *power factor* and can be responsible of large reductions of the power transferred to an electrical component.

In the previous discussion, we focused our attention on a single electrical component. It is easily seen that the result obtained can be extended to sets of linear electrical components connected in series and/or in parallel, as we show in the following example.

Example. Consider a resistance R and an inductance L connected in series and powered by a sinusoidal voltage generator with output $v(t) = V_o\cos\omega t$. This setup can represent an electrical motor connected to the commercial electric power grid, which in Europe supplies effective voltage amplitude of 220 V at frequency of 50 Hz. In this case we have $V_o = 311$ V and $\omega = 314\,\text{s}^{-1}$. Assuming $R = 50\,\Omega$ and $L = 300$ mH (values representative of the electrical motor of a typical home washing machine) we apply the symbolic method to compute the current $i(t) = I_o\cos(\omega t+\phi)$ flowing in the resistance and the inductance. The impedance seen by the generator is $Z = R + j\omega L$, and therefore the complex current is $I = V_o/Z = V_o/(R + j\omega L)$. This result yields $I_o = V_o/\sqrt{R^2+\omega^2L^2} = 310/107 = 2.91$ A and $\phi = \arctan(-\omega L/R) = -1.08$ rad. The power factor $\cos\phi$ is obtained using elementary trigonometry. We get

$$\cos\phi = \frac{R}{\sqrt{R^2 + \omega^2 L^2}} = \frac{50}{\sqrt{(50)^2 + (314 \cdot 0.3)^2}} = 0.47$$

Finally, the average power transferred from the generator to the series of the resistance R and the inductance L is

$$P = \frac{V_o I_o}{2} \cos\phi = \frac{310 \times 2.91}{2} \times 0.47 = 450 \times 0.47 = 212\,\text{W}$$

Note that the phase shift between current and voltage causes a loss of transferred power amounting to 50 % of the power available from the generator. For this reason, operators require that the loads connected to their electrical grid have a limited reactance, see Problem 12 at the end of this chapter.

The attentive reader will have noticed that in Eq. (5.24) we used the real representation of voltage and current, not their symbolic representation. In fact, since the power is a quadratic form, the real and imaginary parts of the symbolic representation would mix providing as a result a complex number with no physical meaning. However, using Eq. (5.25), it is easy to show that the average transferred power P can be expressed in terms of complex amplitudes as

$$P = \frac{1}{2}\Re[V I^*] = \frac{1}{2}\Re[V^* I]$$

5.7 Theorems for AC Linear Networks

We can easily extend the validity of the theorems on linear networks, discussed in Chap. 2, to AC circuits with proofs similar to those we worked out in detail for the case of DC currents. However, it is necessary to formulate them with a terminology more appropriate for AC currents.

- **Superposition theorem**: when a network of linear impedances is powered by more than one generator, the current flowing in a generic branch (or the voltage of a generic node) of the circuit is equal to the sum of the currents (or the voltages) produced by each generator, taken individually, assuming that all the others have been replaced by their internal impedance.
- **Reciprocity theorem**: in a network of reciprocal impedances, if a voltage generator V in the branch AA' produces a current I in the branch BB', the same generator V in the branch BB' produces the same current I in the branch AA'. Similarly, if a current generator I in the branch AA' produces a voltage drop V between nodes B and B', the same the generator I in the branch BB' produces the same voltage drop V between nodes A and A'.
- **Thévenin's theorem**: any network, composed by linear impedances and generators, "seen" between two of its nodes, A and B, is equivalent to an ideal voltage generator V_{eq} in series with an impedance Z_{eq}. The voltage V_{eq} is equal to the

voltage measured between nodes A and B (open-circuit voltage), the impedance Z_{eq} is equal to the impedance "seen" between the two nodes when all the generators in the network have been replaced by their internal impedance.

- **Norton's theorem**: any network, composed by linear impedances and generators, "seen" between two of its nodes, A and B, is equivalent to an ideal current generator I_{eq} in parallel with an impedance Z_{eq}. The current I_{eq} is equal to the *short-circuit current* flowing between nodes A and B; the impedance Z_{eq} is equal to the impedance "seen" between the two nodes when all the generators in the network have been replaced by their internal impedance.

Problems

Problem 1 Show that the sum of a finite number of sinusoidal waveforms with the same frequency and arbitrary phases remains sinusoidal.

Problem 2 Provide the explicit representation of the current described with the following phasors: $I = 3\exp(j\omega t/4)\ A$; $I = 7\exp(-j\omega t/3)\ A$; $I = 5\exp(-j\omega t/6)\ A$

Problem 3 Provide the phasors describing the following voltage waveforms: $v(t) = 10\cos(\omega t - 0.393)\ V$; $v(t) = -3\sin(\omega t + 0.262)\ V$; $v(t) = 5\cos(\omega t + \pi/2)\ V$

Problem 4 We measure the current $5\cos(\omega t + \pi/3)\ A$ by applying to an unknown electrical bipolar component a voltage waveform described by $10\sin(\omega t - \pi/6)\ V$. Assess whether the component is a resistance, an inductance, or a capacitance. [A. Resistance whose value is $2\ \Omega$.]

Problem 5 We measure the current $1.5\sin(200t + 0.33)\ A$ by applying to an unknown electrical bipolar component a voltage waveform described by 4.5 $\cos(200t + 0.33)\ V$. Assess whether the component is a resistance, an inductance, or a capacitance. [A. Inductance whose value is 15 mH.]

Problem 6 We measure the current $3\sin(300t+3\pi/4)\ A$ by applying to an unknown electrical bipolar component a voltage waveform described by $10\cos(300t-\pi/4)\ V$. Assess whether the component is a resistance, an inductance, or a capacitance. [A. Capacitance whose value is 1 μF.]

Problem 7 Compute the equivalent complex impedance of the series connection of a 10 μF capacitor, a 3.0 mH inductance, and a 33 Ω resistance at a frequency of 1.0 kHz. [A. $Z = (33 + 2.92j)\Omega$.]

Problem 8 Compute the equivalent complex impedance of the series connection of a 10 μF capacitor and a 100 Ω resistance at a frequency of 1.0 kHz. [A. $Z = (100 - 15.92j)\ \Omega$.]

Problem 9 Compute the equivalent complex impedance of a 25 Ω resistance in series with the parallel of a 30 nF capacitor with a 3 mH inductance at a frequency of 10 kHz. [A. $Z = (25 + 188.4j)\ \Omega$.]

Problem 10 A sinusoidal voltage generator with output amplitude V_g and angular frequency ω is connected to the series of two capacitances C_1 and C_2. Give the value of the amplitude of the voltage at the interconnection point of the two capacitances (capacitive voltage divider). [A. $V = V_g C_1/(C_1 + C_2)$ where C_2 is grounded.]

Problem 11 A sinusoidal voltage generator with output amplitude V_g and angular frequency ω is connected to the series of two inductances L_1 and L_2. Give the value of the amplitude of the voltage at the interconnection point of the two inductances (inductive voltage divider). [A. $V = V_g L_2/(L_1 + L_2)$ where L_2 is grounded.]

Problem 12 A sinusoidal current generator with output amplitude V_g and angular frequency ω is connected to the parallel of the inductance L and the capacitance C. Find the amplitude of the current flowing in each of these two components. [A. $i_L = I/(1 - \omega^2 LC)$; $i_C = I/(1 - 1/(\omega^2 LC))$.]

Problem 13 An electrical motor with inductance L and resistance R is powered by the public grid at a frequency of 50 Hz. The grid operator requires that for economic reasons the phase shift between voltage and current in the motor is null. To this purpose, it is necessary to connect a capacitor in parallel with the motor. Compute the required value for its capacitance. [A. $C = L/(\omega^2 L^2 + R^2)$.]

References

1. J.D. Jackson, *Classical Electrodynamics*, 3rd edn. (Academic Press, New York, 1998)
2. C. Steinmetz, Phys. Rev. **3**, 335 (1896)

Chapter 6
Alternating Current: Basic Circuits for Applications

6.1 Introduction

In this chapter, we take advantage of the simplicity and effectiveness of the symbolic method to solve some simple circuits under sinusoidal excitation. We address both significant examples for the understanding of physical phenomena and circuits of general interest for practical applications. We start with a closer inspection of circuits consisting of a resistance, an inductance and a capacitance (*RLC* circuits), similar to the circuit used in the previous chapter to introduce the symbolic method. *RLC* circuits have special importance since they show resonant behavior, a phenomenon ubiquitous in physics and of relevance in fields ranging from elementary mechanics to sub nuclear interactions.

We deal in details with these circuits in the following Sect. 6.2 under the assumption of using ideal components. In Sect. 6.3, we start to consider real components, a discussion we conclude in Sect. 6.4 after introducing a technique, the series to parallel transformation, allowing for an easy extension of the results obtained with ideal components. In Sect. 6.5, we discuss reactive bridge circuits and in Sect. 6.6 we describe the properties of two-port circuits, used to transform the input signal into another signal, the output, performing an electrical processing and analysis of information. Sections 6.7 and 6.8 illustrate *RC* filter circuits, obtained with a series connection of a resistance and a capacitance, respectively, in the low-pass and the high-pass configuration, while similar *RL* circuits are the subject of Sect. 6.9. In Sect. 6.10, we show how to use *RC* and *RL* filters to perform some mathematical operations on electrical signals. The static transformer is introduced in Sect. 6.11 where we show how to solve circuits using this important electrical component. Finally Sects. 6.12 and 6.13 are devoted, respectively, to an illustration of the problem of impedance matching and to the electrical networks used to obtain this function in practice.

R. Bartiromo and M. De Vincenzi, *Electrical Measurements in the Laboratory Practice*, Undergraduate Lecture Notes in Physics,
DOI 10.1007/978-3-319-31102-9_6

6.2 Resonant Circuits

In classical physics resonance is a phenomenon observed in systems exhibiting free standing oscillations (oscillating systems) and is obtained subjecting the system to an external perturbation with a specific frequency. Indeed the response[1] of an oscillating system depends upon the frequency of the external perturbation and when it shows a maximum, or a minimum, it is possible that the system is in resonance.

The impedance Z of the series connection of a capacitance C and an inductance L is given by $Z = j\omega L + 1/j\omega C$. It is easy to verify that this impedance *becomes null* when the angular frequency has the value $\omega_o = 1/\sqrt{LC}$. Connecting the same two components in parallel, we have an admittance $Y = 1/j\omega L + j\omega C$ that is null (meaning that the corresponding impedance is infinity) again at the same angular frequency ω_o. We will show in next sections that these circuits at $\omega = \omega_o = 1/\sqrt{LC}$ exhibit a resonant behavior.[2]

6.2.1 *The Circuit RLC Series*

The circuit shown in Fig. 5.2 in the previous chapter, used to introduce solution methods for AC circuits, is referred to as the *RLC series* circuit or series resonant circuit to remind how its reactive components are connected. The (complex) current flowing through it is given by relation (5.18), as obtained in the previous chapter. Using in this relation the two parameters $\omega_0 = 1/\sqrt{LC}$, the resonance angular frequency, and $Q_s = \omega_0 L/R$, known as *quality factor* or simply *Q factor* of the circuit, the amplitude and the phase of this current take the following expressions:

---

[1] By *response* here we mean the manner in which the physical system reacts to an external stress, for example the maximum amplitude of the oscillations of a forced pendulum (the swing of our childhood), the maximum amplitude of the acoustic vibrations in an organ pipe or the maximum charge accumulated on the plates of a capacitor in a *RLC* circuit.

[2] In the published literature, we can come across different definitions of resonance frequency in electric circuits; some of those definitions are: the null reactance of the circuit as seen by the generator (equivalent to null phase shift between voltage and current supplied by the generator); maximum peak voltage or peak current at the generator; excitation frequency equal to the characteristic frequency of free oscillations.

For simple circuits, as those we will discuss in next sections, all definitions are equivalent in the sense that they give the same value of the frequency. However, when the circuits become slightly more complex, for example when the resistance of the inductor is taken into account, different definitions can lead to different expressions for the resonant frequency. The reader can work with Problem 13 at the end of this chapter for more details. In any case, the different determinations of the resonant frequency, although obtained by different mathematical formulas, in general are only slightly different in value.

The interested reader can find a more detailed discussion of the resonance definition in references [1–4].

$$|I| = \frac{|V_g|}{R} \cdot \frac{1}{\sqrt{1 + Q_s^2\left(\frac{\omega}{\omega_0} - \frac{\omega_0}{\omega}\right)^2}} = \frac{|V_g|}{R} \cdot \frac{1}{\sqrt{1 + Q_s^2\left(\frac{\nu}{\nu_0} - \frac{\nu_0}{\nu}\right)^2}}, \tag{6.1}$$

$$\phi_i = \arctan\left[\frac{1}{R}\left(\frac{1}{\omega C} - \omega L\right)\right] = -\arctan\left[Q_s\left(\frac{\omega}{\omega_0} - \frac{\omega_0}{\omega}\right)\right] = -\arctan\left[Q_s\left(\frac{\nu}{\nu_0} - \frac{\nu_0}{\nu}\right)\right] \tag{6.2}$$

where V_g is the output amplitude of the voltage generator while $\nu = \omega/2\pi$ and $\nu_0 = \omega_0/2\pi$ are, respectively, the frequency and its resonance value.

The quality factor of the series resonant circuit Q_s can take different expressions:

$$Q_s = \omega_0 \frac{L}{R} = \frac{1}{\omega_0 RC} = \frac{1}{R}\sqrt{\frac{L}{C}}$$

At the resonant frequency, the impedance of the *LC* series vanishes and consequently the circuit behaves as a resistance connected to a generator. In these conditions, the current flowing in the circuit has a maximum equal to $I_0 = V/R$. In Fig. 6.1, we show a plot of the ratio of the current amplitude *I* to its maximum value I_0 for three different values of the circuit quality factor Q_s ($Q_s = 1, 10, 50$). We can see that the parameter Q_s plays the role of a *scale factor* and determines the width of the *resonance peak* of the current: *the higher the value of* Q_s, *the narrower the bell shaped current plot*. It is evident in this figure that the value of Q_s does not have any influence on the value of the resonant angular frequency ω_0.

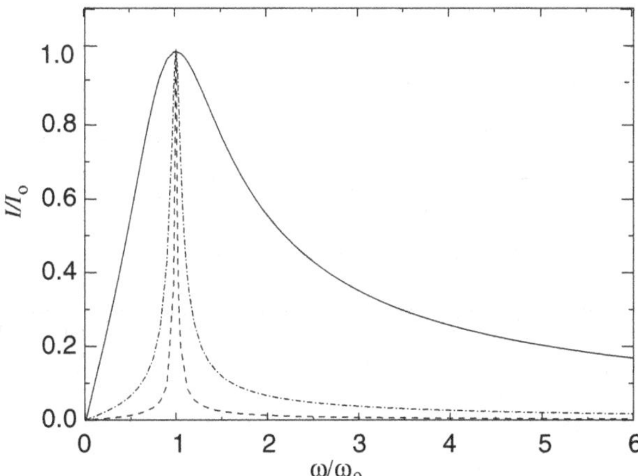

Fig. 6.1 Behavior of the normalized current I/I_0, with $I_0 = V/R$, as a function of the "reduced angular frequency" ω/ω_0 in a series resonant circuit. The three curves refer to three different values of the quality factor Q_s of the circuit (1, 10 and 50 from the wider to the narrower curve). The behavior illustrated in this plot is also valid for a *RLC* parallel circuit, as explained in the text

In summary, the solution of the circuit *RLC series* of Fig. 5.2, powered by a sinusoidal voltage generator with output amplitude $v_g(t) = V_g \cos \omega t$, consists of the following expressions:

Complex current

$$I = \frac{V_g}{R + j\left(\omega L - \dfrac{1}{\omega C}\right)} = \frac{V_g}{R} \cdot \frac{1}{1 + jQ_s\left(\dfrac{\omega}{\omega_0} - \dfrac{\omega_0}{\omega}\right)} \tag{6.3}$$

Complex voltage drop across *R*, *C* and *L*

$$V_R = RI = \frac{RV_g}{R + j\left(\omega L - \dfrac{1}{\omega C}\right)} = V_g \frac{1}{1 + jQ_s\left(\dfrac{\omega}{\omega_0} - \dfrac{\omega_0}{\omega}\right)} \tag{6.4}$$

$$V_C = Z_C I = \frac{V_g}{j\omega RC - \omega^2 LC + 1} = -j\frac{\omega_0}{\omega} \frac{V_g Q_s}{1 + jQ_s\left(\dfrac{\omega}{\omega_0} - \dfrac{\omega_0}{\omega}\right)} \tag{6.5}$$

$$V_L = Z_L I = \frac{-\omega^2 LC V_g}{j\omega RC - \omega^2 LC + 1} = j\frac{\omega}{\omega_0} \frac{V_g Q_s}{1 + jQ_s\left(\dfrac{\omega}{\omega_0} - \dfrac{\omega_0}{\omega}\right)} \tag{6.6}$$

Current

$$i(t) = \frac{V_g}{R} \cdot \frac{1}{\sqrt{1 + Q_s^2\left(\dfrac{\omega}{\omega_0} - \dfrac{\omega_0}{\omega}\right)^2}} \cos\left[\omega t - \arctan Q_s\left(\frac{\omega}{\omega_0} - \frac{\omega_0}{\omega}\right)\right] \tag{6.7}$$

Voltage drop across *R*, *C* and *L*

$$v_R(t) = \frac{V_g}{\sqrt{1 + Q_s^2\left(\dfrac{\omega}{\omega_0} - \dfrac{\omega_0}{\omega}\right)^2}} \cos\left[\omega t - \arctan Q_s\left(\frac{\omega}{\omega_0} - \frac{\omega_0}{\omega}\right)\right] \tag{6.8}$$

$$v_C(t) = \frac{\omega_0}{\omega} \cdot \frac{V_g Q_s}{\sqrt{1 + Q_s^2\left(\dfrac{\omega}{\omega_0} - \dfrac{\omega_0}{\omega}\right)^2}} \cos\left[\omega t - \arctan Q_s\left(\frac{\omega}{\omega_0} - \frac{\omega_0}{\omega}\right) - \frac{\pi}{2}\right] \tag{6.9}$$

$$v_L(t) = \frac{\omega}{\omega_0} \cdot \frac{V_g Q_s}{\sqrt{1 + Q_s^2\left(\dfrac{\omega}{\omega_0} - \dfrac{\omega_0}{\omega}\right)^2}} \cos\left[\omega t - \arctan Q_s\left(\frac{\omega}{\omega_0} - \frac{\omega_0}{\omega}\right) + \frac{\pi}{2}\right] \tag{6.10}$$

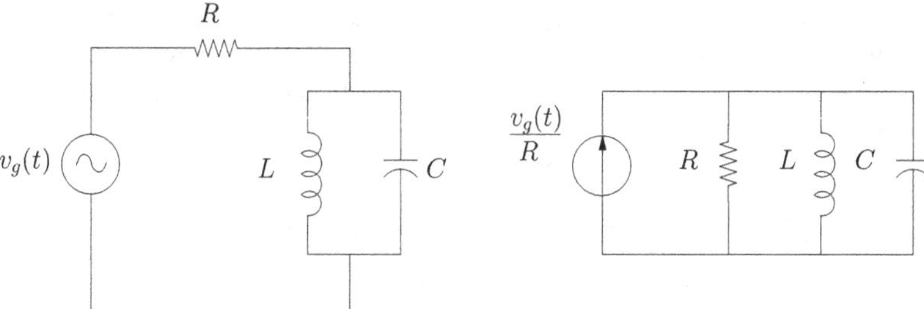

Fig. 6.2 The circuit *RLC* parallel. On the *right*, the equivalent circuit after Norton's theorem is applied to the voltage generator and the resistance R

6.2.2 The Circuit RLC Parallel

In Fig. 6.2, we show the diagram of a circuit referred to as the parallel resonant *RLC* circuit because of the parallel connection between the capacitance C and the inductance L. This circuit is solved rather easily replacing the voltage generator $v_g(t)$ and the resistance R with their Norton's equivalent circuit, as shown on the right hand side of Fig. 6.2. In the transformed circuit, Norton's current generator ($i(t) = v_g(t)/R$) is connected to a load consisting of the parallel of the three components R, L and C. We apply the symbolic method to compute the impedance seen by the current generator:

$$\frac{1}{Z} = \frac{1}{R} + \frac{1}{j\omega L} + j\omega C = \frac{1}{R} + j\left(\omega C - \frac{1}{\omega L}\right)$$

yielding:

$$Z = \left[\frac{1}{R} + j\left(\omega C - \frac{1}{\omega L}\right)\right]^{-1} = \frac{R}{1 + jR\left(\omega C - \frac{1}{\omega L}\right)} = \frac{R}{1 + j\frac{RC}{\sqrt{LC}}\left(\omega\sqrt{LC} - \frac{1}{\omega\sqrt{LC}}\right)}$$

Using the definition of the resonant angular frequency given above $\omega_0 = 1/\sqrt{LC}$ and defining the *quality factor* of the resonant parallel circuit as $Q_p = \omega_0 RC$ (note that with the same components $Q_p = 1/Q_s$), the complex impedance Z becomes:

$$Z = \frac{R}{1 + jQ_p\left(\frac{\omega}{\omega_0} - \frac{\omega_0}{\omega}\right)} \tag{6.11}$$

The module $|Z|$ of the impedance and its phase ϕ are:

$$|Z| = \frac{R}{\sqrt{1 + Q_p^2\left(\frac{\omega}{\omega_0} - \frac{\omega_0}{\omega}\right)^2}} \qquad \phi = -\arctan\left[Q_p\left(\frac{\omega}{\omega_0} - \frac{\omega_0}{\omega}\right)\right] \tag{6.12}$$

The voltage $V(\omega)$, across the impedance Z, is obtained by the product of the Norton's equivalent current $I = V/R$ and the impedance Z. This allows computing the expression for the ratio of the amplitude of $V(\omega)$ to the amplitude of the voltage generator output V_g:

$$\frac{|V(\omega)|}{|V_g|} = \frac{1}{\sqrt{1 + Q_p^2\left(\frac{\omega}{\omega_0} - \frac{\omega_0}{\omega}\right)^2}} \tag{6.13}$$

Note that this equation gives the frequency dependence of the voltage drop across the parallel of L and C and is similar to Eq. (6.1) we found for the frequency dependence of the current flowing in the RLC series circuit. Therefore, the plots shown in Fig. 6.1 can be used to illustrate the behavior of the voltage across the parallel LC, once the label of the y-axis is changed as needed.

In conclusion of our analysis of the RLC parallel circuit, we give in the following the formulas for the current flowing in its components, leaving the proof as an exercise to the reader:

$$i_R(t) = \frac{V_g}{R}\left(1 - \frac{1}{\sqrt{1 + Q_p^2\left(\frac{\omega}{\omega_0} - \frac{\omega_0}{\omega}\right)^2}}\right)\cos\left[\omega t - \arctan Q_p\left(\frac{\omega}{\omega_0} - \frac{\omega_0}{\omega}\right)\right] \tag{6.14}$$

$$i_C(t) = \frac{\omega}{\omega_0}\cdot\frac{V_g Q_p/R}{\sqrt{1 + Q_p^2\left(\frac{\omega}{\omega_0} - \frac{\omega_0}{\omega}\right)^2}}\cos\left[\omega t - \arctan Q_p\left(\frac{\omega}{\omega_0} - \frac{\omega_0}{\omega}\right) + \frac{\pi}{2}\right] \tag{6.15}$$

$$i_L(t) = \frac{\omega_0}{\omega}\cdot\frac{V_g Q_p/R}{\sqrt{1 + Q_p^2\left(\frac{\omega}{\omega_0} - \frac{\omega_0}{\omega}\right)^2}}\cos\left[\omega t - \arctan Q_p\left(\frac{\omega}{\omega_0} - \frac{\omega_0}{\omega}\right) - \frac{\pi}{2}\right] \tag{6.16}$$

6.2.3 The Resonant Behavior

The equations deduced in the previous paragraphs show that in a resonant *RLC series* circuit the voltage drop V_L across the inductance is always in phase opposition to the voltage drop V_C across the capacitance. This means that V_L shows a maximum when V_C has a minimum and vice versa. At the resonant angular frequency ω_0, these two quantities are also equal in amplitude and their sum is identically null for all time values: the circuit behaves as consisting of a resistor connected to a voltage generator.

However, taken separately, both V_L and V_C can have very high amplitudes that at $\omega = \omega_0$ reach a maximum equal to Q_s times the generator output voltage V. In particular, when the value of the resistance R becomes very small we can observe very high values for both V_L and V_C with a very small excitation. In the limit of vanishing resistance, the current flowing in the circuit diverges at the resonance. In practice, this means that we can observe a finite current in the circuit even in absence of excitation. This observation is related to the fact that in these conditions the solution of the homogeneous differential equation, associated to Eq. (5.8) describing the circuit behavior, is stationary and not transient as usual. This is easily seen using the symbolic version of this equation:

$$\left(-\omega^2 L + \frac{1}{C} + j\omega R\right) I = j\omega V$$

and noting that when $V = 0$ we can obtain a non-trivial solution only when the coefficient of I vanishes, that is when

$$\omega = \frac{jR \pm \sqrt{4L/C - R^2}}{2L} = j\frac{R}{2L} \pm \frac{1}{\sqrt{LC}}\sqrt{1 - \frac{R^2 C}{4L}} = j\omega_i \pm \omega_r$$

In general, the presence of an imaginary part of the frequency means that the time evolution of the signal is a damped oscillation. Indeed, since in this case we can write $e^{j\omega t} = e^{j\omega_r t} e^{-\omega_i t}$, the characteristic decay time τ, corresponding to the time interval necessary for the amplitude to decrease of a factor $1/e$, is equal to $1/\omega_i$, while ω_r is the angular frequency of the oscillation. In terms of the circuit parameters, we obtain $\tau = 2L/R$ and when R becomes null the decay time tend to infinity. In this case, there is no more energy dissipation by Joule effect in the resistance and the system can oscillate indefinitely at the resonance angular frequency ω_0.

In physics, every time a system exhibits such a behavior, we say that it has a *normal mode of oscillation*. The dynamics of a *normal mode* is characterized by the time evolution of the energy injected in the system. In general, it consists of an oscillation between different forms. In the present case, we need to consider the magnetic energy stored in the inductance, equal to $LI^2/2$, and the electrostatic energy stored in the capacitance and equal to $CV_C^2/2$. Using the expressions given above for the current flowing in the *RLC series* circuit and for the voltage drop across its capacitance, we can easily show that at the resonant frequency the energy changes from magnetic to electrostatic twice in each oscillation cycle. This is ultimately the same dynamics shown by kinetic and potential energy in the harmonic oscillator, which in fact is the prototype of all resonant physical systems.

However, in practice, as in any real physical system, dissipation is never absent but when R is sufficiently small, oscillation damping is sufficiently small and allows observing a large number of cycles. We can approximate this number to Q_s/π, the product of the decay time τ and the resonance frequency. Obviously, the higher this number, the better we can detect the resonance, hence the name quality factor for the parameter Q_s.

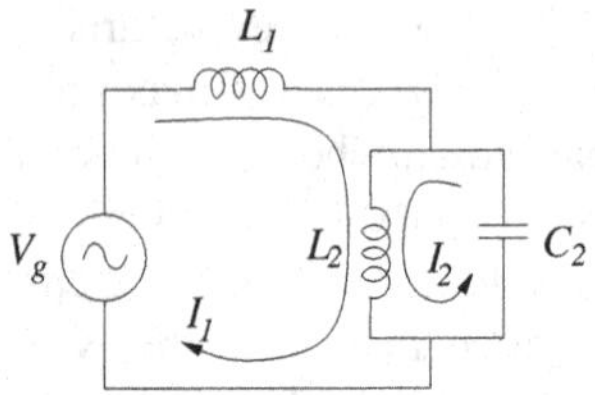

Fig. 6.3 A resonant circuit with two inductances

A different way of quantifying the quality of a resonance consists in comparing the total energy in the system with the energy dissipated in an oscillation cycle. In a resonant *RLC series* circuit, the total energy is equal to $LI_{\max}^2/2 = LI_{rms}^2$ and the dissipated energy is $RI_{rms}^2 2\pi/\omega$ while their ratio is $Q_s/(2\pi)$. At higher quality factor, a smaller amount of energy is dissipated to perform an oscillation cycle.

The behavior illustrated above is quite general: every physical system described by a linear set of differential equations can exhibit normal modes. This happens when the associated homogeneous system shows non-trivial stationary solutions. This analysis is conveniently done exploiting the symbolic method to transform differential equations in a system of algebraic equations. In this case, normal modes can be found looking for values of ω that make the determinant of the coefficient matrix equal to zero.

With this approach, we analyze the circuit of Fig. 6.3. Using the symbolic method and applying the method of loops for the circuit solution, we obtain the following equation system for the two loop currents shown in figure:

$$\begin{cases} (j\omega L_1 + j\omega L_2)I_1 + j\omega L_2 I_2 & = V \\ j\omega L_2 I_1 + \left(j\omega L_2 + \frac{1}{j\omega C_2}\right) I_2 & = 0 \end{cases}$$

Setting to zero it coefficients determinant, we obtain:

$$\omega^2 = \frac{L_1 + L_2}{L_1 L_2} \cdot \frac{1}{C_2} = \frac{1}{L_p C_2}$$

This is the resonance frequency we were looking for, as it can be easily understood by noting that, when the generator is short circuited, the circuit is equivalent to a *LC* parallel where the inductance is obtained with the parallel connection of the two inductors L_1 and L_2.

6.3 Resonant Circuits with Real Components

The study of the series and parallel resonant circuits presented in the previous section has been simplified assuming that all the reactive components involved, inductor and capacitor, were ideal. In reality, however, as discussed in the first chapter of this book, a real inductor has always a series resistance R_L, due to the finite resistivity of the wire used for its coil and possibly to the power dissipation in the ferromagnetic core. Similarly, the power losses in the dielectric of the capacitor can be accounted for with a resistance R_C in series to the capacitance C. With these two modifications, we can solve again the resonant series circuit with the same procedure used in Sect. 6.2.1. We can easily find that the expression for the circuit current is still done by Eq. (6.3) where now we need to add to resistance R both R_L and R_C. The complex voltage across each component can be computed again by the product of the current and the component complex impedance keeping in mind that we need to add reactive and resistive contribution for the inductor and the capacitor.

The resonance frequency turns out the same of the ideal case. At resonance, the current flowing in the circuit remains in phase with the generator voltage and the voltage drops across reactive components are still in phase opposition. However now they are not equal in amplitude and, as a consequence, only a fraction $R/(R+R_L+R_C)$ of the applied voltage is found across the resistance R. On the contrary, the circuit quality factor is affected by the power dissipation in the reactive components. Its expression becomes:

$$Q_s' = \frac{\omega_o L}{R + R_L + R_C}$$

We can obtain a simple representation of nonideal effects considering the inverse of this expression. We obtain:

$$\frac{1}{Q_s'} = \frac{R + R_L + R_C}{\omega_o L} = \frac{R}{\omega_o L} + \frac{R_L}{\omega_o L} + \frac{R_C}{\omega_o L} = \frac{1}{Q_s} + \frac{1}{Q_L} + \frac{1}{Q_C}$$

where $Q_s = \omega_o L/R$ is the ideal quality factor, $Q_L = \omega_o L/R_L$ is the quality factor of the inductor[3] at the angular frequency ω_o and $Q_C = \omega_o L/R_C = 1/\omega_o C R_C$ is the analogous quality factor for the capacitor.

In the parallel resonant circuit, nonideal components have similar effects on the Q factor and, in addition, also affect the value of the resonance frequency.[4] This statement can be more easily proved using a transformation of the impedance of real component, known as the series-parallel transformation, which we illustrate in the next section.

6.4 Series-Parallel Transformation

In this section, we show that the impedance of the series connection of a reactance and a resistance is equivalent to the impedance of the parallel connection of an appropriate reactance with a suitable resistance and vice versa. In the following, we prove

---

[3]This is the quality factor we would observe when the inductor goes in resonance at the angular frequency ω_o with an ideal capacitor and a short circuit replacing R.

[4]For a detailed computation, see Problems 13 and 14 at the end of this chapter.

this statement and we give the expression for the parameters of the transformation. To simplify the discussion, we start our proof considering the impedance Z of a capacitance C_p connected in parallel to a resistance R_p

$$Z = \left(\frac{1}{R_p} + j\omega C_p\right)^{-1} = \frac{R_p}{1 + (\omega R_p C_p)^2} + \frac{1}{j\omega C_p(1 + \frac{1}{(\omega R_p C_p)^2})}$$

This impedance is equivalent to the series of a resistance R_s and a capacitance C_s where

$$R_s = \frac{R_p}{1 + (\omega R_p C_p)^2} \quad \text{and} \quad C_s = C_p\left(1 + \frac{1}{(\omega R_p C_p)^2}\right)$$

The product of these two relations yields $\omega R_s C_s = 1/\omega R_p C_p$, so that we can invert them to obtain:

$$R_p = R_s\left(1 + \frac{1}{(\omega R_s C_s)^2}\right) \quad \text{and} \quad C_p = \frac{C_s}{1 + (\omega R_s C_s)^2}$$

This shows that the impedance of a resistance R_s in series with a capacitance C_s is equivalent to the impedance of the resistance R_p in parallel to the capacitance C_p both derived by using the two previous relations.

With the same line of reasoning followed above, we can generalize the previous result to the combination of a resistance and a generic reactance. We obtain for the impedance of the parallel connection:

$$Z = \left(\frac{1}{R_p} + \frac{1}{jX_p}\right)^{-1} = \frac{R_p}{1 + \left(\frac{R_p}{X_p}\right)^2} + \frac{jX_p}{1 + \left(\frac{X_p}{R_p}\right)^2} = R_s + jX_s$$

that we can use to obtain $(R_p/X_p)^2 = (X_s/R_s)^2$ leading to:

$$R_p = R_s\left[1 + \left(\frac{X_s}{R_s}\right)^2\right] \quad \text{and} \quad jX_p = jX_s\left[1 + \left(\frac{R_s}{X_s}\right)^2\right]$$

We are now in the position to reconsider the case of a parallel *RLC* circuit with real components. By applying a series-parallel transformation to its reactive components, we can obtain a circuit that is formally identical to the ideal case (see Fig. 6.4). Contrary to the case of the series circuit, now the values of both the inductance and the capacitance are frequency dependent. Indeed, we have:

$$L' = L\left[1 + \left(\frac{R_L}{\omega L}\right)^2\right] \quad \text{e} \quad C' = C\frac{1}{1 + (\omega C R_C)^2}$$

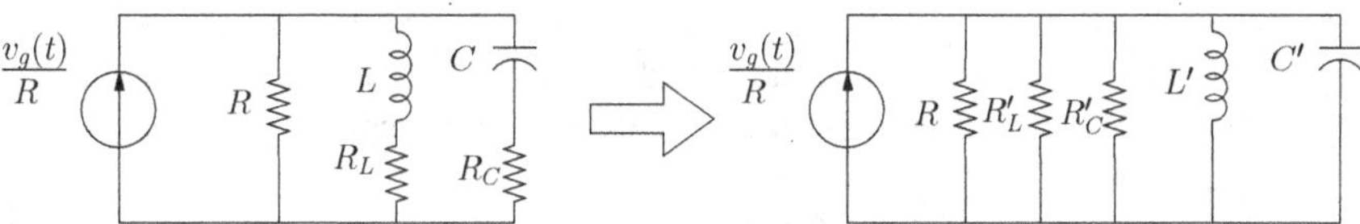

Fig. 6.4 Series-parallel transformation of resonant parallel circuit. The circuit obtained, shown on the *right hand side*, is formally equivalent to the circuit with ideal components

Therefore, the resonance frequency is equal to the ideal case only when $R_L/\omega L \ll 1$ and $\omega CR_C \ll 1$, limits that need to be well satisfied to approximate ideal behavior with real components. For the two parallel resistors, we have $R'_L = R_L[1 + (\omega L/R_L)^2] \simeq \omega^2 L^2/R_L$ and $R'_C = R_C[1 + 1/(\omega CR_C)^2] \simeq 1/\omega^2 C^2 R_C$, where the last two expressions hold in the limits given before. We need to consider the parallel of these two resistances with R to compute the value of the quality factor Q'_p. For its inverse, we easily get:

$$\frac{1}{Q'_p} = \frac{1}{\omega_o C}\left(\frac{1}{R} + \frac{1}{R'_C} + \frac{1}{R'_L}\right) = \frac{1}{Q_p} + \omega_o CR_C + \frac{R_L}{\omega_o L} = \frac{1}{Q_p} + \frac{1}{Q_C} + \frac{1}{Q_L}$$

where Q_p is the ideal value, while Q_C and Q_L are, respectively, the quality factor of the capacitor and of the inductor we already defined in the previous section.

6.5 Bridge Circuits

In Fig. 6.5, we show the AC circuit equivalent to the *Wheatstone's bridge* we already studied for DC currents. In the four branches of the bridge, we have four generic impedances Z_1, Z_2, Z_3 and Z_4; we can solve the circuit with the symbolic method following the procedures used for the DC case. Here, we adopt a simplified approach to find the equilibrium condition of the bridge. With reference to Fig. 6.5, at equilibrium we have $V_A = V_B$. Indicating with I_1 and I_2 the currents flowing, respectively,

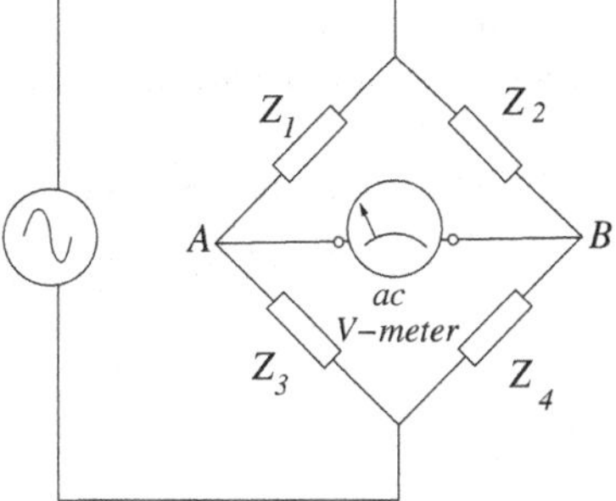

Fig. 6.5 Wheatstone's bridge for alternating currents. When the voltmeter is ideal, the current flowing through it is null

in Z_1 and in Z_2, at equilibrium the currents flowing in Z_3 and in Z_4 are, respectively, equal to I_1 and I_2 independently of the impedance of the AC voltmeter connected to the two nodes. The equilibrium requirement $V_A = V_B$ implies:

$$Z_1 I_1 = Z_2 I_2, \quad \text{and} \quad Z_3 I_1 = Z_4 I_2$$

Taking the ratio of these two relations, currents I_1 and I_2 can be eliminated and we obtain for the *equilibrium condition*:

$$\frac{Z_1}{Z_3} = \frac{Z_2}{Z_4} \tag{6.17}$$

or equivalently:

$$\frac{Z_1}{Z_2} = \frac{Z_3}{Z_4} \tag{6.18}$$

Note that the equality between two complex numbers implies the equality for both their amplitudes and their arguments. For this last quantity, we have:

$$\arg\left[\frac{Z_1}{Z_2}\right] = \arg[Z_1] - \arg[Z_2] = \arg\left[\frac{Z_3}{Z_4}\right] = \arg[Z_3] - \arg[Z_4]$$

and therefore,

$$\arg[Z_1] - \arg[Z_2] = \arg[Z_3] - \arg[Z_4]$$

Bridge circuits take different names depending on the kind of impedance in use. They are exploited mainly for the measure of reactive components, but their use has been much reduced since the introduction of digital instruments. Often, for practical convenience, two of the impedances used in a bridge consist of simple resistances. Let us then assume $Z_1 = R_1$ and $Z_2 = R_2$, pure real quantities. In this case, the bridge equilibrium condition (6.17) is more conveniently written as

$$R_1 Z_4 = R_2 Z_3 \tag{6.19}$$

Supposing that Z_4 is an unknown ideal inductance L_4, this relation shows immediately that we need an inductor of known inductance L_3 to measure L_4. In this case the equilibrium conditions yield $L_4 = R_2 L_3 / R_1$ independently of the generator frequency. It is also evident that to attain the bridge equilibrium we need that at least one of the quantities R_1, R_2 or L_3 should be variable continuously in a known manner. Similarly, when Z_4 represents a capacitance C_4, it is necessary to have a capacitor with known capacitance C_3 and the equilibrium condition yields

$$C_4 = \frac{R_1 C_3}{R_2} \tag{6.20}$$

However, to design an electrical measurement with a bridge circuit, we must keep in mind that very seldom it is sufficient to characterize a real component with just one parameter. For example, in the case of an inductor we need to account also for its series resistance. In this case $Z_4 = R_4 + j\omega L_4$ with both L_4 and R_4 unknown. When we are not interested in the measurement of R_4, a possible strategy consists in using the bridge at a frequency high enough to have $\omega L_4 \gg R_4$ so that we can neglect the inductor resistance and proceed as in the ideal case. However, we must keep in mind that at high frequency the coil inter-turn capacitance becomes important and induces an error in the inductance value. When this happens, or when we are interested to the value of the inductor resistance, we need to introduce a resistance in branch number 3 of the bridge in Fig. 6.5. In this case we have $Z_3 = R_3 + j\omega L_3$ and the complex equilibrium condition yields $R_4 = R_2R_3/R_1$ for the real part, while the imaginary part becomes $L_4 = R_2L_3/R_1$, independently of the generator frequency. Such a bridge can be balanced in two steps. First, we can use a variable R_3 in DC operation to achieve the first of the two conditions above. Then, switching to AC operation, we can balance the imaginary part with a variable L_3 or, alternatively, by changing both R_1 and R_3 but maintaining constant their ratio.

When bridge equilibrium does not depend upon frequency, we can choose this parameter to ease the measurement. This can be achieved by making the voltage drops across the different branches all comparable among themselves to reduce the effect of ambient noise. In this case, we can set the frequency by making ωL_4 roughly equal to R_2.

For the case of a real capacitor, we recall that dielectric losses are the principal nonideal effect. These losses are accounted by a resistor in parallel to the ideal capacitance and the total impedance of the capacitor can be written as $Z_4 = R_4/(1 + j\omega R_4C_4)$. As for the case of the inductive bridge, we need a resistor R_3 in parallel to C_3 so that $Z_3 = R_3/(1 + j\omega R_3C_3)$. In this case, the equilibrium condition becomes:

$$R_1R_4(1 + j\omega R_3C_3) = R_2R_3(1 + j\omega R_4C_4)$$

yielding $R_4 = R_2R_3/R_1$ for the real part while and $C_4 = R_1C_3/R_2$ for the imaginary, again independently of the frequency.

An interesting alternative to the capacitive bridge is the Wien's bridge where the resistor R_3 is connected not in parallel but in series to the capacitor C_3. In this case $Z_3 = R_3 - j/\omega C_3$ and the equilibrium condition becomes

$$\frac{R_1R_4}{1 + j\omega R_4C_4} = R_2\left(R_3 - j\frac{1}{\omega C_3}\right)$$

that yields

$$R_1R_4 = R_2R_3 + R_2R_4C_4/C_3 + jR_2R_3\left(\omega R_4C_4 - \frac{1}{\omega C_3}\right)$$

Its real part requires $R_1/R_2 = R_3/R_4 + C_4/C_3$ while the imaginary part yields a relation dependent upon the frequency: $\omega^2 R_4 R_3 C_4 C_3 = 1$. In this case, it is convenient to use the generator frequency, a quantity that can be easily changed and accurately measured, to balance the imaginary part, and a simple variable resistor for the real part.

6.6 Two-Port Circuits

Two-port circuits are used to transform and/or manipulate electrical signals. Generally speaking, they can be represented as a black box indicating a generic circuit with four terminals giving access to four of its nodes, see Fig. 6.6. The terminals are coupled in *ports*, one being the *input port* where the input signal is connected, the other, named the *output port*, provides the transformed signal for further use. Often, the distinction between input and output is not structural but only functional, as it happens in passive circuits (circuits not including voltage or current generators). In this case, it would be possible to exchange the role of the two ports to obtain a different circuit function.

6.6.1 Two-Port Circuit Characteristics

In this discussion of two-port circuits, we limit ourselves to *linear* networks that we will solve with the symbolic method. This implies that we assume to send a sinusoidal signal to the input port, with a given angular frequency ω and amplitude V_i. Under these assumptions the parameters we need to characterize a linear two-port circuit are:

- *The (complex) input impedance* Z_i, defined as the ratio of the complex voltage drop applied to the input port and the complex current entering the port: $Z_i(\omega) = V_i(\omega)/I_i(\omega)$
- *The (complex) output impedance* Z_u, defined as the ratio of the open circuit voltage and the short circuit current at the output port $Z_u(\omega) = V_{u\,oc}(\omega)/I_{u\,sc}(\omega)$. This relation is a simple corollary to the Thévenin and Norton's theorems.
- *The transfer function* $H(\omega)$, defined as the ratio of the output (complex) voltage V_u to the input (complex) voltage $V_i : H(\omega) = V_u(\omega)/V_i(\omega)$. For passive circuits,

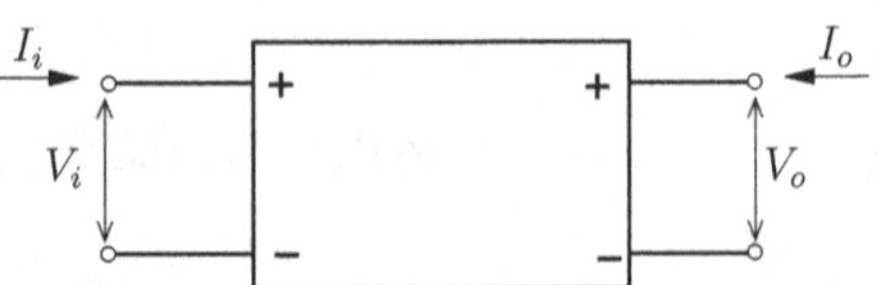

Fig. 6.6 A generic two-port circuit

since typically $|H(\omega)| < 1$, the transfer function is referred to as the *Attenuation* and indicated by $A(\omega)$.

For the time being, we will neglect both the input and the output impedance when discussing the behavior of two-port circuits. This amounts to assuming that the input is connected to an ideal signal generator, whose input impedance is equal to zero, and that the output voltage is measured by an ideal voltmeter having internal impedance equal to infinity. With these approximations, we will evaluate the attenuation of simple but important passive two-port circuits.

6.6.2 The Transfer Function and Bode's Diagrams

The transfer function in general is a complex function of the angular frequency ω (alternatively of the frequency $\nu = \omega/2\pi$). Its properties are usually represented by two diagrams first introduced by Bode.

Bode's diagram for the amplitude. It consists of a plot of the amplitude of the transfer function $|H(\omega)|$ versus the angular frequency ω (or the frequency ν) with logarithmic scales. For the amplitude of the transfer function, the unit of measure adopted is the *decibel* (*dB*). The decibel is a way to represent magnification or attenuation, used in many fields of physics and engineering, whose definition[5] is

$$\text{expression of } |H| \text{ in dB} = 20 \log_{10} |H|$$

Bode's diagram for the phase. It consists in a plot of the argument, or phase, of $H(\omega)$, measured in radians or degrees, versus the angular frequency ω (or the frequency ν) in logarithmic scale.

It can be shown, as first done by Bode (see reference [5]), that the causality principle implies a link between the amplitude and the phase of a transfer function. A detailed discussion of this link requires notions that are usually acquired later during the course of studies.

Bode's diagram for the amplitude is in practice a log–log plot, i.e., a plot of the logarithm of the function versus the logarithm of the independent variable. In this representation a power law, $A(\omega) = \omega^{\alpha}$, becomes a straight line whose angular coefficient is the exponent α. For example, the function $\omega^{\pm 1}$ will be represented by a straight line of slope ± 20 dB/decade,[6] the function $\omega^{\pm 2}$ by a straight line of slope ± 40 dB/decade, and so on. In the study of linear circuits with lumped parameters, functions as those described above are very common and the use of Bode's diagrams

[5]In some application the quantity of interest is the ratio A_p of two *powers* instead of two amplitudes. Since power is proportional to the square of the amplitude, the decibel definition changes to $10 \log_{10} A_p$. Actually this is its original definition, since the decibel was introduced in 1928 by Bell laboratories to measure the power loss of telephone signals.

[6]A decade is a range of numerical values whose end points have a ratio of 10. In a logarithmic scale, all decades have the same length.

is helpful to identify them. In the following, we will come across many examples of Bode's diagrams. To help the reader to become acquainted with the decibels, of common use in electronics and other applications, we report in the following table a detailed list of magnifications/attenuations and the corresponding decibel values.

Ratio	Decibel	Ratio	Decibel
100.00	40.00	0.01	−40.00
10.00	20.00	0.10	−20.00
3.16	10.00	0.316	−10
$\sqrt{2} \simeq 1.41$	3.01	$1/\sqrt{2} \simeq 0.71$	−3.01
2.00	6.02	0.50	−6.02
1.00	0.00		

6.7 The Low-Pass RC Circuit

Consider a circuit consisting of the series of a resistance R and a capacitance C, as shown in Fig. 6.7a. Assume the circuit is powered by a sinusoidal voltage generator with angular frequency ω and amplitude V_i. We are interested in computing the voltage drop across the terminals of the capacitance C, the output voltage of this two-port circuit. Using the symbolic method, the current[7] I is given by:

$$I = \frac{V_i}{R + \dfrac{1}{j\omega C}} \tag{6.21}$$

The output voltage V_u is obtained as the product of this current and the impedance of the capacitance:

$$V_u = IZ_C = \frac{V_i}{R + \dfrac{1}{j\omega C}} \frac{1}{j\omega C} = \frac{V_i}{1 + j\omega RC} = \frac{V_i}{1 + j\dfrac{\omega}{\omega_0}}$$

where in the last expression, we introduced the parameter $\omega_0 = 1/RC$, referred to as the *cutoff angular frequency* ($\nu_0 = 1/2\pi RC$ is the corresponding cutoff frequency). The quantity $\tau = RC = 1/\omega_0$, inverse of the reduced angular frequency, is usually referred to as the circuit *characteristic time* for a reason that will become apparent when we will deal with the same circuit in the time domain, see Chap. 9. It can be easily shown that the product RC has the physical dimension of a time, as required:

[7] More precisely, we should say the complex representation of the current, but from here on this will be understood.

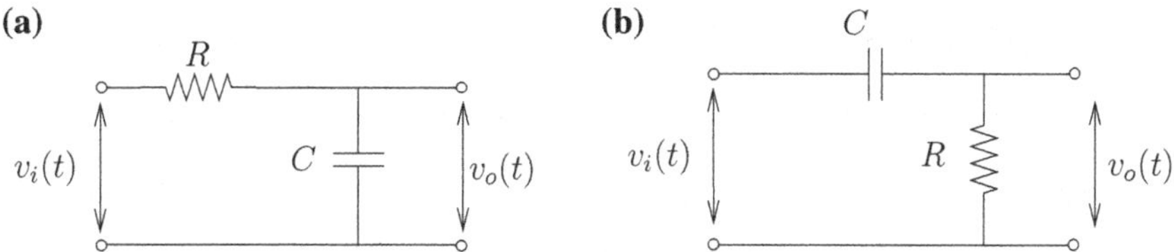

Fig. 6.7 *RC* circuit low-pass (**a**) and high-pass (**b**)

$$[RC] = [R] \cdot [C] = \frac{[\text{Voltage}]}{[\text{Current}]} \cdot \frac{[\text{Charge}]}{[\text{Voltage}]} = [\text{Time}]$$

The transfer function is obtained, by its definition, as the ratio of the complex representations of the output to the input voltage:

$$A(\omega) = \frac{V_u}{V_i} = \frac{1}{1 + j\dfrac{\omega}{\omega_0}} \tag{6.22}$$

Therefore, the amplitude and the phase of the attenuation of the *low-pass RC* circuit are:

$$|A(\omega)| = \frac{1}{\sqrt{1 + \left(\dfrac{\omega}{\omega_0}\right)^2}} = \frac{\omega_0}{\sqrt{\omega_0^2 + \omega^2}}; \quad \phi = -\arctan\frac{\omega}{\omega_0} \tag{6.23}$$

Note that in the two previous expressions, the values of the components (R and C) appear only through their product RC, and, therefore, the behavior of the circuit is defined by only one parameter, in this case the characteristic time RC. Later in this chapter, we will find the same result for *RL* circuits, which together with *RC* circuits are referred to as *one-parameter circuits*.

When the angular frequency is equal to its cutoff value, $\omega = \omega_0$, we have $|A(\omega)| = 1/\sqrt{2} \simeq -3.0\,\text{dB}$. Figure 6.8 shows the frequency plot of $|A(\omega)|$, expressed both in absolute and in decibel units, and of the phase. Inspection of the behavior of $|A(\omega)|$, in Fig. 6.8 explains why the circuit is qualified as *low-pass*. Indeed, signals of frequency lower than the cutoff frequency suffer of little attenuation while above the cutoff, the amplitude of the output signal decreases in proportion of the inverse of the frequency. Circuits acting selectively on signals, allowing only the passage of some frequencies are said filter circuits. For this reason, the *RC* circuit we are dealing with is a *low-pass filter* since it allows the passage of low-frequency signals from its input to its output while blocking the rest of the frequency spectrum.

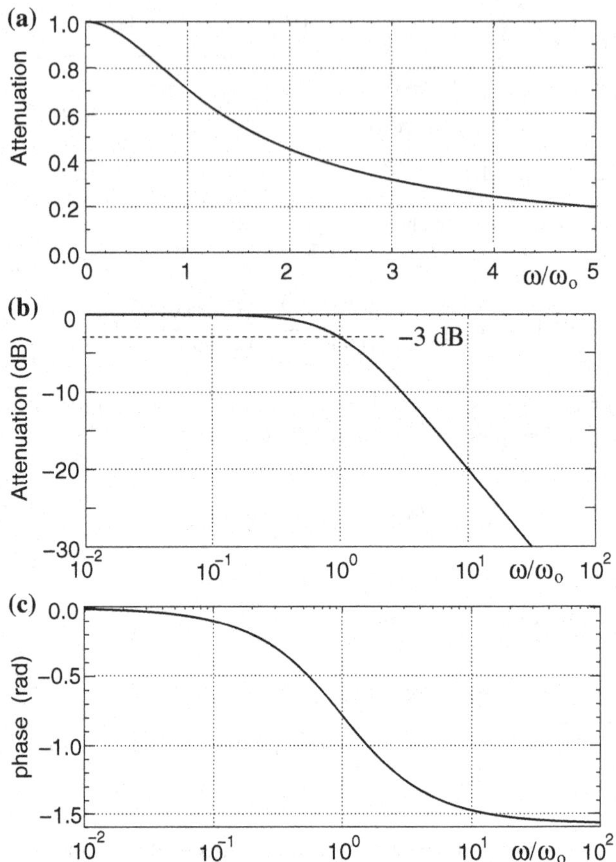

Fig. 6.8 Plots pertaining to the *low-pass RC circuit.* **a** linear plot of the attenuation ($V_u(\omega)/V_i$). In abscissa, we use the "reduced angular frequency" $\omega\tau = \omega/\omega_0$. **b** *Bode's* diagram of the *low-pass RC circuit.* In abscissa, we use the reduced angular frequency, on the vertical axis the attenuation is reported in decibel. The horizontal line at −3.01 dB corresponds to the attenuation of $1/\sqrt{2}$ obtained for $\omega = \omega_0$, **c** log-plot of the phase shift introduced by the circuit as a function of the reduced angular frequency

6.8 The High-Pass RC Circuit

The high-pass *RC* circuit is obtained from the low-pass by exchanging resistance and capacitance among themselves, as shown in Fig. 6.7b. Obviously the expression for the current is still given by the Eq. (6.21) and the output voltage in this case is obtained as the product of the current and the resistance *R*:

$$V_u = IR = \frac{V_i}{R + \dfrac{1}{j\omega C}} R = V_i \frac{j\omega RC}{1 + j\omega RC} = V_i \frac{j\omega\tau}{1 + j\omega\tau} = V_i \frac{j\omega/\omega_0}{1 + j\omega/\omega_0}$$

It follows that the attenuation of the high-pass RC circuit is:

$$A(\omega) = \frac{j\omega/\omega_0}{1 + j\omega/\omega_0} \tag{6.24}$$

while its amplitude and phase are:

$$|A(\omega)| = \frac{\dfrac{\omega}{\omega_0}}{\sqrt{1 + \left(\dfrac{\omega}{\omega_0}\right)^2}} = \frac{\omega}{\sqrt{\omega_0^2 + \omega^2}}; \quad \phi = \arctan\frac{\omega_0}{\omega} \tag{6.25}$$

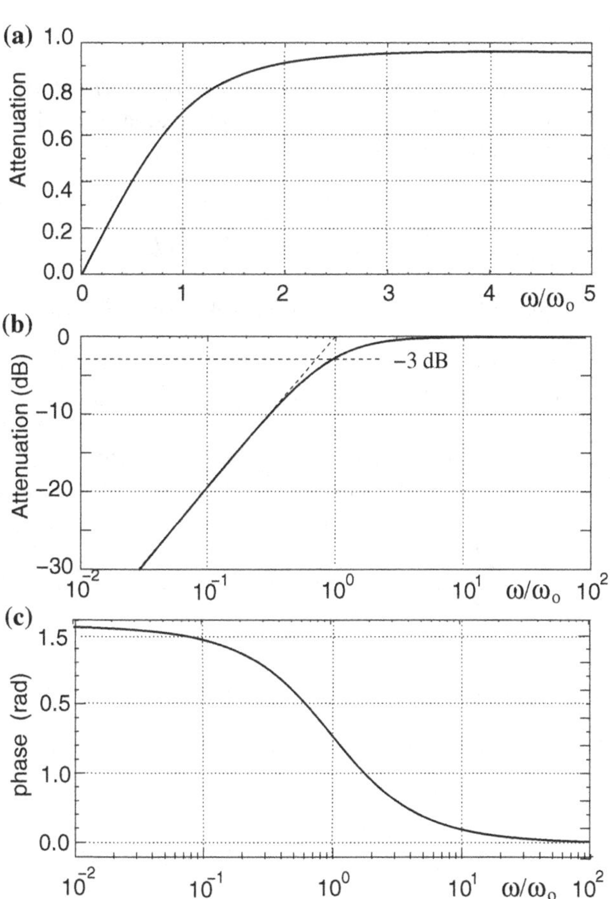

Fig. 6.9 Plots pertaining to the *high-pass RC circuit*. **a** linear plot of the attenuation $(V_u(\omega)/V_i)$. In abscissa, we use the "reduced angular frequency" $\omega\tau = \omega/\omega_0$. **b** *Bode's* diagram of the *high-pass RC circuit*. In abscissa, we use the reduced angular frequency, on the vertical axis the attenuation is reported in decibel. The horizontal line at -3.01 dB corresponds to the attenuation of $1/\sqrt{2}$ obtained for $\omega = \omega_0$, **c** log-plot of the phase shift introduced by the circuit as a function of the reduced angular frequency

When the angular frequency is equal to its cutoff value, $\omega = \omega_0$, once again we have $|A(\omega)| = 1/\sqrt{2} \simeq -3.0$ dB. Figure 6.9 show the frequency plot of $|A(\omega)|$, expressed both in linear and in decibel units, and of the phase. Inspection of the behavior of $|A(\omega)|$, in Fig. 6.9 explains why the circuit is qualified as *high-pass*. Indeed, signals with frequency higher than the cutoff frequency suffer little attenuation while below the cutoff, the amplitude of the output signal decreases in proportion of the frequency. Therefore, this circuit is also known as a *high-pass filter* since it allows the passage of high frequency signals from its input to its output while blocking the rest of the frequency spectrum.

6.9 RL Circuits

In this section, we apply the symbolic method to analyze the *RL* circuits whose diagrams are shown in Fig. 6.10. For the *low-pass* configuration, Fig. 6.10a, the current flowing in the circuit is given by:

$$I = \frac{V_i}{Z_{tot}} = \frac{V_i}{Z_R + Z_L} = \frac{V_i}{R + j\omega L}$$

Noting that the output voltage is $V_u = RI$, the attenuation of this circuit is:

$$A(\omega) = \frac{V_u}{V_i} = \frac{RI}{V_i} = \frac{R}{R + j\omega L} = \frac{1}{1 + j\omega L/R} = \frac{1}{1 + j\omega/\omega_0} \tag{6.26}$$

where in the last step we introduced the expression of the cutoff angular frequency for *RL* circuits, $\omega_0 = R/L$. For the amplitude and the phase of the attenuation, we get:

$$|A(\omega)| = \frac{1}{\sqrt{1 + \left(\dfrac{\omega}{\omega_0}\right)^2}} = \frac{\omega_0}{\sqrt{\omega_0^2 + \omega^2}} \qquad \phi = -\arctan\frac{\omega}{\omega_o} \tag{6.27}$$

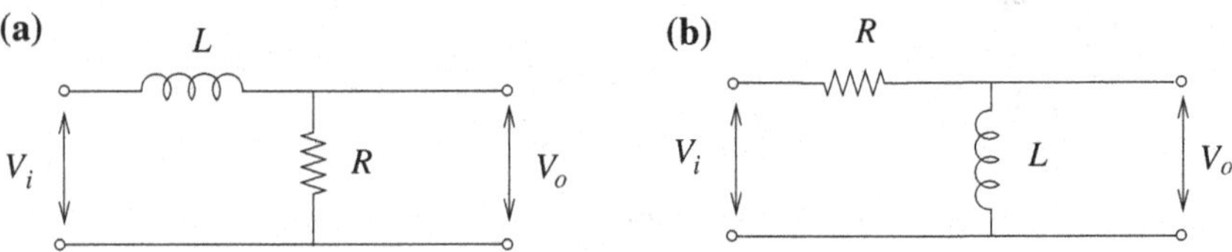

Fig. 6.10 RL circuit: **a** low-pass and **b** high-pass

Note that expressions (6.26) and (6.27) are identical to their equivalent obtained for the *RC low-pass* circuit, the only difference being the definition of ω_0.

Moving to the *high-pass RL* circuit, shown in Fig. 6.10b, we first remark that the expression for the current I flowing in the circuit is the same of the low-pass circuit (the current in a series connection does not depend upon the sequence order). In this case, we can obtain the output voltage as $V_u = Z_L I = j\omega L I$. Therefore, the attenuation of the circuit is:

$$A(\omega) = \frac{V_u}{V_i} = \frac{j\omega L I}{V_i} = \frac{j\omega L}{R + j\omega L} = \frac{j\omega L/R}{1 + j\omega L/R} \tag{6.28}$$

Using the definition $\omega_0 = R/L$, amplitude and phase of $A(\omega)$ are:

$$|A(\omega)| = \frac{\dfrac{\omega}{\omega_0}}{\sqrt{1 + \left(\dfrac{\omega}{\omega_0}\right)^2}} = \frac{\omega}{\sqrt{\omega_0^2 + \omega^2}}, \quad \phi = \arctan\frac{\omega_0}{\omega} \tag{6.29}$$

As in the case of the low-pass, we note that the expression of the attenuation of the high-pass *RL* circuit is equal to that of the high-pass *RC* circuit, as it is easily seen comparing expressions (6.29) and (6.25).

RL circuits are one-parameter circuits. Using the definition of their cutoff frequency, we obtain the characteristic time as $\tau = 1/\omega_0 = L/R$.

6.10 RL and RC Circuits as Differentiators and Integrators

In this section, we show that *RC* and *RL* circuits we discussed in the previous sections, under suitable conditions, are able to supply in output a signal proportional to the derivative or the integral of the input signal.

We begin by making qualitative remarks on the similitude of the mathematical operations of differentiation and integration with the signal transformations performed by low-pass and high-pass filters.

The mathematical operation of differentiation highlights the variations of the function operated upon: the faster the variation of the dependent variable for given change in the independent variable, the higher the value of the derivative. In terms of the function Fourier expansion, we see that the higher frequency components give the larger contribution to the derivative while the contribution of low frequency components is not very important. Therefore, we are induced to expect that the output of a high-pass circuit could show some resemblance with the derivative of the input signal.

On the contrary, the operation of integration tends to smooth fast variations of the integrand function; the integral attenuates the high frequency components of the signal while magnifying those at low frequency. Therefore, we can expect that the output of a low-pass circuit could show some resemblance to the integral of the input signal.

We show now that a low-pass circuit works as an integrator for sinusoidal signals with an angular frequency $\omega \gg \omega_0$. The transfer function of this kind of circuits is given by (6.22) or equivalently by (6.26). In the limit of high frequencies ($\omega \gg \omega_0$), we have:

$$A(\omega) = \frac{V_u}{V_i} = \frac{1}{1 + j\dfrac{\omega}{\omega_0}} \xrightarrow{\omega \gg \omega_0} \frac{1}{j\dfrac{\omega}{\omega_0}} = \frac{\omega_0}{j\omega} \propto \frac{1}{j\omega}$$

Recalling the expression (5.16) for the integral of a complex exponential function, this result shows that the output signal of a low-pass filter is proportional to the integral of the input signal.

Similarly, to show that a high-pass circuit works as a differentiator for sinusoidal signals, consider its transfer function as given by expression (6.24) or equivalently by (6.28). In the low frequency limit, ($\omega \ll \omega_0$), we have:

$$A(\omega) = \frac{V_u}{V_i} = \frac{j\omega/\omega_0}{1 + j\omega/\omega_0} \xrightarrow{\omega \ll \omega_0} \frac{j\omega}{\omega_0} \propto j\omega$$

Recalling once again the expression (5.16) for the derivative of a complex exponential function, this result shows that in this case the output signal is proportional to the derivative of the input signal.

The results of this section can be also obtained by analyzing the circuits directly in the *time domain* as we will discuss in Sects. 9.4 and 9.6.

6.11 Mutual Magnetic Induction and Transformers

Electromagnetic induction, described by the law of Faraday–Newman–Lenz, in the presence of a variation of the magnetic flux $\Phi(B)$ linked to a circuit, produces an electromotive force, equal in module to $d\Phi(B)/dt$, whose sign is such as to induce in the circuit a current generating a magnetic field opposing the flux variation. As already discussed in the first chapter of this book (see Sect. 1.2.4), if the flux $\Phi(B)$ is due to the magnetic field produced by the current i_1 flowing in an adjacent circuit, it can be written as $\Phi(B) = Mi_1$, where the *mutual induction* coefficient M depends only upon the geometry of the two circuits. Consequently, a time dependent current flowing in a circuit can induce an electromotive force in a separate circuit. The intensity of this coupling can be quantified by a *coupling coefficient* K defined by:

$$K = \frac{M}{\sqrt{L_1 L_2}}$$

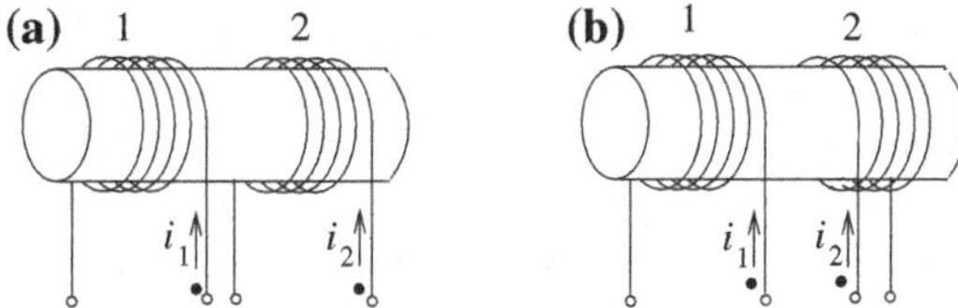

Fig. 6.11 Mutual induction. In **a** the two coils "*1*" and "2" have the same *helicity* while in **b** they have opposite helicity. As explained in the text, currents entering the coil from the terminal marked with the dot produce magnetic field with the same sense. In particular, with this convention, the magnetic field generated by all coils in the figure points from *right* to *left*

where M is the *mutual induction coefficient*, L_1 and L_2 are the inductance, respectively, of the first and the second circuit. The coupling coefficient K is always lower than unity and reaches this value only in ideal conditions, when all the magnetic field lines generated by the first circuit are linked to the second and vice versa.[8] Electromagnetic induction is exploited in a very important electric component, the static[9] transformer that allows manipulating the amplitude of sinusoidal signals.

A static transformer consists of two coils in one of the configurations shown in Fig. 6.11. Denoting with $i_1(t)$ and $i_2(t)$ the currents flowing, respectively, in coil 1 and 2, the absolute values of the mutually induced *e.m.f.* are:

$$\begin{array}{ll} |v_1(t)| = |M\frac{di_2(t)}{dt}| & \text{in coil 1} \\ |v_2(t)| = |M\frac{di_1(t)}{dt}| & \text{in coil 2} \end{array}$$

Since the direction of the magnetic field depends upon the *helicity* of the coil, in a transformer the same symbol, usually a dot, marks the terminals that produce the same direction of the magnetic field for the same current signal, see Fig. 6.11.

This information allows identifying the polarity of an *e.m.f.* due to mutual induction. Assuming that the coil 1 in Fig. 6.11a is open while in coil 2 flows the current i_2. Lenz's law implies that the *e.m.f.* $\mathcal{E}_1$ induced in coil 1 should drive a current opposing any magnetic flux variation. Therefore, when $di_2/dt > 0$ $\mathcal{E}_1$ drives a current flowing toward the terminal identified by the black dot in coil 1. This implies that $\mathcal{E}_1 < 0$ and, assuming $M > 0$:

$$\mathcal{E}_1 = -M\frac{di_2}{dt}$$

Real transformer are typically built by winding its coils around a core of ferromagnetic material with a shape equivalent to a toroid; the ferromagnetic material

---

[8]The reader can find the proof of this statement in a good textbook of electromagnetism as for example in Ref. [6].

[9]The qualifier *static* means that this device exploits the electromagnetic induction in circuits at rest, implying that the flux variation is entirely due to the time derivative of a current, not to circuit motion.

increases the inductance of the coils and their coupling is maximized by the toroidal shape that helps to convey the flux lines from one circuit to the other minimizing the magnetic flux leakage.

The symbols used in circuit diagrams to represent a static transformer are shown in Fig. 6.12a, b, respectively, for an air and a ferromagnetic core transformer. These symbols represent *ideals* transformers, a concept that we will clarify in the following of this section. Usually, the winding connected to an electric power source is identified as the primary while the other as the secondary.

Solution of a transformer circuit. In the diagram of Fig. 6.13 we show a transformer whose primary coil is connected to a voltage generator v_p while its secondary is closed on a load R_L. For the analysis of this circuit, we assume that the primary and the secondary coil consist, respectively, of N_p and N_s turns with an ohmic resistance R_p and R_s, respectively. We denote with L_p, L_s and M the inductance of the primary coil, of the secondary coil and their mutual induction coefficient. We also recall that for solenoidal coils self-inductance is proportional to the square of their turns:

$$L_p \propto N_p^2 \qquad L_s \propto N_s^2 \tag{6.30}$$

Using Kirchhoff's law of voltages, we can write the two following equations for the two unknown currents i_p and i_s flowing, respectively, in the primary and the secondary winding:

Fig. 6.12 Symbols representing static transformers: **a** air core transformer, **b** ferromagnetic core transformer

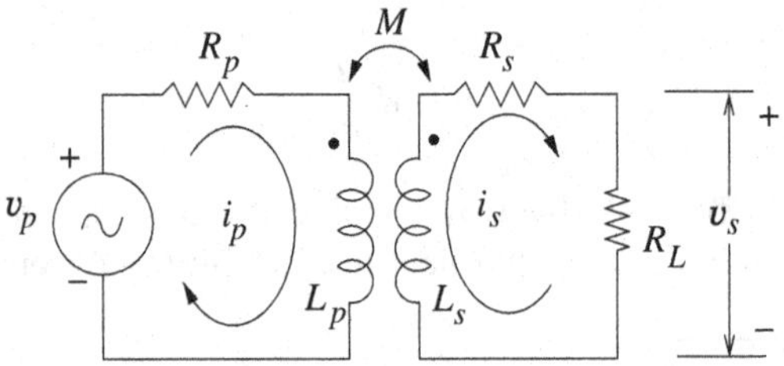

Fig. 6.13 Diagram of a transformer with a voltage generator in the primary circuit and a resistive load in the secondary. The reader is invited to verify that in this scheme, an increase (reduction) of i_p causes an increase (reduction) in i_s

$$\begin{cases} v_p = R_p i_p + L_p \dfrac{di_p}{dt} - M\dfrac{di_s}{dt} \\ 0 \ = -M\dfrac{di_p}{dt} + L_s\dfrac{di_s}{dt} + i_s(R_s + R_L) \end{cases} \tag{6.31}$$

Note that in these two equations the term proportional to M is preceded by a minus sign because of the Lenz's law, of the helicity of the windings, as marked by the dots, and of the choice of the positive direction made for the two currents i_p and i_s. When v_p has a sinusoidal waveform, we can apply the symbolic method and the Eq. (6.31) become:

$$\begin{cases} V_p = R_p I_p + j\omega L_p I_p - j\omega M I_s \\ 0 \ = -j\omega M I_p + j\omega L_s I_s + I_s(R_s + R_L) \end{cases} \tag{6.32}$$

whose solutions are easily obtained as:

$$I_p = V_p \frac{R_L + R_s + j\omega L_s}{R_L(R_p + j\omega L_p) + R_p(R_s + j\omega L_s) + j\omega[L_p(R_s + j\omega L_s) - j\omega M^2]} \tag{6.33}$$

$$I_s = V_p \frac{j\omega M}{R_L(R_p + j\omega L_p) + R_p(R_s + j\omega L_s) + j\omega[L_p(R_s + j\omega L_s) - j\omega M^2]} \tag{6.34}$$

We can simplify these two rather complicated expressions exploiting a few approximations usually very well verified in practical applications. The first of these approximations assumes that the leaked magnetic flux is negligible, allowing to set $K = 1$ or equivalently $M = \sqrt{L_p L_s}$. Equations (6.33) and (6.34) become:

$$I_p \simeq V_p \frac{R_L + R_s + j\omega L_s}{R_L(R_p + j\omega L_p) + R_p(R_s + j\omega L_s) + j\omega L_p R_s}$$

$$I_s \simeq V_p \frac{j\omega\sqrt{L_p L_s}}{R_L(R_p + j\omega L_p) + R_p(R_s + j\omega L_s) + j\omega L_p R_s}$$

Next, we assume that the angular frequency ω is high enough to make reactance much higher than resistance in both the primary and the secondary circuit: $R_p \ll \omega L_p$, $R_s \ll \omega L_s$. Furthermore, the resistance of the secondary winding must be negligible with respect to the load: $R_s \ll R_L$. With these approximations, the previous two equations become:

$$I_p \simeq V_p \frac{R_L + j\omega L_s}{j\omega(R_L L_p + L_s R_p)}$$

$$I_s \simeq \frac{V_p}{R_L + R_p \frac{L_s}{L_p}} \sqrt{\frac{L_s}{L_p}}$$

The third approximation assumes that the ohmic resistance R_p is negligible compared to the load resistance R_L multiplied by the ratio of the primary to the secondary inductance (we shall see in the following that this is the impedance of the load as "seen" by the primary winding). We obtain:

$$I_p \simeq V_p \frac{R_L + j\omega L_s}{j\omega L_p R_L}$$

$$I_s \simeq \frac{V_p}{R_L} \sqrt{\frac{L_s}{L_p}}$$

With these three approximations, we can compute the output voltage of the transformer as $V_s = I_s R_L$ and its transfer function $H(\omega) = V_s/V_p$:

$$H(\omega) = \frac{V_s}{V_p} = \sqrt{\frac{L_s}{L_p}} = \frac{N_s}{N_p} \tag{6.35}$$

Note that this transfer function does not depend upon frequency. However, this is only true when we can neglect the ohmic resistance of the two windings, as postulated with the second approximation above.

A quantity of important use in the following is the ratio of the currents in the two windings:

$$\frac{I_s}{I_p} = \frac{j\omega\sqrt{L_p L_s}}{R_L + j\omega L_s} = \sqrt{\frac{L_p}{L_s}} \cdot \frac{1}{1 + \dfrac{R_L}{j\omega L_s}} \tag{6.36}$$

In addition to the three approximation described above, we now assume also that the load resistance is negligible compared with the reactance of the secondary winding $R_L/\omega L_s \ll 1$. Under this hypothesis, the Eq. (6.36) becomes:

$$\frac{I_s}{I_p} = \sqrt{\frac{L_p}{L_s}} = \frac{N_p}{N_s} \tag{6.37}$$

Ideal transformer. Since the four approximations adopted above are well satisfied in many circumstances, it is useful to define the ideal transformer as a two-port electrical component with the following characteristic property:

$$\frac{V_p}{V_s} = \frac{I_s}{I_p} = \sqrt{\frac{L_p}{L_s}} = \frac{N_p}{N_s} \equiv n \tag{6.38}$$

the parameter n is referred to as the *transformation ratio* of the ideal transformer. Transformers are widely used in commercial electric power grid to change the voltage along the lines to optimize its efficiency and safety.

An important property of the ideal transformer is its capability of manipulating load impedance. We compute the equivalent impedance of the primary winding as $Z_p = V_p/I_p$ and similarly for the secondary winding as $Z_L = V_s/I_s = R_L$ with the last identity valid only in the ideal case. Using Eq. (6.38) we obtain:

$$Z_p = \frac{V_p}{I_p} = \frac{V_p}{I_s}\frac{I_s}{I_p} = R_L\frac{L_p}{L_s} = n^2 R_L \tag{6.39}$$

This relation shows that the load applied to the secondary winding is seen as multiplied by the squared transformation ratio by the primary winding.

An important consequence of Eq. (6.38) is the power balance relation

$$V_p I_p = V_s I_s \tag{6.40}$$

This shows that in an ideal transformer the power input from the voltage generator to the primary winding is equal to the power dissipated by the resistive load in the secondary circuit, as expected since we neglected all other dissipative elements.

6.12 Impedance Matching

In the previous section, we have learned that by interposing an ideal transformer we can manipulate the impedance of a load connected to a voltage generator, see Eq. (6.39). In many electrical and electronic applications, it is necessary to match the impedances of different circuits performing different functions before interconnecting them. Consider, for example, a voltage generator with output amplitude V_g and purely resistive output impedance R_g, connected to a resistive load of impedance R_L. The power P_L transferred from the generator to the load is easily computed as the product of the voltage drop across the load and the current flowing through it:

$$P_L = \frac{V_g^2 R_L}{(R_g + R_L)^2}$$

Often in applications, it is required to extract the maximum possible amount of power from a generator. Using the previous results, we can show that this requires that the two resistances are equal, $R_g = R_L$, see Sect. 2.5. If the generator has a complex impedance Z_g and the load has a complex impedance Z_L, it can be easily shown that

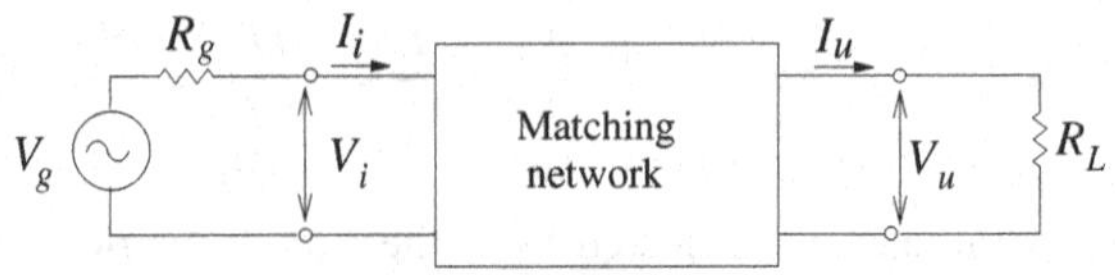

Fig. 6.14 Matching network

the maximum power transfer is obtained when $Z_g = Z_L^*$. Similar considerations apply when we try to obtain the highest possible signal from an electrical sensor, a device that transforms a physical quantity into a voltage or a current, such as a microphone, a photodiode or a semiconductor thermometer. All these devices can be schematized according to Thévenin or Norton and therefore, in addition to the electrical signal, we need to take into account their output impedance too. Finally, we shall see in the last chapter of this book that impedance matching becomes mandatory any time we need to use a long connection cable at high frequency.[10]

Impedance matching between two different circuits is obtained via the interposition of a suitable matching network, as shown schematically in Fig. 6.14. This network should consist of purely reactive components, to avoid losses due to dissipative components, and therefore, impedance matching becomes only practically feasible in AC circuits. This is the reason why commercial electric power distribution grids are operated with sinusoidal currents at a frequency sufficiently high, to allow impedance matching, and sufficiently low, to minimize the power loss due to eddy currents induced in the surrounding environment.

It is always possible to assume that the impedances to be adapted are real. Consider the case of a generator having a reactance jX_g to be connected to a load with reactance jX_L. We can first compensate these two reactances by connecting the generator is series with a reactance $-jX_g$ and similarly the load with a reactance $-jX_L$. In this way, we have reduced the problem to the matching of two resistive components. In this hypothesis, the complex amplitudes of voltages and currents can be expressed with real numbers and, with reference to Fig. 6.14, using the definition of impedance we obtain:

$$R_{in} = \frac{V_i}{I_i} \quad \text{and} \quad R_L = \frac{V_u}{I_u}$$

Dissipation being absent in a purely reactive network, the input power $V_i I_i/2$ is equal to the output power $V_u I_u/2$, and we can write

$$R_{in} = \frac{V_i}{I_i} = \frac{V_i^2}{V_i I_i} = \frac{V_i^2}{V_u I_u} = \frac{V_i^2}{V_u^2} R_L = \frac{R_L}{h^2} \tag{6.41}$$

[10] A similar problem of impedance matching is encountered when we need to connect a sensor to a signal amplifier: to minimize the noise generated at the input it is often necessary that the impedance of the sensor is equal to an optimal value which depends by the characteristics of the amplifier input stage.

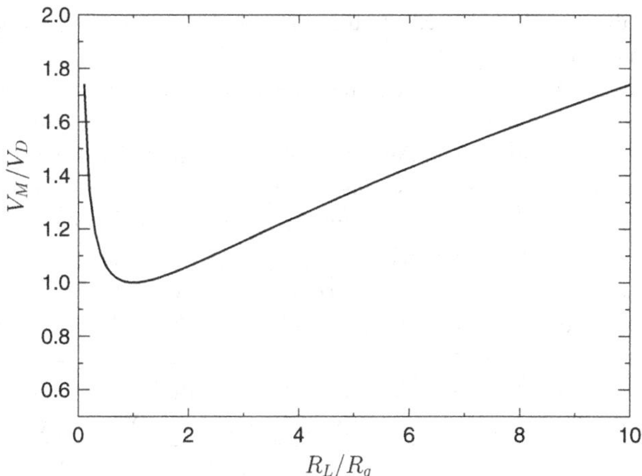

Fig. 6.15 Plot of V_M/V_D as a function of the ratio R_L/R_g

where $h = V_u/V_i$ is the output attenuation of the matching network. This result, valid only when the input impedance to the network is purely real, shows that impedance matching can be obtained through network attenuation equal to the square root of the ratio of the load resistance to the output resistance of the voltage generator. We shall see in the next section how we can design the needed network, but the result obtained above already allows computing the voltage on the matched load.

In this case, due to the matching, the voltage at the network input is $V_i = V_g/2$. Taking into account its attenuation, we have on the load $V_M = (V_g/2)\sqrt{R_L/R_g}$. This voltage is always higher than the voltage obtained with a direct connection of the load to the generator $V_D = V_g R_L/(R_g + R_L)$. Figure 6.15 shows that the ratio V_M/V_D reaches its minimum value, equal to unity, only for $R_L = R_g$, i.e., when the matching network is not needed.

6.13 Matching Networks

Equation (6.41) shows that impedance matching requires a reactive network with attenuation $h = \sqrt{R_L/R_g}$. An ideal transformer can perform this function since its attenuation between primary and secondary winding does not depend upon the working frequency but only on the number of turns in the windings, see Eq. (6.39). However, it must be noted that it is nearly impossible to obtain an ideal behavior with a real transformer at high frequencies because of the parasitic inter-turn capacitance and of hysteresis losses in the ferromagnetic core. Furthermore, from a practical point of view, a transformer is a rather bulky component of difficult deployment. Finally, very often, the matching network is also used to filter out noise at frequency different from the signal of interest. In this case, it is possible to match impedances using only capacitors and/or inductors.

Following Eq. (6.41), when the load resistance is higher than the generator resistance, the matching network needs to amplify its input voltage; this can only be obtained using a resonant circuit. On the contrary, when the load has a lower resistance than the generator, a simple reactive voltage divider can be deployed. Both these situations can be handled by the generic circuit shown in Fig. 6.16. Using the series-parallel transformation described in Sect. 6.4, the parallel of jX_2 and R_L is equivalent to the series of jX_2' and R_L' as shown on the right side of the figure. Choosing jX_1 equal to the opposite of the transformed reactance $jX_2' = jX_2/[1 + (X_2/R_L)^2]$, the network impedance becomes equal to the transformed resistance $R_L' = R_L/[1 + (R_L/X_2)^2]$, which can be made equal to the generator resistance with a suitable choice of X_2. In this way, we have a resonance between the two reactances jX_1 and jX_2' and it is easy to verify that the attenuation is equal to $\sqrt{1 + (R_L/X_2)^2}$. The quality factor of this resonant circuit, $Q = X_2'/R_L'$, is also equal to R_L/X_2 because of the properties of the series to parallel transformation. The input impedance of the network can be written as $R_L/(1 + Q^2)$.

In conclusion, when the load has a higher resistance than the generator, the matching can be obtained at the operation frequency with a resonant circuit whose quality factor is $Q = \sqrt{R_L/R_g - 1}$, which works also as a filter.

When in Fig. 6.16, the two reactances are taken with the same sign, the circuit becomes a reactive voltage divider and therefore its attenuation is always lower than unit. This circuit can be used to match a load with a resistance lower than the generator. However, it is necessary to use a third reactance, opposite in sign to those already in the circuit, to make real the input impedance. The detailed calculation of this reactance is left to the reader as an exercise, see Problem 11 at the end of this chapter.

When $R_L < R_g$ we can obtain a simpler division of the input voltage with the generic circuit shown in Fig. 6.17. We first transform the series of R_L and jX_2 in a parallel obtaining the equivalent circuit, shown on the right side of the figure, where $X_2' = X_2[1 + (R_L/X_2)^2]$ and $R_L' = R_L[1 + (X_2/R_L)^2]$. Choosing $1/X_2' = -1/X_1$, the conductance of the parallel of the two reactive elements is null and the network input impedance is only given by the resistance R_L'. In this case, we have a parallel resonance between jX_1 and jX_2' whose quality factor $Q = R_L'/X_2'$ is equal to X_2/R_L because of the property of the series to parallel transformation. For this matching

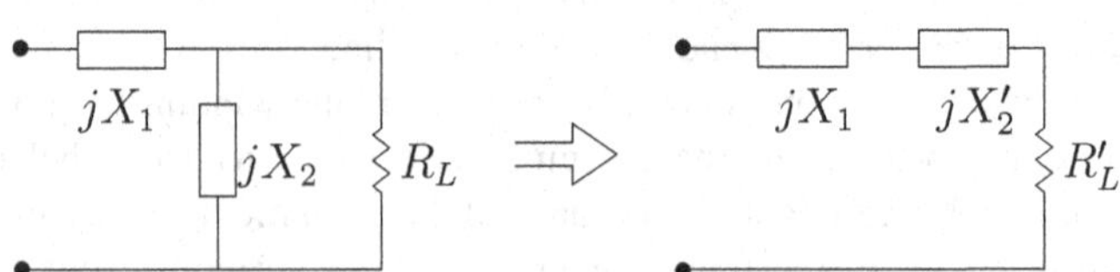

Fig. 6.16 Matching network with a series resonance and its parallel to series transformation

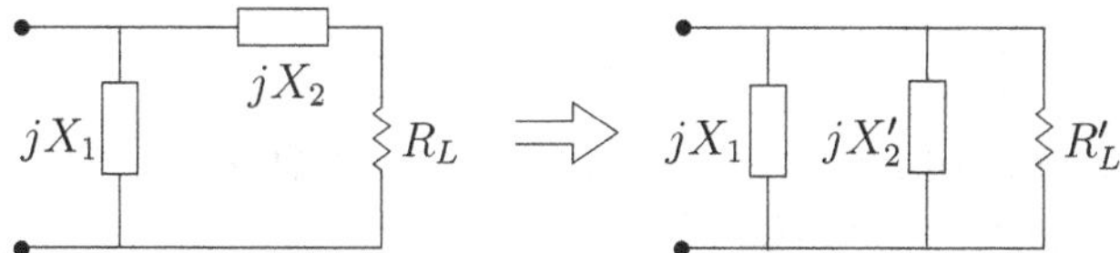

Fig. 6.17 Matching network with a parallel resonance and its series to parallel transformation

network too, the input impedance can be expressed as a function of the quality factor as $R_L[1+Q^2]$ and the matching to the generator is obtained for $Q = \sqrt{(R_g/R_L)-1}$.

Example. To observe the phenomenon of nuclear magnetic resonance [7] we need to place a sample in an magnetic field oscillating at the resonance frequency. This is obtained with the sample inside a linear solenoid powered by a sinusoidal voltage generator. Obviously, the higher the oscillating magnetic field, the better the signal to noise ratio achieved in the experiment. To proceed with a quantitative analysis, we assume that the solenoid has an inductance $L = 10\,\mu$ H and a resistance $R_L = 10$ ohm. We also assume that the available generator has a maximum voltage output $V_g = 12$ V and an output impedance $R_g = 50$ ohm.

With a direct connection to the generator, the maximum peak current in the solenoid would be equal to $I = V_g/\sqrt{(\omega L)^2 + (R_g + R_L)^2}$. At a working frequency of 1 MHz we would get $I = 138$ mA. In a first step, we compensate the inductive reactance of the solenoid with a series capacitance C_1 equal to 2.53 nF and the peak current becomes $I_D = V_g/(R_g + R_L) = 200$ mA.

In the second step, we match the two resistances. Since $R_g > R_L$ we can adopt the approach illustrated in Fig. 6.17. We need a value of the quality factor $Q = \sqrt{R_g/R_L - 1} = 2$. The value of the reactance X_2 is given by $X_2 = QR_L = 20\,\Omega$. At the working frequency, this reactance can be obtained with a capacitance $C_2 = 7.96$ nF. This implies that for the reactance $X_1 = X_2[1+(1/Q)^2]$ we need an inductance $L_1 = 3.98\,\mu$ H. When the matching condition is satisfied, the input voltage to the matching network is equal to $V_i = V_g/2$. Then, we obtain the voltage applied to the load taking into account the attenuation of the matching network: $V_M = (V_g/2)\sqrt{R_L/R_g}$. In these conditions, the current in the solenoid is equal to $I_M = V_g/(2\sqrt{R_L R_g}) = 268$ mA, a factor 1.3 higher than the value obtained in the preceding step.

In Fig. 6.18, we show a practical implementation of the circuitry discussed above. Note that we have integrated the two capacitances C_1 and C_2 in a single component of capacitance equal to 10.49 nF. The reader is invited to calculate the current in the solenoid by solving this circuit for example with the method of the nodes.

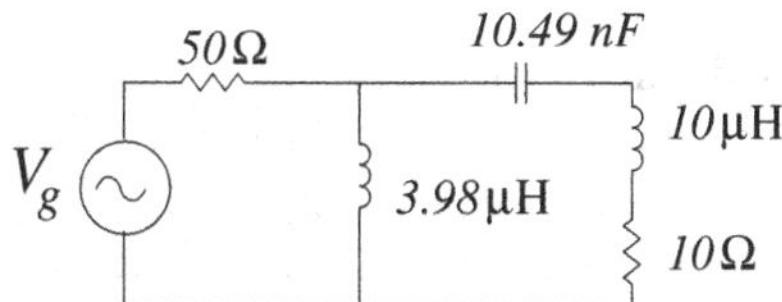

Fig. 6.18 Matching a solenoid to a real voltage generator

Problems

Problem 1 Compute the transfer function of a *low-pass RL* filter taking into account the inductor ohmic resistance R_L.

Problem 2 Compute the transfer function of a *high-pass RL* filter taking into account the inductor ohmic resistance R_L.

Problem 3 Derive relations (6.14), (6.15) and (6.16).

Problem 4 Compute the parameters of the Thévenin's equivalent circuit of a series *RLC* circuit seen from the inductor terminals at the resonance frequency. [A. $Z_{eq} = Q_s^2 R + j\omega_0 L$; $V_{eq} = Q_s V_g$.]

Problem 5 Compute the equilibrium conditions for the bridge circuit shown in the figure. [A. $R_1 R_4 = (R_3 + R) R_2$; $R_1 L_4 = R_2 L_3$.]

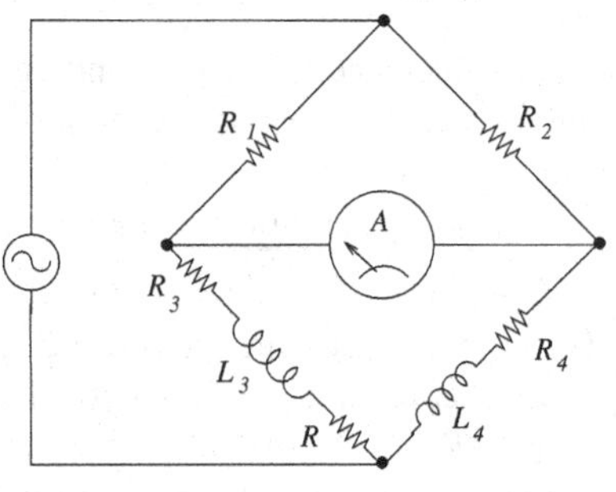

Problem 5

Problem 6 Using the series–parallel transformation, compute the equilibrium conditions for the bridge circuit shown in the figure. [A. $R_2/R_4 = R_4/R_3 + L_4/L_3$; $\omega^2 = R_4 R_3 / L_4 L_3$.]

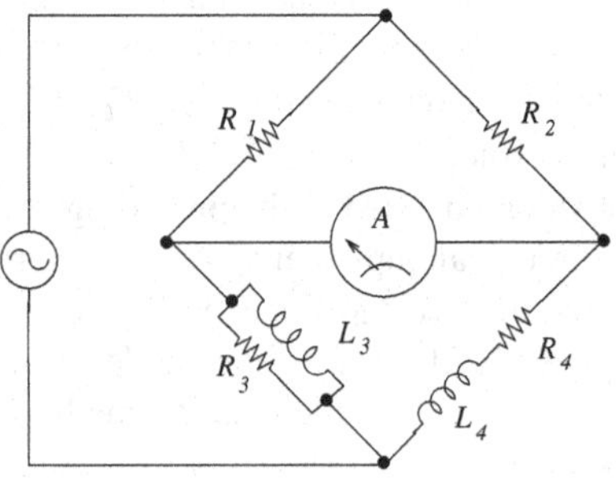

Problem 6

Problem 7 The Heaviside bridge, shown in the figure, can be used to measure the mutual induction coefficient between the two windings of a transformer. Show that at the equilibrium the following relation holds: $M = (R_1 L_3 - R_2 L_4)/(R_1 + R_2)$, independently of the generator frequency.

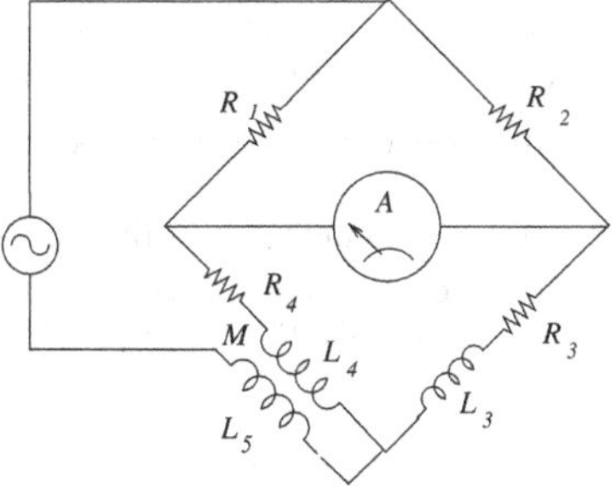

Problem 7

Problem 8 Compute the complex attenuation of the circuit in the figure, using for example Thévenin's theorem. Show that it is a bandpass filter with transmission maximum at $\omega_0 = 1/RC$. Compute the angular frequencies corresponding to the attenuation equal to its peak value divided by $\sqrt{2}$. [A. $A(\omega) = 1/(3 + j(\omega RC - 1/\omega RC))$, $\omega_{1,2} = \omega_0(\sqrt{13} \pm 3)/2$.]

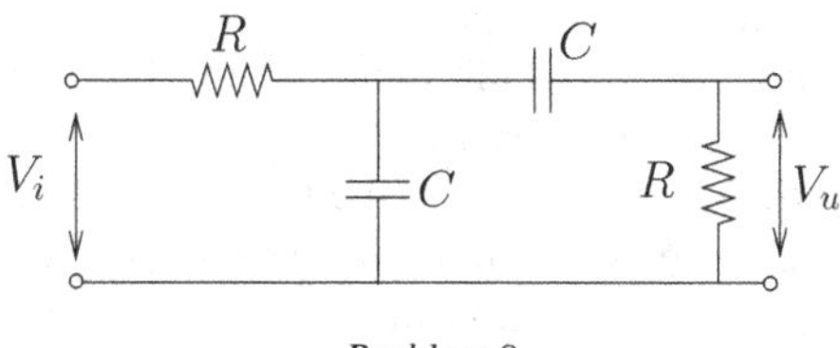

Problem 8

Problem 9 Use the loop method to obtain the equation system describing the circuit shown in the figure. Perform the analysis of its coefficient determinant to show that the circuit is resonant for two values of the angular frequency ω. Show that for these frequencies the impedance seen by the generator is null. [A. $\omega_{1,2}^2 = (3\pm\sqrt{5})/(2LC)$.]

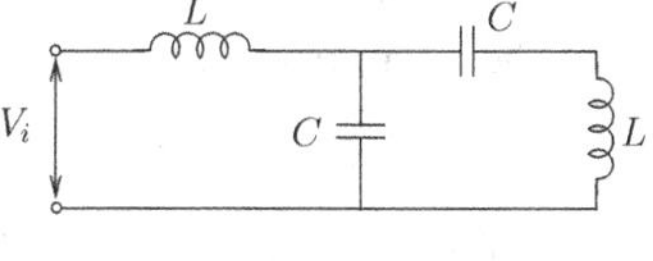

Problem 9

Problem 10 Compute the attenuation $A(\omega)$ of the two-port circuit shown in the figure and show that it is always real but divergent at two frequency values. [A. $A(\omega) = 1/[3 - (\omega^2 LC + \frac{1}{\omega^2 LC})]$, $\omega_{1,2}^2 = (3 \pm \sqrt{5})/(2LC)$.]

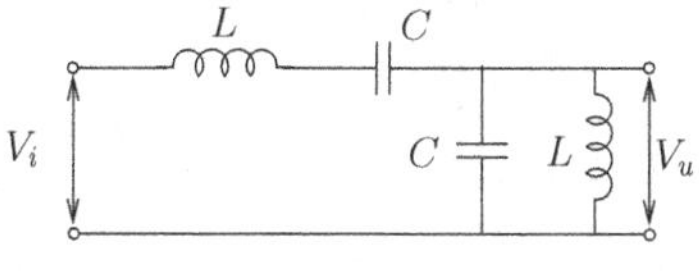

Problem 10

Problem 11 Show that the circuit of previous problem has two resonances at the angular frequencies $\omega_{1,2}^2 = (3 \pm \sqrt{5})/(2LC)$ for which the impedance seen at the input is null. Show also that this impedance diverges at $\omega = 1/\sqrt{LC}$.

Problem 12 Compute the input impedance Z_{eq} of the network shown in the figure choosing the value of the inductance L to cancel its reactance under the hypothesis $R_L \ll 1/\omega C_2$. [A. $Z_{eq} = R_L(1 + C_2/C_1)^2$; $L = (1/C_1 + 1/C_2)/\omega^2$.]

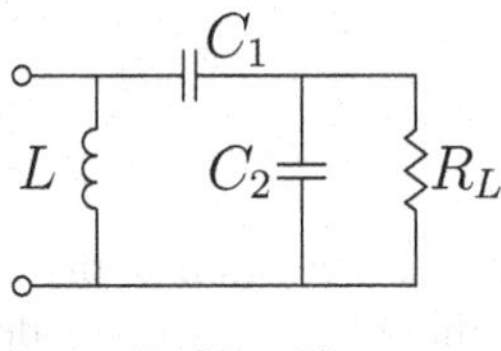

Problem 12

Problem 13* In this problem, the reader can verify that different definitions of resonance given in the literature are not equivalent.

Solve the *RLC parallel* circuit taking into account the ohmic resistance of the inductor. The diagram of the circuit is obtained by adding a resistance r in series to the inductance L in the circuit shown in the text in Fig. 6.2. Then, compute the transfer function $H(\omega)$ given by the ratio of the output voltage, measured across the capacitor, and the generator voltage. Show that the amplitude of the transfer function has a maximum at the angular frequency $\omega_{\max}$ given by:

$$\omega_{\max} = \omega_o \sqrt{\sqrt{1 + 2\frac{r}{R} + 2\left(\frac{r}{\omega_o L}\right)^2} - \left(\frac{r}{\omega_o L}\right)^2}$$

where $\omega_o = 1/\sqrt{LC}$ is the resonance frequency we obtained with $r = 0$. Adopting a McLaurin series expansion in the small parameter r, to second order the previous expression becomes:

$$\omega_{\max} \simeq \omega_o \left(1 + \frac{r}{2R} - \frac{3}{8}\frac{r^2}{R^2} + \cdots\right)$$

The reader is warned that the calculation of $\omega_{\max}$ is laborious and requires an appropriate amount of attention in the various steps that lead to the solution.

Show that the admittance of the circuit, as seen by the voltage generator, vanishes at an angular frequency ω_X, different by $\omega_{\max}$:

$$\omega_X = \sqrt{\frac{1}{LC} - \left(\frac{r}{L}\right)^2} = \sqrt{\omega_o^2 - \left(\frac{r}{L}\right)^2} \simeq \omega_o \left(1 - \frac{1}{2}\frac{C}{L}r^2 + \cdots\right)$$

Finally show that the circuit has a normal mode of oscillation at the angular frequency ω_n:

$$\omega_n = \omega_o \sqrt{1 - \frac{L}{4CR^2} + \frac{r}{2R} - \frac{Cr^2}{4L}}$$

Problem 14* With reference to the previous problem, evaluate the effect of the resistance r of the inductor on the value of the merit factor Q of the circuit from the width of the attenuation peak. Show that Q is approximatively given by:

$$Q = \frac{1}{\omega_o}\frac{R}{L} \simeq \frac{1}{\omega_o}\frac{R}{L + RCr}$$

Hint: replace the series of r and L with the parallel of r_{eq} and L_{eq} adopting a series-parallel transformation. Both r_{eq} and L_{eq} are frequency dependent:

$$r_{eq} = r\left(1 + \frac{\omega^2 L^2}{r^2}\right) \quad L_{eq} = L\left(1 + \frac{r^2}{\omega^2 L^2}\right)$$

Compute Q in the approximations $r_{eq} \simeq \omega_o^2 L^2/r$ and $L_{eq} \simeq L$, both verified at sufficiently high frequency.

References

1. E.J. Burge, Am. J. Phys. **29**, 19 (1961)
2. E.J. Burge, Am. J. Phys. **29**, 251 (1961)
3. J.D. Dudley, W.J. Strong, Am. J. Phys. **55**, 610 (1987)
4. J.A. Stuller, Am. J. Phys. **4**, 296 (1988)
5. H.W. Bode, Bell Syst. Tech. J. **19**, 421–450 (1950)
6. R.C. Dorf, J.A. Svoboda, *Introduction to Electric Circuits* (Wiley, Hoboken, 2010)
7. C. Slichter, *Principles of Magnetic Resonance* (Springer, Berlin, 1996)

Chapter 7
Measurement of Alternating Electrical Signals

7.1 Introduction

The most comprehensive information on alternating signals is obtained by observing their temporal evolution, i.e., by measuring continuously in time the value of the signal (current or voltage as required by the problem under consideration). This kind of measurement is generally carried out with devices capable of displaying the time development of the waveform as the oscilloscopes, instruments that we will deal with extensively in the following chapter. However, it is not always necessary to know waveform details. This is the case, for example, when the waveform is known a priori, as for a sinusoidal signals of given frequency, or when we need high-accuracy measures that visual techniques could not be able to provide, or when the presence of high-frequency electrical noise makes difficult and ambiguous the measurement of the instantaneous value of the signal, or, finally, when we require an accurate evaluation of the electrical noise present in a circuit.

In all these cases it is possible to characterize the alternating signal with a single parameter, related to its *amplitude*, currently identified in its root mean square (*rms*) value, also called the *effective value*.

The measure of AC (short for alternating current) signals, contrary to the DC case, poses a set of problems related to the very nature of the alternating signal. One problem, partly solved by the introduction of integrated circuits, stems from the difficulty of building devices able to measure correctly the *effective* value. Even the most modern tools can provide information not corresponding to the required quantity in the case of AC measures. The experimenter should then always know the principles of operation of the instrument in use to avoid gross errors in measurement evaluations.

Technical specifications of professional instruments report whether they measure the *true-rms* value, meaning that the instrument is able to process the signal through the necessary mathematical operations to obtain its effective value. Cheaper instruments, although widely used, do not measure the *true-rms* and their users must be

R. Bartiromo and M. De Vincenzi, *Electrical Measurements in the Laboratory Practice*, Undergraduate Lecture Notes in Physics,
DOI 10.1007/978-3-319-31102-9_7

careful to avoid aforementioned errors. Furthermore, it must be taken into account that the impedance of real electrical components (resistors, capacitors, inductors, and active components) used to build measuring devices is always frequency dependent. This implies that outside a given frequency range, referred to as the *passband*, the instrument produces erroneous results. Moreover, with increasing frequency, parasitic series inductance and parallel capacity of the cables connecting the circuit with the instrument can become important and may degrade the accuracy of AC measurements for a considerable amount. These observations emphasize that the measurement of alternating signals requires a careful analysis of the operational circumstances and more attention from the experimenter than in the case of DC signals.

In the following Sect. 7.2, we introduce the parameters used to characterize time-dependent signals and we describe the waveforms most commonly encountered in laboratory experiments and related instrumentation. In Sect. 7.3, we discuss the coupling technique usually adopted to connect AC signals to instruments. Section 7.4 is devoted to the illustration of problems posed by analog instruments when they are used for the measurement of AC signals. Finally in Sect. 7.5, we illustrate the operational principles of a number of modern digital instruments for AC signals and we discuss their use for the measurement of amplitude and phase. A brief introduction to the problem of impedance measure concludes the chapter.

7.2 Characteristics of Alternating Signals

This chapter only deals with *stationary* periodic signals.[1] In other words, we will not attempt here to give a description of waveforms with finite time duration, commonly referred to as *transients*,[2] an important subject in the field of electronics that will be dealt in Chap. 9.

As defined in Sect. 5.2, a generic periodic signal $x(t)$ is qualified as alternating when its time-averaged value is it null. Alternating signals can be characterized by the set of parameters shown in the Fig. 7.1. Their meaning is given below as follows:

- T: signal period. $T = 1/\nu = 2\pi/\omega$ where ν is the frequency quoted in Hz and ω the pulsation or angular frequency, measured in rad s^{-1}.
- X_{pp}: peak-to-peak amplitude. The difference between the maximum and the minimum value of the signal.
- X_p: peak amplitude. Maximum value in the positive half wave. This parameter is useful only for signal symmetric with respect to the time axis.
- X_{rms}: effective or rms amplitude. Root mean square signal value.

[1] In our context the term *stationary* does not refer to quantities that *do not depend on time*, we would use the term static instead. The alternating signals are time dependent but periodic and the infinitely repeated waveform justifies the term *stationary* to qualify them.

[2] To have an example of transient waveform, consider a capacitor being connected to a battery through a resistance. The current flowing to charge the capacitor has a *transient* waveform.

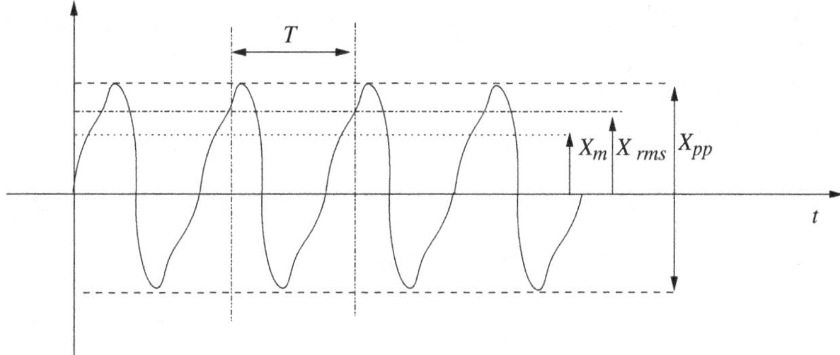

Fig. 7.1 Time evolution of a generic alternate signal. Signal characteristics illustrated in the text are shown. The average level of the signal is null

- X_m: average absolute amplitude: obtained using the average value of the rectified signal, as we shall see below. The average can be carried out for the single or the double half wave.

As we discussed already in the introduction, the effective, or *rms*, amplitude is the quantity more commonly used to characterize an alternating signal and this value is usually measured by AC instruments. Its exact accepted definition is given as follows:

The effective, or rms, amplitude is the value of the continuous voltage (or current) producing the same Joule effect as the considered alternating signal in a single period, in a generic resistance R.

Consider for example an alternating voltage signal $v(t)$, with period T. When this voltage is applied at the terminals of a resistance R, the energy dissipated by Joule effect in the resistance, in a period, is

$$\mathscr{E} = \frac{1}{R}\int_0^T v(t)^2\,dt \tag{7.1}$$

The DC voltage V_{rms} (or V_{eff}) producing the same net effect is given by the relation as follows:

$$\frac{V_{rms}^2}{R}T = \frac{1}{R}\int_0^T v(t)^2\,dt \tag{7.2}$$

that yields

$$V_{rms} = \sqrt{\frac{1}{T}\int_0^T v(t)^2\,dt} \tag{7.3}$$

A similar expression can be found for the value of the effective current I_{rms}. In conclusion, the effective, or *rms*, value is given by the positive determination of the square root of the mean quadratic value of the alternating signal.

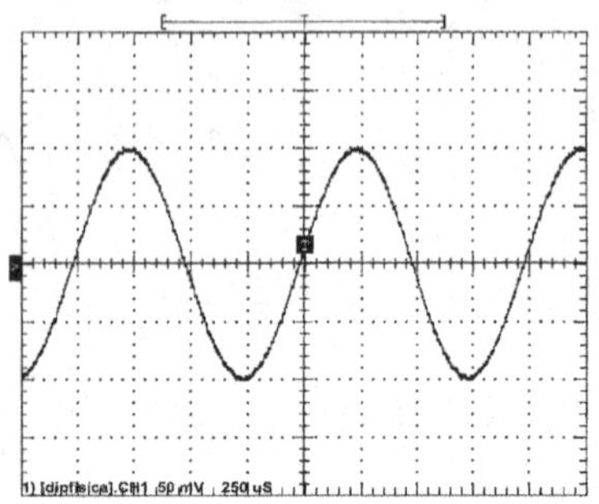

Fig. 7.2 Sinusoidal signal. Plot obtained on the screen of a digital oscilloscope with vertical sensitivity 50 mV per division, and horizontal sensitivity 250 μs per division

In the following, we apply this definition to a number of specific cases. In the examples, we will parameterize the amplitudes in terms of the peak-to-peak amplitude, since this is the most immediate value that can be measured with an oscilloscope, as explained in next Chap. 8.

7.2.1 Sinusoidal Signal

The sinusoidal signal with period T, represented in Fig. 7.2, is given by the following mathematical expression:

$$x(t) = x_0 \cos(\omega t) = \frac{X_{pp}}{2} \cos\left(\frac{2\pi}{T} t\right) \tag{7.4}$$

For this kind of signal we have $X_p = x_0$ and $X_{pp} = 2x_0$. Using the definition (7.3) of effective value, we obtain the following for the sinusoidal signal:

$$X_{rms} = \sqrt{\frac{1}{T} \int_{-T/2}^{T/2} \left[x_0 \cos\left(\frac{2\pi}{T} t\right)\right]^2 dt} = \frac{X_{pp}}{2\sqrt{2}} \simeq 0.354 X_{pp} \tag{7.5}$$

The effective value of a sinusoidal signal is given by the peak-to-peak amplitude of the waveform divided by $2\sqrt{2}$.

7.2.2 Triangular Signal

The triangular signal with period T, represented in Fig. 7.3, is given by the following mathematical expression:

$$x(t) = \begin{cases} \frac{2X_{pp}}{T}\left(t + \frac{T}{4}\right) & \text{for } -\frac{T}{2} < t < 0 \\ -\frac{2X_{pp}}{T}\left(t - \frac{T}{4}\right) & \text{for } 0 < t < \frac{T}{2} \end{cases} \tag{7.6}$$

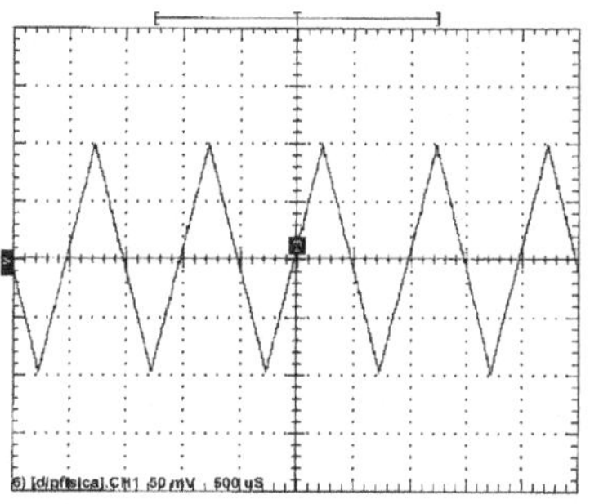

Fig. 7.3 Triangular signal. Plot obtained on the screen of a digital oscilloscope with vertical sensitivity 50 mV per division, and horizontal sensitivity 500 μs per division. Note that time origin is shifted with respect to definition 7.6

where X_{pp} is *the peak-to-peak* amplitude of the signal. The origin of the time axis ($t = 0$) has been located at the signal maximum to simplify its mathematical expression.

Using the definition (7.3) of effective value, we obtain the following for the triangular signal:

$$X_{rms} = \left[\frac{1}{T}\left(\int_{-T/2}^{0} \frac{4X_{pp}^2}{T^2}\left(t+\frac{T}{4}\right)^2 dt + \int_{0}^{T/2} \frac{4X_{pp}^2}{T^2}\left(t-\frac{T}{4}\right)^2 dt\right)\right]^{1/2}$$
$$= \frac{X_{pp}}{\sqrt{12}} \simeq \frac{X_{pp}}{3.46} = 0.289X_{pp}$$

The Fourier series expansion of the triangular signal, as defined by relation (7.6), is given by the following expression:

$$x(t) = \frac{2X_{pp}}{\pi^2}\sum_{k=1}^{\infty}\frac{1-(-1)^k}{k^2}\cos\frac{2\pi k}{T}t = \frac{2X_{pp}}{\pi^2}\left(\cos\frac{2\pi}{T}t + \frac{1}{3^2}\cos\frac{2\pi}{T}3t + \dots\right). \tag{7.7}$$

Note that all even harmonics of the triangular signal are null.

7.2.3 Rectangular Signal

The rectangular signal is described by a waveform taking just two different values during the period T, see Fig. 7.4. It is characterized by a parameter δ ($0 < \delta < 1$), referred to as the *duty factor*, which indicates the period fraction corresponding to the upper level. Therefore, the time interval corresponding to the higher value is δT while the lower value corresponds to a time interval $(1 - \delta)T$. If X_{pp} is the peak-to-peak amplitude, the mathematical representation of the rectangular alternating signal is as follows:

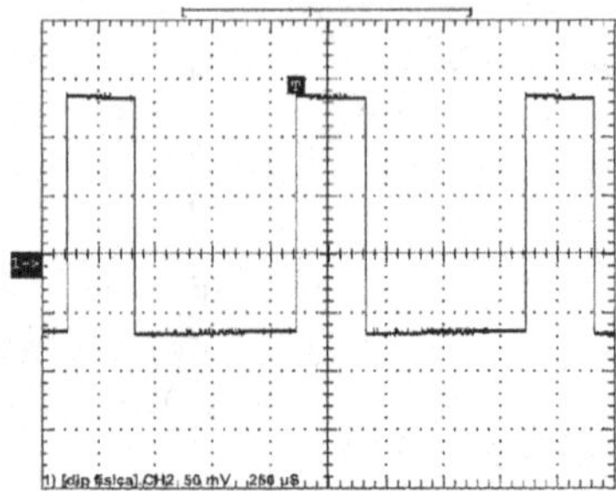

Fig. 7.4 Rectangular signal with 30 % *duty cycle*. Plot obtained on the screen of a digital oscilloscope with vertical sensitivity 50 mV per division, and horizontal sensitivity 250 μs per division. It can be noticed in this experimental plot that the two signal levels are not exactly constant. This is caused by the low-frequency cutoff introduced when using the oscilloscope in AC coupling (see Sect. 7.3)

$$x(t) = \begin{cases} X_{pp}(1-\delta) & \text{for} \quad |t| < \frac{\delta T}{2} \\ -X_{pp}\delta & \text{for} \quad \frac{\delta T}{2} < |t| < \frac{T}{2} \end{cases} \tag{7.8}$$

When $\delta = 0.5$ the rectangular waveform defined by (7.8) becomes a *square wave* since the signal assumes its two values for the same duration $T/2$. Using the definition (7.3) of effective value, we obtain the following for the rectangular signal:

$$\begin{aligned} X_{rms} &= \left[\frac{1}{T}\left(\int_{-T/2}^{-\delta T/2}(X_{pp}\delta)^2\,dt + \int_{-\delta T/2}^{\delta T/2}(-X_{pp}(1-\delta))^2\,dt + \int_{\delta T/2}^{T/2}(-X_{pp}\delta)^2\,dt\right)\right]^{1/2} \\ &= X_{pp}\delta(1-\delta) \end{aligned}$$

This relation shows that the *rms* value of a rectangular signal, besides depending on the peak-to-peak amplitude, changes with its *duty factor* δ. In particular for a *square wave* ($\delta = 0.5$) we get the following:

$$X_{rms} = \frac{X_{pp}}{4} = 0.25 X_{pp}$$

The Fourier series expansion of the rectangular signal, as defined by relation (7.8), is given by the following expression:

$$x(t) = X_{pp}\delta \sum_{k=1}^{\infty} \frac{\sin k\pi\delta}{k\pi\delta} \cos\frac{2\pi k}{T} t \tag{7.9}$$

Note that for the particular case of the square wave ($\delta = 1/2$), all even harmonics vanish.

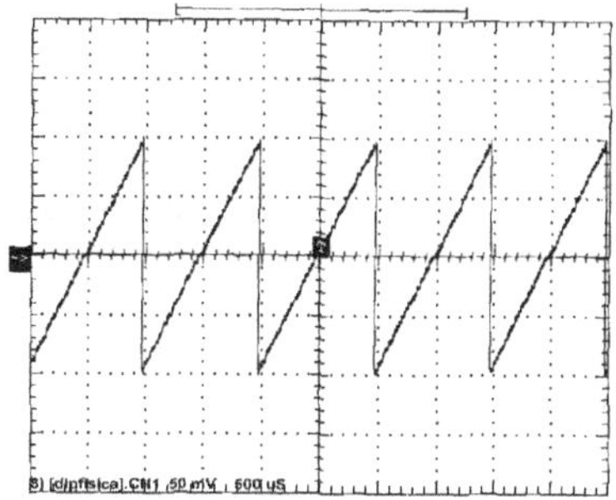

Fig. 7.5 Sawtooth signal. Plot obtained on the screen of a digital oscilloscope with vertical sensitivity 50 mV per division, and horizontal sensitivity 500 μs per division

7.2.4 Sawtooth Signal

A sawtooth signal of period T and peak-to-peak amplitude X_{pp}, represented in Fig. 7.5, is given by the mathematical expression:

$$x(t) = \frac{X_{pp}}{T}t \qquad \text{for} \quad -\frac{T}{2} < t < \frac{T}{2} \tag{7.10}$$

This kind of waveform was used to drive the sweep of the electron beam in cathode ray tubes once used in analog oscilloscopes (see next chapter) and home television sets.

Applying the definition (7.3) of effective, or rms, value, we obtain the following for the sawtooth signal:

$$X_{rms} = \left[\frac{1}{T}\int_{-T/2}^{T/2} \frac{X_{pp}^2}{T^2} t^2 \, dt\right]^{1/2} = \frac{X_{pp}}{\sqrt{12}} \simeq \frac{X_{pp}}{3.46} = 0.289 X_{pp}$$

Note that the triangular and the sawtooth signals are characterized by the same conversion factor between the peak-to-peak and the effective amplitudes.

The Fourier series expansion of the sawtooth signal, as defined by relation (7.10), is given by the following expression:

$$x(t) = -\frac{X_{pp}}{\pi}\sum_{k=1}^{\infty}\frac{(-1)^k}{k}\sin\frac{2\pi k}{T}t \tag{7.11}$$

It contains both even and odd harmonics.

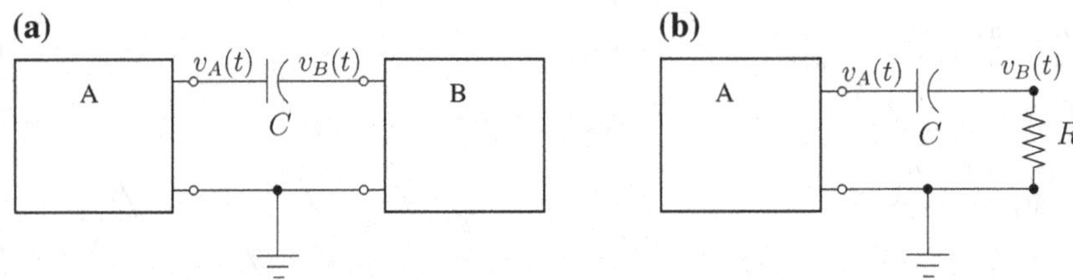

Fig. 7.6 AC coupling of two different circuits (**a**) and its equivalent representation (**b**)

7.3 AC Coupling

Before discussing the operational principles of instruments used to measure alternating signals, we describe the *AC coupling* technique typically adopted to connect these instruments to the circuit under test. This technique is used specially by the most accurate and expensive devices, for the safety of circuits and for a better definition of the measurand.

With the expression "AC coupling," we intend that two circuits are connected through a capacity, as shown in Fig. 7.6a. Usually one of this two circuits is a signal generator (indicated by the letter A in Fig. 7.6) while the other is a receiver (indicated by the letter B in the same figure). Often the receiver has purely resistive input impedance and the circuit can be schematized as in Fig. 7.6b.

We now proceed to evaluate the effect of the capacity C by comparing the characteristics of the signal at its two terminals. Consider the Fourier expansion of the periodic signal $v_A(t)$ generated by the circuit A, as shown in Fig. 7.6b,

$$v_A(t) = \frac{a_0}{2} + \sum_{n=1}^{\infty} A_n \cos(\omega_1 nt + \phi_n)$$

where $\omega_1 = 2\pi/T$ is the signal angular frequency. This expression consists of a time-independent contribution $a_0/2$ and a sum of oscillating terms. Using the superposition principle, the signal $v_B(t)$, measured at the receiver input $v_B(t)$, can be obtained by summing the effects of the single terms of the Fourier expansion of $v_A(t)$. Since the capacity has infinite impedance for DC signals, the continuous component of the signal v_A is blocked.[3] From the inspection of the circuit of Fig. 7.6b and using the symbolic method, we obtain for a generic term $V_B = V_A R/Z_{tot} = V_A j\omega RC/(1 + j\omega RC)$ and for the complete signal we can write as follows:

$$v_B(t) = \sum_{n=1}^{\infty} A_n \cdot \frac{j\omega_1 nRC}{1 + j\omega_1 nRC} \cos(\omega_1 nt + \phi_n)$$

[3] A similar analysis can be carried out for the case of a *block capacitor* used to separate the continuous voltage component of different sections in the same circuit.

When $\omega_1 n \gg 1/RC$, the previous expression is correctly approximated by

$$v_B(t) \simeq \sum_{n=1}^{\infty} A_n \cos(\omega_1 nt + \phi_n) = v_a(t) - \frac{a_0}{2}$$

Under these circumstances the two signals v_B and v_A differ only by the constant amount $(a_0/2)$ and the average value of v_b is null. In conclusion, the AC coupling has the following effects:

1. it blocks the continuous component of a time-dependent periodic signal transforming it in an alternating signal with null time-averaged value: $\int_{-\infty}^{+\infty} v(t)dt = 0$
2. as a side effect, it modifies the low frequency range (where the relation $\omega_1 n \gg 1/RC$ is not valid) of the signal spectrum, consequently distorting the signal shape (see example in Fig. 7.4).

7.4 Analog Instruments

In Sect. 4.3.1 of Chap. 4 we described in detail the properties of the moving coil ammeter and its use to measure a DC current. We discuss now the response of this instrument when an alternating current circulates in the coil. The solution to this problem is still provided by the differential equation (4.4) where now the alternating current $i(t)$ is used instead of the constant current I.

Qualitatively, the motion of the moving coil is determined by the difference between the angular frequency of the alternating current and the characteristic frequency ω_M of the coil mechanical system.[4] When $\omega \ll \omega_M$ the moving coil replicates the time evolution of the signal while on the contrary when $\omega \gg \omega_M$ the coil will only oscillate with very small amplitude around its equilibrium position.

To obtain the quantitative solution to this problem, we start by writing the modified version of Eq. (4.4) as follows:

$$\mathscr{I}\frac{d^2\theta}{dt^2} + \beta\frac{d\theta}{dt} + C\theta = m(t) \tag{7.12}$$

where $\mathscr{I}$ is the inertia moment of the moving coil with respect to its rotation axis, β is the coefficient describing the viscous torque, and C is the coefficient describing the return torque of the spiral spring. The term $m(t)$ is the electromagnetic torque caused by the alternating current flowing in the coil. When this current is sinusoidal we can write $m(t) = M \cos \omega t$, where M is the maximum torque value. Since Eq. (7.12) is

[4]Usually the moving coil motion is critically damped, as described in Sect. 4.3.2, and its characteristic frequency is $\omega_M \simeq \sqrt{C/\mathscr{I}}$.

linear with time-independent coefficients, we can use the symbolic method to obtain its steady-state solution, the evolution of the *transient* phase being irrelevant here. After denoting the complex amplitude of the coil rotation angle with Θ_c, we have

$$(-\omega^2 \mathscr{I} + j\omega\beta + C)\Theta_c = M$$

$$\Theta_c = \frac{M}{-\omega^2 \mathscr{I} + j\omega\beta + C} \tag{7.13}$$

Indicating the modulus of Θ_c with Θ, we can write

$$\Theta = \frac{M}{\sqrt{(C - \omega^2 \mathscr{I})^2 + \omega^2\beta^2}} = \frac{M}{\sqrt{C^2 - 2\omega^2 \mathscr{I} C + \omega^4 \mathscr{I}^2 + \omega^2\beta^2}}$$

This allows obtaining the following solution of Eq. (7.12):

$$\Theta(t) = \Theta \cos(\omega t + \phi)$$

As in the ammeter the motion of the coil is *critically damped* ($\beta^2 = 4\mathscr{I} C$), the oscillation amplitude becomes as follows:

$$\Theta = \frac{M}{C + \omega^2 \mathscr{I}} \tag{7.14}$$

Identifying the characteristic angular frequency of the moving coil with $\omega_M = \sqrt{C/\mathscr{I}}$, we can write

$$\Theta = \frac{M}{C\left[1 + \left(\frac{\omega}{\omega_M}\right)^2\right]} \tag{7.15}$$

This equation confirms the qualitative picture on the coil motion presented at the beginning of this section and shows the following:

- when $\omega \gg \omega_M$ then $\Theta \simeq 0$ the index of the instrument "*vibrates*" around the zero value
- when $\omega \ll \omega_M$ then $\Theta \simeq M/C$ and the index follows the time evolution of the input current.

It is important to remark that the characteristic angular frequency is determined by mechanical parameters and its value cannot exceed a few cycles per second. Given these limits, the moving coil ammeter is of little use for measuring alternating currents.

The use of diodes to rectify the AC current entering the ammeter permits to obtain a useful response from these instruments. Figure 7.7 shows two possible setups exploiting this idea. In both cases, diodes rectify the current before it enters the coil in such a way to make its average value proportional to the effective amplitude of

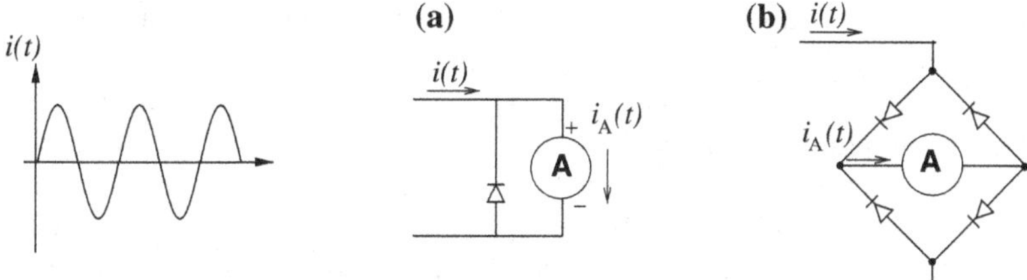

Fig. 7.7 Principle of operation of AC ammeters with rectification: (**a**) single half-wave rectification and (**b**) full-wave rectification. The input current is shown in the *left plot* while the currents flowing in the coil in the two cases are shown in Fig. 7.8

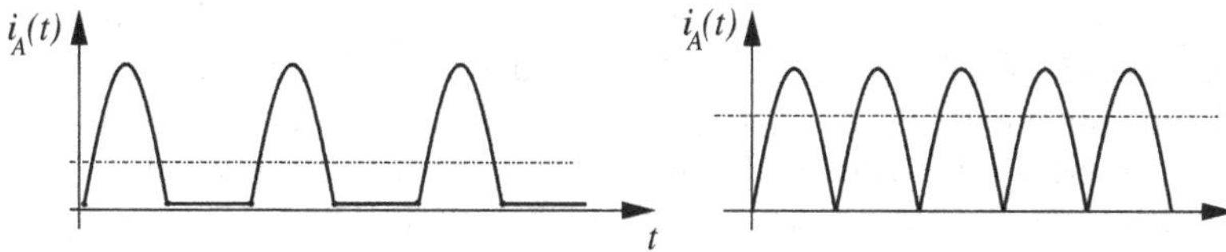

Fig. 7.8 Time evolution of the currents flowing in the ammeter of Fig. 7.7: on the *left* single half wave, on the *right* full-wave rectification. The *dashed line* represents the average current value of each waveform

the original alternating waveform. The diodes used in the circuits shown in Fig. 7.7a ideally behave in the following way: for current flowing in the direction shown in the symbol, their impedance is null and, vice versa, for current flowing in the opposite direction this impedance becomes infinity. Therefore, when the signal is $i(t) = i_0 \sin \omega t$, this current flows in the ammeter only during the positive half wave with the waveform shown to the left of Fig. 7.8. When expressed as a Fourier series, this waveform is made of a continuous component, equal to its average value, and an infinite number of oscillating components with frequency given by integer multiples of the signal frequency ω and with amplitudes decreasing as their frequency increases. Since the response of the ammeter is linear, the effect of each component can be obtained using Eq. (7.15). Consequently, provided $\omega \gg \omega_M$, only the continuous component of the current produces a measurable displacement of the instrument index and the measured value coincides with the time average of the current $i_A(t)$. In the case of a single half wave, this amounts to the following:

$$< I >= \frac{1}{T}\int_0^T i_A(t)dt = \frac{1}{T}\int_0^{T/2} i(t)dt = \frac{1}{T}\int_0^{T/2} i_0 \sin(\frac{2\pi}{T}t)dt = \frac{i_0}{\pi} = \frac{I_{rms}}{2.22} \tag{7.16}$$

Obviously, for the scheme of Fig. 7.7b where the full wave is rectified, this value doubles and we obtain $< I >= I_{rms}/1.11$.

The results obtained show that it is possible to exploit a moving coil ammeter to measure alternating sinusoidal currents adopting one of the two options illustrated above and calibrating accordingly the scale of the instrument to indicate directly the effective amplitude (*rms*) of the current. Similarly, it is possible to modify a DC voltmeter to measure alternating voltage signals. We remark once again that the scale of ammeters exploiting rectifying circuits for alternating signals is calibrated for sinusoidal waveforms and, consequently, gives wrong values if used to measure the effective value of non-sinusoidal signals.

Besides the rectified moving coil ammeter, there are others analog instruments useful to perform AC measurements, all of them now superseded by the developments of digital electronics. Among them, we deem useful to mention the *electrodynamic instruments* that exploit the force between two coils with the same flowing current. This force is quadratic in the current amplitude and it is therefore suitable for the measure of alternating currents. Similarly, the *electrostatic voltmeters* have a quadratic response, see Sect. 4.4.2, and can be used to measure the effective value of alternating voltage signals. Note however that, contrary to the DC case, a current flows in electrostatic voltmeters when used with AC voltages. Indeed they consist of a capacitor and therefore their impedance, very high for continuous voltages, becomes progressively lower with increasing signal frequency.

7.5 Digital Instruments

Nowadays the scientific electronic market offers digital instruments for AC signals based on many different designs. The diagram in Fig. 7.9 is a valid representation for most of them. The central element determining the measurement principle is the AC/DC converter used to transform the AC signal (voltage or current) in a DC voltage level. This continuous voltage is digitized and transformed in a binary signal by an analog-to-digital converter, ADC. Finally, a display element transforms the ADC binary output in a decimal number and shows it to the user. In the following, we describe the most important approaches to the conversion of AC signal in DC levels.

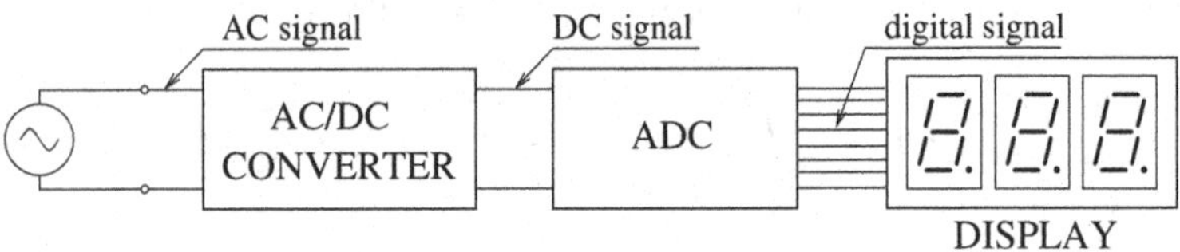

Fig. 7.9 Block diagram of a digital instrument for measuring AC signals. The circuit in the box labeled *AC/DC CONVERTER* converts the alternating signal input into a DC voltage, which is digitized by the device shown in the figure with the acronym ADC (analog-to-digital converter), and finally the device indicated with DISPLAY shows the voltage value

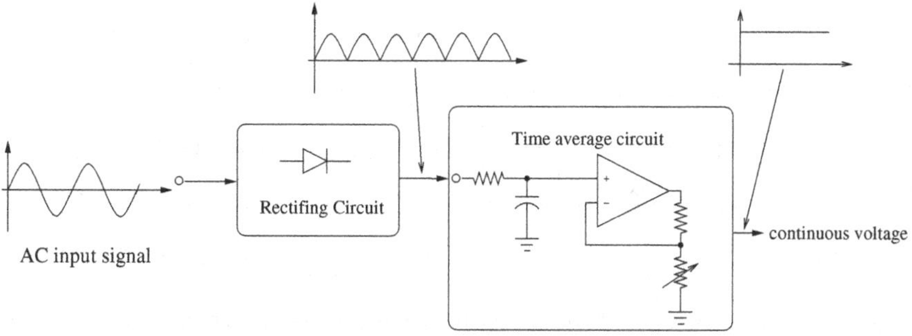

Fig. 7.10 Diagram showing the working principle of an instrument converting the average amplitude of an AC signal in a DC voltage level. The circuit performing the time average of the signal exploits an *ideal operational amplifier*

7.5.1 Peak or Average Converter

Low cost digital multimeters, and among them especially portable devices, compute the signal *rms* value from the measure of the peak amplitude or from the average value of the rectified signal, see Fig. 7.10. These instruments are calibrated under the assumption that the input signal has a sinusoidal waveform. Therefore, similarly to the case of analog ammeters, they produce inaccurate results if used with non-sinusoidal signals, the measured value depending on the signal waveform.

Instruments based on the conversion of the peak value or of the average value should not be used to measure non-sinusoidal signals.

7.5.2 Analog AC/DC Converters

To obtain the *rms* value of a voltage (or of a current) it is necessary to compute the expression (7.3), i.e., the instrument first needs to square the signal, then perform the time average of this new signal and finally to take the square root of the result. These tasks become possible with the use of *operational amplifiers*, active electronic components designed to perform mathematical operations, such as the sum, the logarithm, the product, the quotient, the integral, and the derivative, of electrical signals connected to their inputs.

In Fig. 7.11 we show the block diagram of a voltmeter based on this kind of component: the first operational amplifier performs the product of the signal with itself and divides this signal by the output V_0 of the complete circuit through a *feedback* connection.[5] The *RC* circuit connected to the output of the first operational

[5] The *feedback* is a feature of operational amplifiers whereby the circuit output is fed back to one of the inputs of the device. A detailed description of the design of an operational amplifier requires notions that are usually learned at a successive stage of the course of studies.

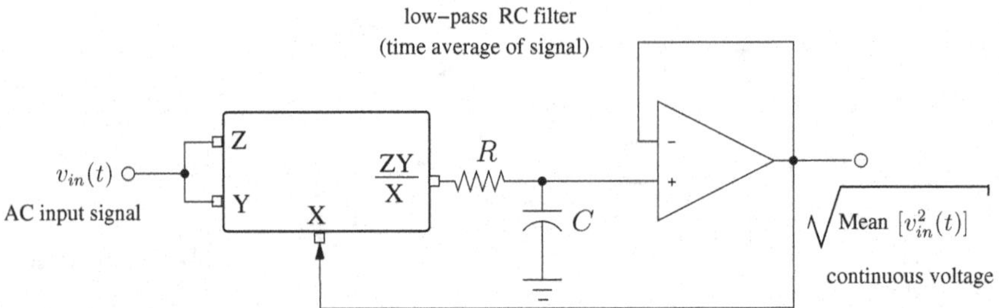

Fig. 7.11 Block diagram of an instrument exploiting an analog converter. The *rectangle* represents a circuit capable of executing the indicated operations on inputs *X*, *Y*, and *Z*. The figure is based on a schematic published on Ref. [1]. The understanding of the implementation of this scheme requires the knowledge of the elementary principles of operational amplifiers

amplifier works as an integrator while the second amplifier has unitary gain and is used for its high impedance to obtain the feedback signal with minimal disturbance of the preceding circuits. Therefore, for the output V_0 of the complete circuit the following relation holds:

$$V_o \propto \int \frac{v_{in}^2(t)}{V_o}\, dt$$

that yields $V_o \propto \sqrt{\int v_{in}^2(t)\, dt}$.

This kind of voltmeter is free from the systematic errors plaguing instruments based on rectifier. However, we remark that in operational amplifiers the product of two signals is obtained by summing their logarithms and this reduces the dynamic range of the instrument.

7.5.3 Thermal AC/DC Converters

The operation of thermal AC/DC converters is based on the definition of the effective value of a time-dependent signal as the DC level with equivalent capability of heating a resistive load by Joule effect. By construction, instruments using this kind of converter are not sensitive to the waveform of the input signal. Moreover, all Fourier components of the signal are equally weighted making this technique one with the widest frequency *passband*. In Fig. 7.12 we show schematically the working principle of a *thermal AC/DC converter*. The device consists of two identical temperature sensors and of an operational amplifier. The signal under test heats the input resistor by Joule effect. The sensor measuring its temperature drives the positive input of the operational amplifier whose output is used to heat an identical resistor. The sensor measuring the temperature of this last resistor is connected to the negative input of the operational amplifier. The output voltage of the operational amplifier decreases

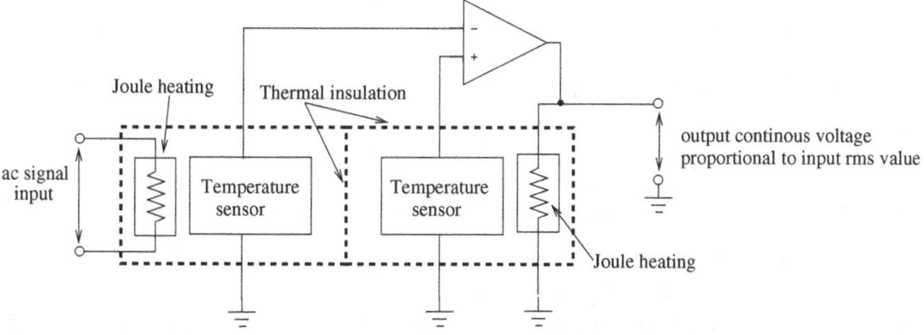

Fig. 7.12 Block diagram of an AC–DC converter based on thermal effect. The AC signal heats a resistor R and the circuit measures a DC voltage that causes the same thermal effect in an identical resistor. See text for more details

until its two inputs have practically equal voltage values. After a transient phase due to the finite thermal capacity of resistors and sensors, the temperatures of the two resistors become equal and this output voltage stabilizes to a constant value. This value, producing the *same thermal effect* in an identical resistor, is equal *by definition* to the effective value of the AC signal.

7.5.4 Digital Sampling

The availability of fast ADC circuits makes it possible to obtain the effective value of a time-dependent signal by measuring and storing its instantaneous value at a sampling frequency sufficiently higher than its frequency. Denoting with V_k, $(k = 1, \ldots, N)$ the digitized sampled values, we have

$$V_{rms} \simeq \sqrt{\frac{1}{N}\sum_{k=1}^{N}(V_k - < V >)^2} \quad \text{where} \quad < V > = \frac{1}{N}\sum_{k=1}^{N} V_k$$

provided the sampled values cover an integer number of signal periods. In this expression, the subtraction of the average value is in principle redundant when the signal is AC coupled to the ADC. However, it is advisable to keep it to correct for the systematic effects due to the finite ADC offset. The statistical uncertainty affecting the value of V_{rms}, due to the ADC finite resolution and to ambient electrical noise superimposed to the signal, can be easily computed. As expected, it is inversely proportional to the square root of the samples number N. This algorithm is implemented with digital electronics and is used to characterize AC signals in all modern digital oscilloscopes, as discussed in detail in next chapter.

7.5.5 Phase Measurements

Consider the two waveforms: $v_1(t) = V_1 \cos \omega t$ and $v_2(t) = V_2 \cos \omega(t + \phi)$. Their product can be written as follows:

$$V_1 V_2 \cos(\omega t) \cos(\omega t + \phi) = \frac{V_1 V_2}{2}[\cos(\phi) + \cos(2\omega t + \phi)]$$

It consists of a continuous component and an alternating signal with frequency equal to the double of the frequency of the signals in input.This suggests to measure $\cos \phi$ using the ratio of the amplitude of these two components. Therefore, we can use operational amplifiers to obtain the measure of the phase difference between two sinusoidal signals.

It must be noted, however, that the mere knowledge of the phase cosine does not allow determining its sign. This ambiguity can be solved introducing a phase lag of $\pi/2$ in one of the two signals and using the same setup to measure the sine of the original phase difference. An instrument capable of measuring both the amplitude and the phase difference of sinusoidal signals is usually referred to as a *vectorial instrument*.

Digital sampling. Similarly to the effective amplitude, the availability of high-frequency digital sampling allows for high-accuracy measurement of the phase difference of sinusoidal signals using ad hoc algorithms yielding an estimate of its value with accuracy better than a thousandth of a radian. A detailed presentation of these techniques falls beyond the scope of this book. The interested reader can find more details on this subject in Ref. [2].

7.5.6 Impedance Measurement

We can obtain the value of unknown impedances using a vectorial voltmeter, an instrument that can measure both the amplitude and the phase difference of sinusoidal waveforms. A simple circuit allowing to perform this task is shown in Fig. 7.13. It

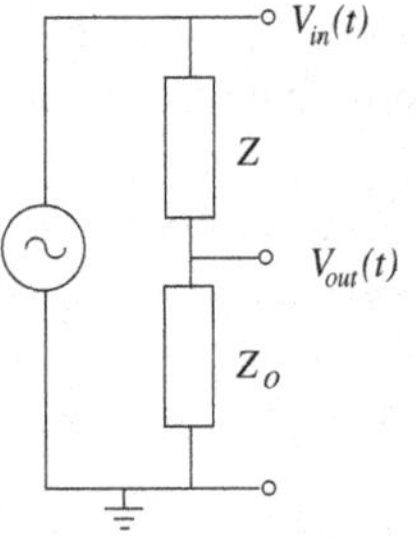

Fig. 7.13 Voltage divider for impedance measure. See text for details

consists of a simple voltage divider whose output signal is given by

$$V_{out} = V_{in}\frac{Z_o}{Z + Z_o}$$

From the values of the amplitude of input and output voltage and from their phase difference we can evaluate the unknown impedance value Z as

$$Z = Z_o\left(\frac{V_{in}}{V_{out}} - 1\right) = Z_o\left(\frac{|V_{in}|}{|V_{out}|}e^{-j\phi_o} - 1\right) = Z_o\left[\frac{\cos\phi_o}{A_o} - 1 - j\frac{\sin\phi_o}{A_o}\right]$$

where $A_o = |V_{out}|/|V_{in}|$ is the output attenuation and ϕ_o the phase difference. This equation allows obtaining both resistive and reactive components of the impedance. This method can be more convenient with respect to the use of a bridge circuit, which requires the availability of an accurately measured inductance and/or of a capacitance as reference impedances. In this case, the reference impedance Z_o can be made of a simple resistance that can be accurately measured with a simple digital multimeter.

Problems

Problem 1 An ADC digitizes a sinusoidal voltage signal taking N samples at constant time intervals covering a span equal to the signal period. Sampled values are affected by quantization errors, uncorrelated and identically distributed with variance equal to σ^2. Data are used to identify their maximum and minimum from which the peak-to-peak Xpp is computed. Evaluate the standard uncertainty on this last quantity. [A. $u_{pp} = \sqrt{2}\sigma$.]

Problem 2 An ADC digitizes a sinusoidal voltage signal taking N samples at constant time intervals covering a span equal to the signal period. Sampled values are affected by quantization errors, uncorrelated and identically distributed with variance equal to σ^2. Show that the estimate of the effective value, obtained as the root mean square of the sampled values, is affected by a bias related to the variance of quantization errors. [A. $\hat{V}_{rms} = \sqrt{V_{rms}^2 + \sigma^2}$.]

Problem 3 An ADC digitizes a sinusoidal voltage signal taking N samples at constant time intervals covering a span equal to the signal period. Sampled values are affected by quantization errors, uncorrelated and identically distributed with variance equal to σ^2. Show that the estimate of the effective value, obtained as the root mean square of the sampled values, besides being biased (see previous problem) is affected by a statistical uncertainty related to the variance of quantization errors. [A. $u_V = \sigma/\sqrt{N}$.]

Problem 4 An ADC, with negligible quantization error, digitizes a sinusoidal voltage signal taking N samples at constant time intervals covering a span equal to the signal period. Sampled values are affected by an ambient noise, uncorrelated and identically distributed with variance equal to σ_n^2. Show that the estimate of the effective value, obtained as the root mean square of the sampled values, is affected by a bias related to the variance of ambient noise. [A. $\hat{V}_{rms} = \sqrt{V_{rms}^2 + \sigma_n^2}$.]

Problem 5 An ADC, with negligible quantization error, digitizes a sinusoidal voltage signal taking N samples at constant time intervals covering a span equal to the signal period. Sampled values are affected by an ambient noise, uncorrelated and identically distributed with variance equal to σ_n^2. Show that the estimate of the effective value, obtained as the root mean square of the sampled values, besides being biased (see previous problem) is affected by a statistical uncertainty related to the noise variance. [A. $u_V = \sigma_n/\sqrt{N}$.]

Problem 6 A resistor R is heated by Joule dissipation. To compute its temperature, assume that the resistor is cooled by thermal conduction and that the power lost to the environment is given by $Q = -k(T - T_0)$. The resistor is in equilibrium with the environment at temperature T_0 before being connected to an ideal voltage generator with output equal to V_0. Denoting with C its thermal capacity, calculate the time evolution of the resistor temperature after the connection to the generator. [A. $T = T_0 + (P\tau/C)(1 - \mathrm{e}^{-t/\tau}$ where $P = V_0^2/R$ and $\tau = C/k$.]

References

1. D.H. Sheinglod, *Non Linear Circuits Handbook* (Analog Devices Inc., Norwood, 1976)
2. R. Bartiromo, M. De Vincenzi, AJP **82**, 1067–1076 (2014)

Chapter 8
The Oscilloscope

8.1 Introduction

The oscilloscope is an instrument which allows the observation of the time evolution of electrical signals. Nowadays most physical quantities can be measured through specialized sensors, also referred to as transducers, which convert them in to electrical signals. Therefore, the oscilloscope is an indispensable tool for engineers and physicists (and not only) and can be found in every measurement laboratory.

The first embodiment of an instrument capable of detecting fast electrical signal dates back to the late 19th century. The oscilloscopes became of widespread use immediately after the Second World War. They were analog instruments using a cathode ray tube and a fluorescent screen capable of displaying the point of electron impact. Thanks to the development of digital electronics, more recently digital oscilloscopes have become available that are more versatile to use and offer a wide range of possible processing of the signals acquired.

Although the analog oscilloscope is nowadays out of date, nevertheless we start this chapter in the next Sect. 8.2 with a detailed discussion of its operational principles. This is an effective introduction to the understanding of the functions needed in an instrument capable of following the time evolution of fast electric signals. We will discuss the problem of synchronizing the instrument with the experiment, then we will illustrate a block diagram of its components and functions and finally we will evaluate a realistic estimate of its response time. In Sect. 8.3, we will introduce the components of a digital oscilloscope and illustrate its most important characteristics and operational features. In Sect. 8.4, we will show how to obtain the measure of the phase difference between sinusoidal signals, and finally in Sect. 8.5 we will discuss some practical aspects to be kept well in mind when using an oscilloscope in the laboratory practice.

R. Bartiromo and M. De Vincenzi, *Electrical Measurements in the Laboratory Practice*, Undergraduate Lecture Notes in Physics,
DOI 10.1007/978-3-319-31102-9_8

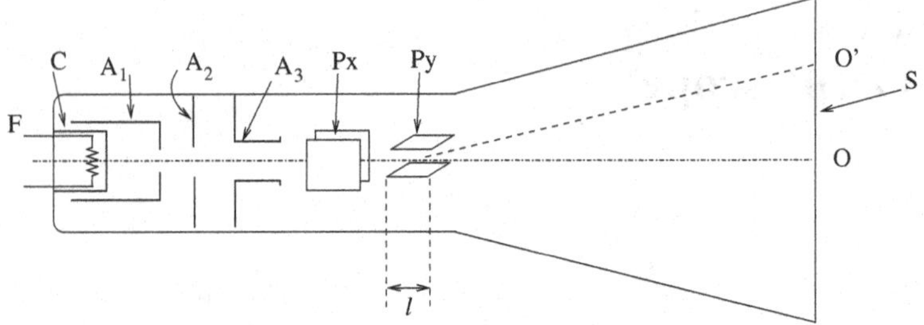

Fig. 8.1 Schematic representation of an analog oscilloscope

8.2 The Analog Oscilloscope

A cathode ray tube, schematically represented in the Fig. 8.1, consists of a glass bulb hosting an electron gun in vacuum to produce a collimated electron beam. The electrons are emitted by thermionic effect from a cathode C, indirectly heated by thermal conduction to temperatures that can reach around 800 K by the filament F in its turn heated by the Joule effect of an electric current flowing through it.

The electrons exit the cathode with an average energy of about 0.1 eV[1] corresponding to its temperature. The anode A_1 accelerates them toward the fluorescent screen S up to energy of the order of 10^3 eV. Two more anodes, A_2 and A_3, are used to regulate the beam intensity and to focus it on the center of the screen. This screen consists of a glass which serves to close the vacuum tube and on which a grid is engraved for the measurement of the impact point of the electrons. Indeed, a fluorescent substance deposited on the inner side of the screen, produces the emission of visible light, typically in the green, when stricken by fast electrons.

In the space between the gun exit and the screen, the electron beam travels through two pairs of flat electrodes used to deflect the impact point across the screen. A similar system was used until the end of last century to produce the image in a home television set. It is easy to show that the displacement of the beam impact point on the screen is proportional to the voltage difference applied to the electrodes. To obtain this result, we use a reference system with the z-axis passing through the centers of the cathode and of the screen and the x and y axes in the plane of the screen, see Fig. 8.2.

First consider the pair of electrodes P_y, in the following also referred to as the vertical deflecting plates. When a voltage difference V_y is established across them, the electric field E_y in the space region between these electrodes is nearly constant in amplitude and directed along the vertical y-axis on the screen. The electrons come from the cathode along the z-axis and at the P_y entrance, they have velocity v_0 along the z direction as given by the accelerating potential V_0. This is equal to the voltage difference between the cathode and the anode A_3. Taking the time origin at the instant

[1]The electron-volt (eV) is an energy unit: 1 eV corresponds to the energy gained by an electron when moving between two space points with a voltage difference of 1 V.

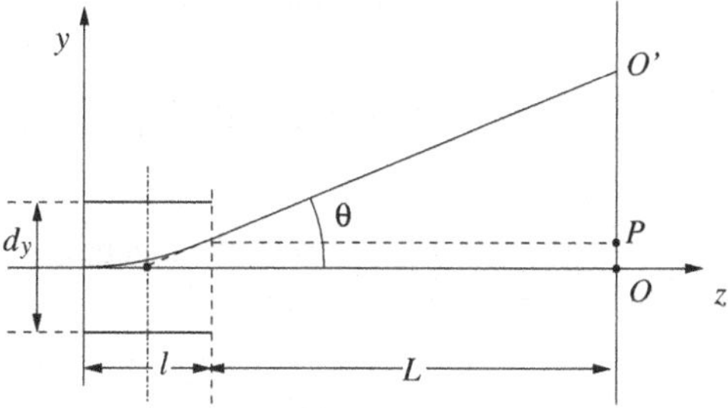

Fig. 8.2 Deviation of the electron beam within an analog oscilloscope

when the electrons enter the space between the electrodes and start feeling the action of the E_y field, their motion is described by the following equations:

$$v_z = \sqrt{\frac{2eV_0}{m_e}} = v_0 = \text{constant}, \quad v_y = -\frac{eE_y}{m_e}t = \frac{eV_y}{m_e d_y}t \text{ and } v_x = 0$$

where d_y is the distance separating the electrodes P_y, e is the electron charge, and m_e is the electron mass. The electron trajectory lays in the plane y–z and is described by the parametric equations

$$z = v_0 t \text{ and } y = \frac{1}{2} \cdot \frac{eV_y}{m_e d_y} t^2$$

Inserting the value of the time t obtained from the first equation in the second we get the trajectory equation:

$$y = \frac{z^2}{4d_y} \frac{V_y}{V_0}$$

The electrons follow a parabolic trajectory while traveling between the electrodes but they will resume their straight uniform motion once they exit the region where the E_y field is present. Their new flight direction will be given by the tangent to the parabolic trajectory at the exit of the electrodes.

With the help of Fig. 8.2 we can now evaluate the displacement of the beam impact point on the fluorescent screen. It is composed of the segment $\overline{OP}$ due to the deflection inside the electrodes and of the segment $\overline{PO'}$ due to the deviation of the flight direction at the electrodes exit.

Denoting with l the extension of the electrodes along the z direction and using the last of previous equations we obtain

$$\overline{OP} = \frac{l^2}{4d_y} \frac{V_y}{V_0}$$

while for the other segment we have

$$\overline{PO'} = L \tan\theta = L\left(\frac{\partial y}{\partial z}\right)_{z=l} = L\frac{l}{2d_y}\frac{V_y}{V_0}$$

where L is the distance between the electrodes exit point and the fluorescent screen. Adding these two terms, we obtain

$$\overline{OO'} = \frac{l}{2d_y}\left(\frac{l}{2} + L\right)\frac{V_y}{V_0} = kV_y$$

This shows that the displacement of the beam impact point along the y-axis is proportional to the voltage difference V_y applied to the vertical deflecting plates P_y.

The deflection factor, $k = l(l/2+L)/V_0 d_y$, depends on the geometrical parameters of the cathode tube and on its accelerating voltage. A typical value is obtained assuming $d_y = 1\,\text{cm}, l = 4\,\text{cm}, L = 20\,\text{cm}$ and $V_0 = 1000\,\text{V}$ that yield $k = 4.4 \times 10^{-2}\,\text{cm/V}$. This means that signal amplitude of 1 V produces a deflection of about half a millimeter, showing that it is necessary to use a signal amplifier to connect it to the electrodes and obtain an easily measurable deflection. After an adequate calibration of the instrument, this deflection can be used to measure the voltage difference across the electrodes by means of a ruler engraved on the screen.

It should be noted at this point that contrary, for example, to the case of the indicator in the moving coil ammeter, the fluorescent screen is a two-dimensional object where x and y directions are completely equivalent. Therefore, it is possible to use the P_x electrodes to measure a potential difference V_x simultaneously to V_y, and consequently to obtain directly on the screen a plot displaying one quantity in function of the other.

Let us now see how to use an oscilloscope to measure the time evolution of the voltage drop across a capacitor C while it is charged through a resistance R to the output potential V supplied by a voltage generator. With reference to the Fig. 8.3, we will connect the points A and B to the electrodes P_y while using the electrodes P_x as it follows. Starting from the beginning of the experiment, i.e., from the instant when the switch T is closed, we let the voltage V_x applied to these electrodes to increase linearly in time. This moves the x coordinate of the bright spot on the screen proportionally to the charging time while the coordinate y changes proportionally to the voltage across the terminals of the capacitor C. In this way, we obtain a plot of the time evolution of this voltage during the charge.

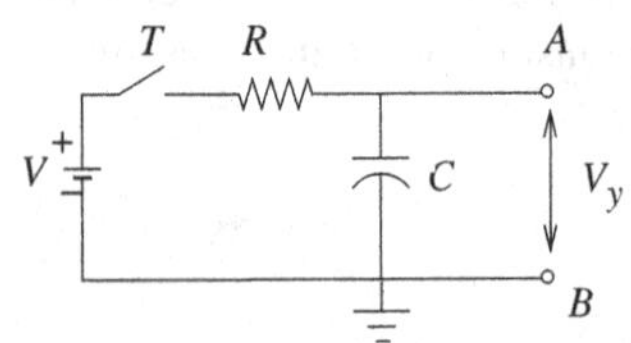

Fig. 8.3 Experimental setup to study the charge of a capacitor

We see from such example that in order to measure the time evolution of electric signals the oscilloscope needs an auxiliary circuit to generate the voltage ramps to apply to the P_x electrodes. By changing the rate at which this ramp changes in time, it will be possible to change the time scale of the instrument.

8.2.1 Synchronization

It is important to note at this point that to start the horizontal V_x voltage ramp in the experiment just discussed we used the knowledge of the instant at which the phenomenon begins. For a practical implementation the most effective approach consists of using a switch driven by an electric pulse with the same pulse also providing the start to the V_x voltage circuit. In this case, we would say that the oscilloscope is used with an external synchronization pulse, commonly referred to as the *trigger* pulse.

This is not always feasible and often it is necessary to obtain the *trigger* pulse directly from the signal under observation. In our previous experiment, assuming a charging voltage of 10 V, we could have built the pulse with a threshold discriminator. For example, when the voltage across the capacitor becomes higher than 0.1 V, the discriminator fires the pulse that starts the horizontal ramp. In this case, we say that the *trigger* pulse is *internal*, since an ancillary circuit built in the oscilloscope generates it.

Since the persistence time of the track on the fluorescent screen is very short, an analog oscilloscope is of little use for the study of transient phenomena unless we use a photographic camera to record and analyze results. However, it becomes extremely useful for the analysis of periodic waveforms that allow a continuous display on the fluorescent screen exploiting the finite persistence time of the human retina. In these conditions, we must be periodically restarting the voltage ramp V_x whose waveform takes a characteristic sawtooth shape. It must be noted that in these circumstances too the *trigger* circuit plays an important role since it allows starting the track always at the same point of the periodic signal avoiding confusion of subsequent sweeps on the screen.

8.2.2 The Analog Oscilloscope: Scheme of the Instrument

The layout of an oscilloscope is well represented by the block diagram of Fig. 8.4. The cathode ray tube (*CRT*) is powered with the voltages needed for the heather and the electron gun. The signal under test is applied to the vertical deflecting plates after being amplified by an appropriate circuit to obtain on the screen an easily measurable displacement. A switch (not shown) allows inserting an input capacitor for cutting the DC component of the signal if necessary to enhance the component that varies over time (*AC coupling*). A sawtooth voltage is applied to the horizontal deflecting plates and its slope can be adjusted through an amplifier to scan the request time

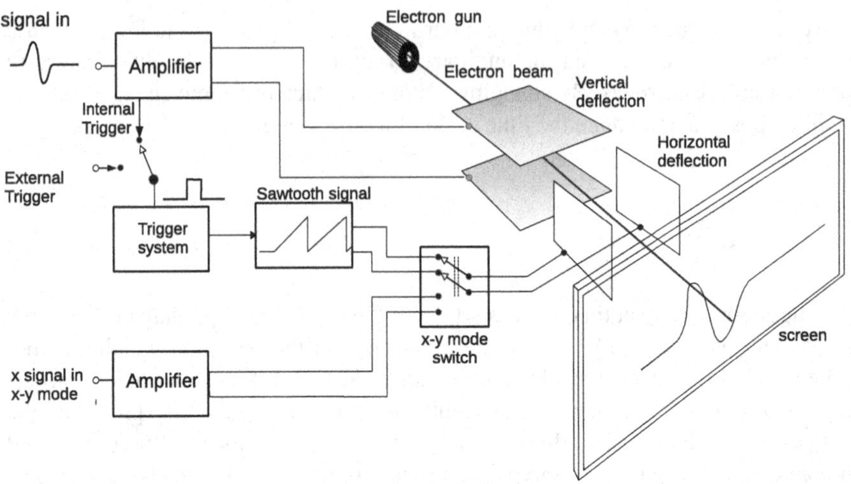

Fig. 8.4 Functional scheme of a CRT oscilloscope

interval. The circuit that generates the horizontal ramp can be controlled through an appropriate processing of the signal under measurement, internal *trigger*, or through an auxiliary pulse coming from the experiment, the *external trigger*, selectable via a switch contact.

Often, a probe, essentially consisting of a compensated resistive divider, which enables to attenuate the input signal to extend the measurement range to higher amplitudes, accompanies laboratory instruments. The use of a suitable probe is highly recommended when measuring high voltages also for safety reasons. It should be remembered that in general, the probe increases the impedance seen by the circuit under test but can reduce the bandwidth of the instrument. For a deeper understanding of working principles of an oscilloscope probe, we recommend the study of the compensated voltage divider presented in Sect. 9.10.

It is important to remark again that the use of the oscilloscope is not limited only to the study of the time evolution of electrical signals. Indeed, it is possible to connect the horizontal deflecting plates to any kind of signal to study its influence on the quantity being measured on the y-axis. To this purpose, the instrument is equipped with an external input for the x-axis and with a switch to address it in place of the horizontal ramp. When using this feature, we say that the oscilloscope is operated in the x–y configuration.

As an example, we show how to use the oscilloscope to measure the voltage–current characteristic of a diode. Following the scheme in Fig. 8.5, a bipolar oscillator supplies a time-dependent voltage, for example with a sinusoidal or triangular waveform, ranging from a negative minimum to a positive maximum. Adjusting the amplitude of this signal, the voltage drop across the diode terminals can change in the requested interval. The vertical input of the oscilloscope is connected in parallel to the diode terminals to obtain the measure of the diode voltage. The voltage drop

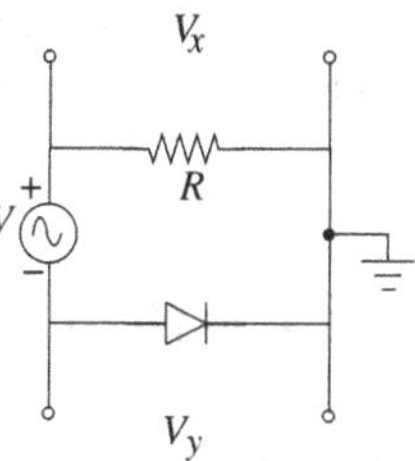

Fig. 8.5 Circuit to measure the voltage-current characteristic of a diode with an oscilloscope

across the resistor is proportional to the current flowing in the diode and can be connected to the horizontal input of the oscilloscope to obtain directly on its screen a representation of the diode characteristic.

It is important to note in Fig. 8.5 the solution adopted for the ground connections. Since usually all inputs to the oscilloscope are referred to the same ground, attention must be paid to avoid short circuits. This solution requires that the oscillator output must be floating, i.e., it should be disconnected from the ground connection of its power supply. Otherwise, a voltage transformer must be used to obtain a floating voltage to use in our measurement, see Chap. 6. This complication can be avoided if the oscilloscope has two independent vertical channels to measure the voltage across the diode as the difference between the oscillator signal and the voltage drop across the resistor. We will see in the following that this function is easily implemented in a digital oscilloscope.

8.2.3 *Response Time*

In general oscilloscopes are rather fast instruments since they exploit the motion of electrons, the lightest stable particles present in nature. Nevertheless, there is a limit to the time duration of a signal that can be accurately measured. This limit arises from the finite time that electrons employ to cross the space between the deflection electrodes. In fact, it is clear from the derivation presented above that, if the voltage varies appreciably while the electrons are deflected, its measurement will be distorted. Denoting with l the length of the deflecting plates in the electron flight direction, see Fig. 8.2, the transit time is given by

$$\tau_r = \frac{l}{v_0} = \sqrt{\frac{l^2 m_e}{2eV_0}}$$

With typical parameters, $l = 4$ cm and $V_0 = 1000$ V, this time turns out of the order of 10^{-9} s. This implies that pulses with a duration longer than 10 ns can be reliably measured, corresponding to a high-frequency cutoff of the order of 100 MHz. This limit is due only by the properties of the cathode ray tube. However, in practice the amplifier used for the input signal sets the bandpass of the oscilloscope. Indeed, in a well-designed instrument the bandpass of this amplifier is sharp and narrower than the bandpass of the tube to guarantee that all measured signals are not distorted.

8.3 The Digital Oscilloscope

In the analog instrument, the cathode ray tube performs a dual function since it is used both to carry out the measurement through the deflection of the electron motion and to show the result through the fluorescent screen. The advances of modern digital electronics allow separating these two functions to get more accurate, flexible and versatile instruments.

A digital oscilloscope exploits an analog-to-digital converter (ADC) to measure the input voltage at fixed time intervals, a microprocessor to store the data files and to perform additional data analysis, if required, and a *display* to represent the results for the user. The first embodiment of a digital oscilloscope, which still did use a cathode ray tube for the display, dates to about 40 years ago. The rapid progress in the production of flat liquid crystal screens of the last 20 years has caused the complete disappearance of the cathode ray tube from the laboratories (and from our homes).

A digital oscilloscope, in addition to performing all the functions of an analog one, offers a better screen and a number of modes of use exploiting the digital nature of the instrument. The main ones are:

- Digital direct reading of voltage and time
- Recording of transient phenomena
- Simultaneous measure of multiple voltages (multi-channel instrument)
- Accurate time base
- Electronic processing of the acquired signals: sum, difference, spectral analysis
- Average of repeated measurements
- Interface for data transfer to the memory of a computer for data storage and further analysis

Lastly, it is worth noting that a digital oscilloscope offers the opportunity to exploit the electronic memory to record events in the past with respect to the *trigger* pulse (post-*trigger* operation). This is achieved by filling the memory sequentially in time in a cyclical manner and stopping the acquisition of data only upon arrival of the *trigger* therefore keeping in memory the values measured before.

The theoretical limit to the frequency bandwidth of a digital oscilloscope is due to the sampling frequency of the analog-to-digital converter that can be easily made of the order of one *gigasample* per second, i.e., 10^9 samples every second. Indeed, as we show in Fig. 8.6, when the signal to be measured has a frequency higher than the sampling frequency, the reconstructed signal is strongly distorted both in frequency and in shape (aliasing). It is possible to show that the maximum frequency that can be reliably measured is limited to half the sampling frequency. This limit is usually referred to as Nyquist frequency. Therefore, in a digital oscilloscope the bandpass of

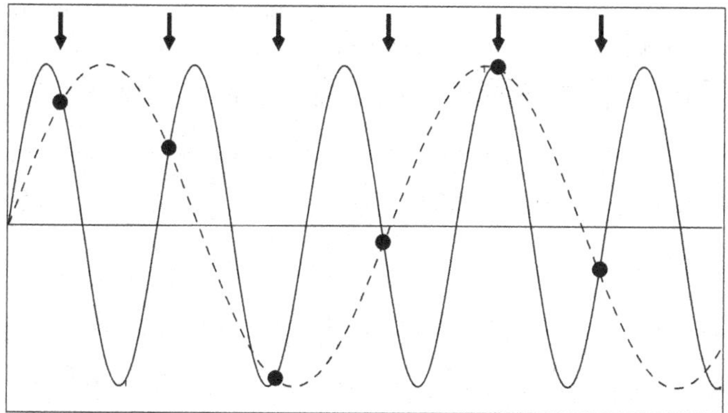

Fig. 8.6 The figure illustrates the phenomenon known as "aliasing". A high-frequency signal (continuous line) is reconstructed erroneously due to an insufficient sampling. The points, sampled at the time indicated by the *arrows*, are shown by the *dots* and the *dashed line* represents the reconstructed signal

the input amplifier is limited below half the sampling frequency to avoid distorting the measurements (anti-aliasing filter)

The readout of data from a digital oscilloscope can be done through the grating on its display, in which case the measurement accuracy will be determined by the division amplitude. However, the potential of the instrument is fully exploited only when the data are used in full digital form. In the digital measurement of a voltage, the uncertainty is given by the value of the last significant bit of the analog-to-digital converter and by the accuracy of its calibration, see Sect. 4.4.1. In general, for economic reasons these two contributions are made of the same order of magnitude. As an example, an eight bit converter has a last significant bit equivalent to 1/256 of its full range and it does not make much sense trying to calibrate it with accuracy much better than half a percent of the full range.

As for the accuracy of time measurements, it must be emphasized that it is in general very high, of the order of one hundred parts per million, thanks to the use of quartz-stabilized oscillators for timing the conversion in ADCs. This accuracy, however, is accessible only by transferring data to a computer through a special interface. In fact, the capacity of representation of a liquid crystal display is limited: the horizontal scan is divided into a finite number of points, typically a few thousand. Therefore, the accuracy of time measurements obtained by reading from the display is in the order of one part in a thousand, about an order of magnitude worse than the intrinsic accuracy. The full use of the features of a digital oscilloscope in general can only be achieved by transferring the data to a computer through an interface.

By way of example, we report in the Table 8.1 the most important parameters of a Tektronics TDS1012 oscilloscope shown in Fig. 8.7.

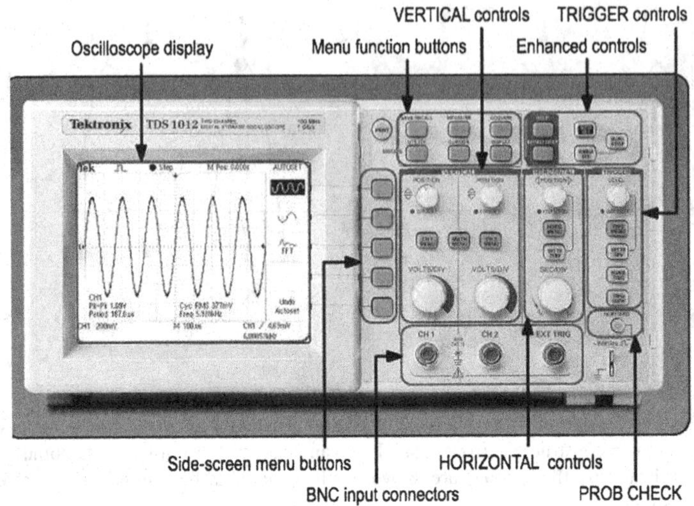

Fig. 8.7 Tektronics TDS1012 digital oscilloscope (Copyright ©Tektronix. Reprinted with permission. All Rights Reserved)

Table 8.1 Principal parameters of the TDS1012 digital oscilloscope

Sampling frequency	1 GSample/s
Frequency bandpass	100 MHz
Vertical sensitivity	from 2 mV/div–5 V/div
Vertical resolution	8 bit
Voltage accuracy	±3 %
Time base accuracy	50 ppm
Buffer memory size	2500 data

8.4 Phase Measurements

The measurement of signal amplitude with a digital oscilloscope is easily done with simple manipulation of the voltage data supplied by the instrument. In particular, most instruments nowadays offer a direct reading both of the peak-to-peak amplitude and of its root mean square (*rms*) value. On the contrary, phase measurements require special arrangements. Two different methods can be used when data are obtained directly from the display. The first only uses amplitude data and it can be implemented with a single channel analog instrument. The second method requires the measure of time intervals and is more easily implementable with an instrument having two independent channels. A very accurate measure of the phase difference between two sinusoidal signals can be obtained by exporting the data from a two channel digital oscilloscope to a personal computer. A detailed description of the algorithms needed to process the data is beyond the scope of this book. The interested readers can find an introduction to this argument in reference [1].

Phase difference. Method of the ellipse. This method makes use of the oscilloscope in the configuration x–y, whereby two voltage signals are sent respectively to the x and to the y axes. To begin the analysis, suppose that the same sinusoidal signal is sent to the two inputs. On the screen of the instrument, we will observe the plot of a curve parametrically described by the two following equations:

$$\begin{cases} x(t) = A\cos(\omega t + \phi_0) = A\cos\phi(t) \\ y(t) = A\cos(\omega t + \phi_0) = A\cos\phi(t) = x(t) \end{cases}$$

with $\phi(t) = \omega t + \phi_0$ e $-\pi < \phi(t) < +\pi$. This is a segment of the straight line inclined 45° with respect to the x-axis. When the signal sent to the y-axis has different amplitude, the last equation becomes

$$y(t) = B\cos(\omega t + \phi_0) = B\cos\phi(t) = Bx(t)/A$$

and the curve is transformed in a segment of the straight line with inclination equal to $\arctan(B/A)$.

In case the two previous signals have a phase difference of $-\pi/2$, the two equations become

$$\begin{cases} x(t) = A\cos(\omega t + \phi_0) = A\cos\phi(t) \\ y(t) = B\cos(\omega t + \phi_0 - \pi/2) = B\sin\phi(t) \end{cases}$$

and yield the parametric description of an ellipse centered in the origin and with axes aligned to the reference system: $(x/A)^2 + (y/B)^2 = 1$.

Generalizing the value of the phase difference, we have

$$\begin{cases} x(t) = A\cos(\omega t + \phi_0) \\ y(t) = B\cos(\omega t + \phi_0 + \Delta\phi) \end{cases}$$

This is the equation of an ellipse centered in the origin but inclined with respect to the reference system. This can be shown by obtaining from the first equation $\cos(\omega t + \phi_0) = x/A$ and using it in the second to write

$$\frac{y}{B} = \frac{x}{A}\cos\Delta\phi - \sqrt{1 - \left(\frac{x}{A}\right)^2}\sin\Delta\phi$$

Isolating the square root on the left member and squaring the resulting equation, we get:

$$\sin^2\Delta\phi\left[1 - \left(\frac{x}{A}\right)^2\right] = \left(\frac{x}{A}\right)^2\cos^2\Delta\phi + \left(\frac{y}{B}\right)^2 - 2xy\frac{\cos\Delta\phi}{AB}$$

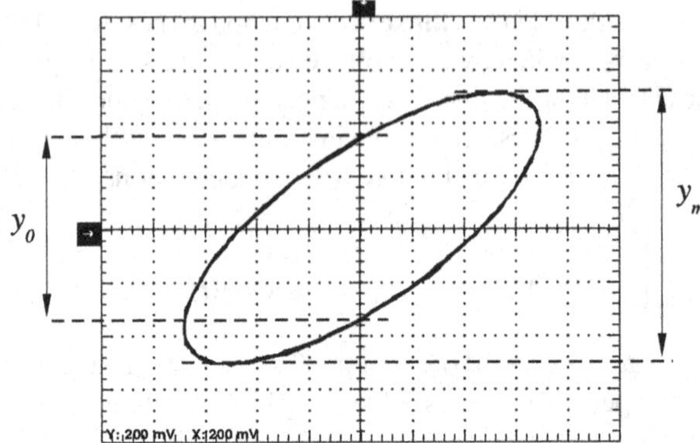

Fig. 8.8 An ellipse can be seen on the screen of an oscilloscope using two sinusoidal input signals with the same frequency displayed in the x–y mode

that with simple algebra finally yields

$$\left(\frac{x}{A\sin\Delta\phi}\right)^2+\left(\frac{y}{B\sin\Delta\phi}\right)^2-2xy\frac{\cos\Delta\phi}{AB\sin^2\Delta\phi}=1$$

that is the equation of an ellipse. The absence of terms linear in x and y implies that this ellipse is centered in the origin. Its axes are rotated with respect to the reference system of the screen, as shown in the Fig. 8.8.

Usually with this method the phase difference is obtained comparing the value y_0 of the signal on the y channel, measured when the amplitude of the signal on the x channel is null, to its maximum value y_m, see Fig. 8.8. When $x = 0$ we have $\omega t + \phi_0 = \pm\pi/2$ and this implies that $y_0 = B\cos(\pm\pi/2 + \Delta\phi) = \pm B\sin(\Delta\phi)$. Since $y_m = B$ we obtain $\Delta\phi = \arcsin(y_0/y_m)$.

Phase difference. Method of the time shift. The second method for determining the phase difference between two signals requires the measurement of the time interval that elapses between the passages of the two signals across a point corresponding to an assigned phase (refer to Fig. 8.9). In practice, we should always choose a point that the signal crosses with the maximum slope to minimize the uncertainty on the determination of the time. For sinusoidal signals, this corresponds to the zero crossing. To increase the accuracy of measurement of the times, it is helpful in any case to expand both the vertical and horizontal scale of the instrument to the maximum feasible.

In conclusion, to obtain the phase difference between two sinusoidal signals it will be sufficient to measure the time interval that elapses between the passages through the zero level, with positive (or negative) slope, of the two signals. Denoting with t_1 and t_2 the two time points corresponding to these crossings, we have $\omega t_1 = \omega t_2 + \Delta\phi$ yielding $\Delta\phi = 2\pi(t_1 - t_2)/T$ where T is the period of the two sinusoidal signals and can be measured with the same oscilloscope.

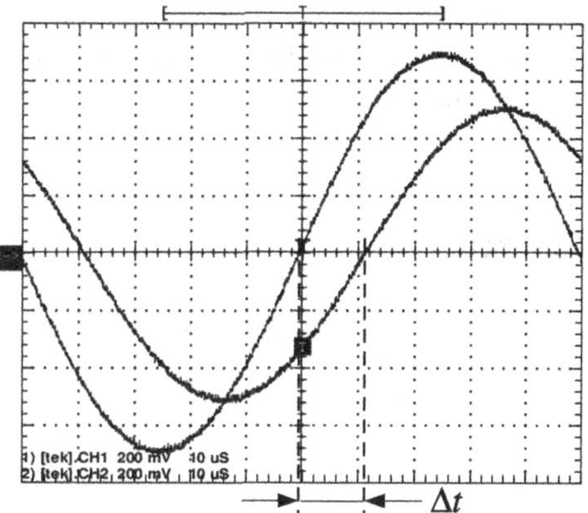

Fig. 8.9 The phase difference measurement with the method of time delay between sinusoidal signals

Although this measurement is more easily performed with a double channel oscilloscope, nevertheless using the external trigger it is possible to implement it in a single channel instrument. In this case, t_1 and t_2 can be obtained with two consecutive measures but one of the two signals must be always connected to the trigger input to guarantee that the time axis does not shift between the two measurements.

8.5 A Few Practical Considerations

When using an oscilloscope, we must always keep in mind that the input amplifiers have finite input impedance Z_{in}, well represented in general by a resistor R_{in}, typically of the order of 1 MΩ, in parallel to a capacity C_{in}, of the order of 10 pF. Therefore, if a circuit with a Thévenin equivalent impedance Z_s provides the input signal with amplitude V_s, the signal measured by the oscilloscope will have amplitude equal to $V_s Z_{in}/(Z_s + Z_{in})$. Usually $Z_{in} \gg Z_s$, and often the correction to obtain the unperturbed value is not important. However, the reactance of the capacity C_{in} decreases with increasing signal frequency and can become comparable to Z_s. When this last impedance is purely resistive, $Z_s = R_s$, the net effect will be the spurious presence of a low-pass filter with a frequency cutoff $\nu = 1/(2\pi R_s C_{in})$. Therefore, special attention must be paid when measuring circuits with high output impedance.

The problem is often complicated by presence of cables to connect signals to the oscilloscope. Cables will be dealt with in details in Chap. 10 of this book. Referring to the *RG58C/U* type of cable, of widespread use in laboratories, it must be stressed that the use of a meter of this cable for the input connection implies that and additional capacity of about 100 pF is added in parallel to C_{in}. Furthermore, the same length of cable has a series impedance of about 0.25 μ H. Assuming for example that the circuit under test has an output capacitive reactance corresponding to 1 nF, it is easy to calculate that the presence of a few meters of cable can lead to the spurious presence of resonant frequencies of the order of a few *MHz*. In these cases, it is advisable to connect the signal to the input with a compensated probe, a device that we discuss in details in Sect. 9.10.

Reference

1. R. Bartiromo, M. De Vincenzi, AJP **82**, 1067–1076 (2014)

Chapter 9
Pulsed Circuits

9.1 Introduction

In previous chapters, we exploited the *symbolic method* to study the behavior of electrical circuits with sinusoidal voltages and currents. The *symbolic method* assigns resistance and reactance to circuit components and allows solving circuits for sinusoidal signals at given frequency using the two Kirchhoff's laws. This approach to the solution of a circuit is known as the analysis in the *frequency domain*.

Since a sinusoidal signal retains its shape going through a linear circuit (only its amplitude and phase can change), there is no more information to be gained studying it as a function of time. Preserving the waveform of the input signal is a characteristic unique to the sinusoidal excitation; in general, even a linear circuit will deform waveforms other than sinusoidal.[1] In principle, using the Fourier expansion, it would be possible to study signals of any shape in the frequency domain. However, to analyze in this way pulsed signals, typically characterized by rapid rising and/or falling edges, we should use a considerable number of harmonics in the Fourier expansion, with a resulting increase in complexity of computations and with the risk of missing the salient features of the waveform. In these cases, it is convenient to analyze the circuit response directly as a function of time. Of course, the two approaches must obtain the same result and the choice between them is only a matter of convenience.

In general, the analysis of circuits with non-sinusoidal signals is addressed by solving the differential equations describing them directly in the time variable.[2] This approach to solving circuits is known as the *time domain* analysis and is the subject of the present chapter. In particular, in this chapter we study pulsed signals that highlight

---

[1] A generic signal shape is only preserved in the particular case of circuit consisting uniquely of resistances.

[2] We have already seen in Sect. 5.3.2 that circuits with capacitance and inductance are described by differential equations.

R. Bartiromo and M. De Vincenzi, *Electrical Measurements in the Laboratory Practice*, Undergraduate Lecture Notes in Physics,
DOI 10.1007/978-3-319-31102-9_9

interesting properties of electrical components and are of fundamental importance for the understanding of the operation of the digital circuits underpinning most of modern electronic devices.

We start in Sect. 9.2 introducing the voltage step, a signal playing in the time domain a role analogous to the sinusoidal signal in the frequency domain. In Sect. 9.3, we study the response of the high-pass RC circuit, already studied in Chap. 6 in the frequency domain, to different pulsed waveforms and in Sect. 9.4, we show that it is possible to use this circuit to obtain in output the time derivative of the input signal. In Sect. 9.5, we study the response of the low-pass RC circuit while in Sect. 9.6 we show how to obtain with this circuit the time integral of the input signal. After briefly addressing in Sect. 9.7 similar RL circuits, we discuss the response to the voltage step of the parallel and the series RLC circuits, respectively, in Sects. 9.8 and 9.9. Finally, we address in Sect. 9.10 the subject of signal attenuation with an instructive comparative discussion of the compensated voltage divider in both the frequency and the time domains.

9.2 The Step Signal

Among impulsive signals, the *step signal* plays a special role. This waveform has zero amplitude before a given time instant t_0 (often it is possible to set $t_0 = 0$ without loss of generality) and a constant value thereafter. The Heaviside θ function is usually adopted for its mathematical representation and, as an example, for a voltage signal starting at $t = 0$ we can write $v(t) = V\theta(t)$.

From a mathematical point of view the step signal has a *discontinuity* at $t = 0$; in fact, the limit of the function at the point $t = 0$ is 0 when reached from $t < 0$, while it is V when reached from $t > 0$. We indicate the time $t = 0$ reached from negative values with $t = 0^-$, and with $t = 0^+$ when reached from positive values. Therefore, for the voltage step we can write: $v_i(0^-) = 0$ and $v_i(0^+) = V$.

The special role of the step function stems from the possibility to express any signal as a superposition of step functions. In fact, given a generic function $v(t)$, for $t > 0$ the following relation holds:

$$v(t) = v(0) + \int_0^t \frac{dv}{dt'}dt' = v(0)\theta(t) + \int_0^\infty \frac{dv}{dt'}\theta(t - t')dt'$$

where the key point is that, with the use of the step function, we moved the independent variable t from the integration limit to the argument of the integrand function.

Once we know the response $G(t)$ of a linear circuit to the unit step $v(t) = \theta(t)$, using the superposition theorem we can compute the output signal v_o of the circuit to any input waveform $v_i(t)$. It is easy to verify that the following relation holds:

$$v_o(t) = v_i(0)G(t) + \int_0^\infty \frac{dv_i}{dt'}G(t - t')dt' \tag{9.1}$$

In the following, we will see several applications of this equation, which shows that the importance of the step function for the analysis in the time domain is somewhat similar to that of the sine function for the analysis in the frequency domain.

9.3 High-Pass RC Circuit

Figure 9.1 shows a high-pass RC circuit, already studied[3] using the symbolic method in the frequency domain. The analysis of this circuit in the time domain is performed solving the differential equation describing the time evolution of its output voltage.

Referring to Fig. 9.1, we denote with $v_i(t)$ the input voltage and $i(t)$ the circuit current. Applying Kirchhoff's law of voltages, we can write

$$v_i(t) = v_C(t) + v_R(t) = \frac{1}{C}\int i(t)dt + Ri(t) \quad \text{or} \quad \frac{dv_i(t)}{dt} = \frac{i(t)}{C} + R\frac{di(t)}{dt} \tag{9.2}$$

The output signal $v_o(t)$ is the voltage difference across the resistance R, $v_o(t) = Ri(t)$, and using this relation Eq. (9.2) becomes

$$\frac{dv_i(t)}{dt} = \frac{1}{RC}v_o(t) + \frac{dv_o(t)}{dt} \tag{9.3}$$

This is a first-order differential equation with constant coefficients whose solution gives the output voltage signal $v_o(t)$ as a function of time. It is worth to note that, as in the frequency domain analysis of this circuit, R and C only appear in Eq. (9.3) through the characteristic time $\tau = RC$. We shall see in the following that the solution of this equation is relevant for all circuits, both RC and RL, characterized by a single parameter. We can solve Eq. (9.3) exploiting the solution formula of a generic first-order differential equations.[4] However, when its left-hand side is constant, a straightforward integration shows that its solution reduces to

$$v_o(t) = A_1 + A_2 e^{-t/RC} \tag{9.4}$$

where A_1 and A_2 are constants linked to initial and final values of the output voltage.

---

[3] See Sect. 6.8.

[4] The first-order linear differential equation $y'(t) = \alpha(t)y(t) + \beta(t)$ has for general solution:

$$y(t) = \exp\left(\int_0^t \alpha(\tau)\,d\tau\right)\left[c + \int_0^t d\tau\,\beta(\tau)\exp\left(-\int_0^\tau \alpha(\tau')\,d\tau'\right)\right]$$

where c is the integration constant. If, as of interest in our case, $\alpha(t)$ and $\beta(t)$ are constants ($\alpha(t) = \alpha_o$ and $\beta(t) = \beta_o$), the previous relation simplifies to

$$y(t) = (c + \beta_o/\alpha_o)\exp(\alpha_o t) - \beta_o/\alpha_o$$

For Eq. (9.3) we have $\alpha_o = -1/RC$ and, requiring $y(\infty) = V_f$ and $y(0) = V_i$, we get $\beta_o/\alpha_o = V_f$ and $c = (V_i - V_f)$.

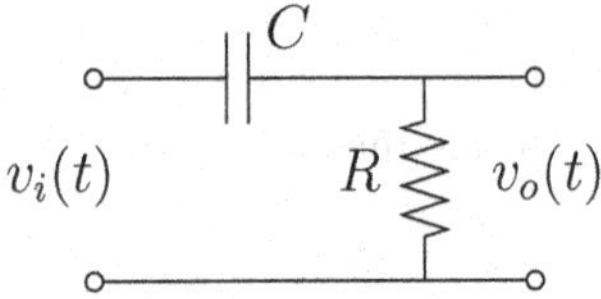

Fig. 9.1 High-pass RC circuit

The constant A_1 is the level reached at $t = \infty$, i.e., the *final* value of the output voltage. We denote it with the mnemonic symbol V_f. The constant A_2 is determined by the initial value of the output voltage that we denote by V_i. For $t = 0$ Eq. (9.4) gives $v_o(0) \equiv V_i = A_1 + A_2$ yielding $A_2 = V_i - V_f$. Substituting these values, we obtain

$$v_o(t) = V_f + (V_i - V_f)e^{-t/RC} \tag{9.5}$$

This equation will be used several times in the following as it represents the general solution for circuits with a single characteristic time constant having V_i and V_f, respectively, as initial and final values of the output signal.

9.3.1 Response to Voltage Step

When the input to the *high-pass RC* circuit of Fig. 9.1 is a voltage step, $v_i = V\theta(t)$, then the left-hand side in Eq. (9.3) is zero *almost everywhere*[5] and we can find the output signal using Eq. (9.5). To compute the constants V_i and V_f we note that, at $t = 0$, the input voltage v_i steps up from 0 to V. The output signal must behave in same way: indeed, the voltage drop across the capacitor is $(1/C)\int_0^t i(t')dt'$ and for $t \to 0$ the integral tends to zero when $i(0)$ is finite.

This leads us to state the following important rule for the analysis of capacitors in pulsed circuits:

the voltage drop across a capacitor cannot change instantaneously if the current flowing through it remains finite

In the present case, since the input voltage abruptly steps up from 0 to V and the resistance R limits the current in the capacitor, the output voltage signal shows the same discontinuity as the input, since the voltage across a capacitor cannot change abruptly. Therefore, $v_o(0^+) = V$, and in Eq. (9.5) we set $V_i = V$.

The final value of the output voltage, at $t = \infty$, is zero, because the current in R becomes null once the capacitor C is charged at the voltage V. Seen in another way, we can look at the capacitor between input and output of the circuit as a component that blocks the direct current.[6] Therefore, we have $V_f = 0$.

---

[5]This expression means that the property enunciated is not verified only for a set of points of zero measure.

[6]The capacitors used to block the direct current flow are called *blocking capacitors*.

Substituting the values of V_i and V_f in Eq. (9.5) we get the output voltage signal for the high-pass RC circuit as

$$v_o(t) = \begin{cases} V e^{-t/RC} & \text{for } t > 0 \\ 0 & \text{for } t < 0 \end{cases} \tag{9.6}$$

Using the Heaviside $\theta(t)$ function, we can write this solution in a more compact way as

$$v_o(t) = \theta(t) V e^{-t/RC} \tag{9.7}$$

9.3.2 Response to Voltage Pulse

An ideal pulse signal or, briefly, a pulse, see Fig. 9.2, of time duration T can be mathematically written as $V[\theta(t) - \theta(t - T)]$. Therefore, a pulse is the sum of two step signals, with opposite amplitudes, separated by a time interval T. Since the RC circuit is linear, we can apply the superposition theorem to find its response to an ideal pulse from relationship (9.7). It is a simple exercise to get the solution

$$v_o(t) = V \left[\theta(t) e^{-t/RC} - \theta(t - T) e^{-(t-T)/RC}\right] \tag{9.8}$$

It is instructive to justify the result in (9.8) analyzing step by step what happens in the circuit. In the time interval $0 < t < T$, the circuit response is the same as for the step signal $v_o(t) = V e^{-t/RC}$. At time $t = T$, the input signal steps down from V to 0 in a zero time interval; since the voltage across a capacitor cannot change instantaneously, the output voltage will have the same "jump" as the input, as shown in Fig. 9.2. In other words, the capacitor for "fast" voltage variations behaves like a short circuit, so that at time $t = T^+$ the output voltage will be $V(e^{-T/RC} - 1)$. Note that this value is *negative*. For $t > T$, the capacitor C discharges on the resistance R and the output voltage tends to zero with the known exponential law (see Fig. 9.2).

The qualitative features of the output signal depend on the ratio of the pulse duration T to the characteristic time RC of the circuit. For the signal shown in Fig. 9.2, the two times are of the same order of magnitude. If $T \ll RC$ the output is

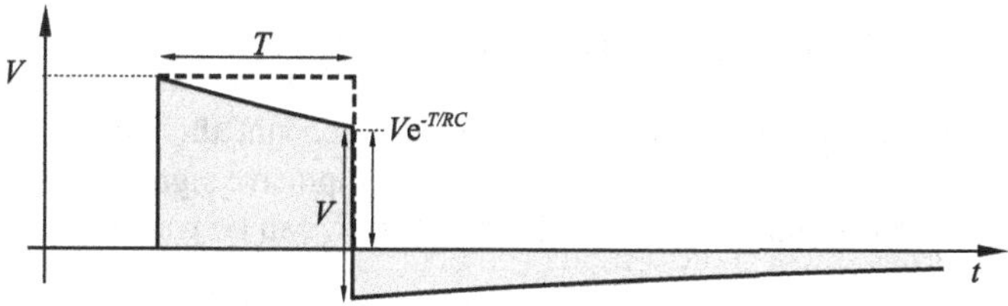

Fig. 9.2 Response of a high-pass RC circuit to a pulse of time duration T (*dashed line*). In this plot $RC = 3T$. Shaded areas for positive and negative voltages are equal

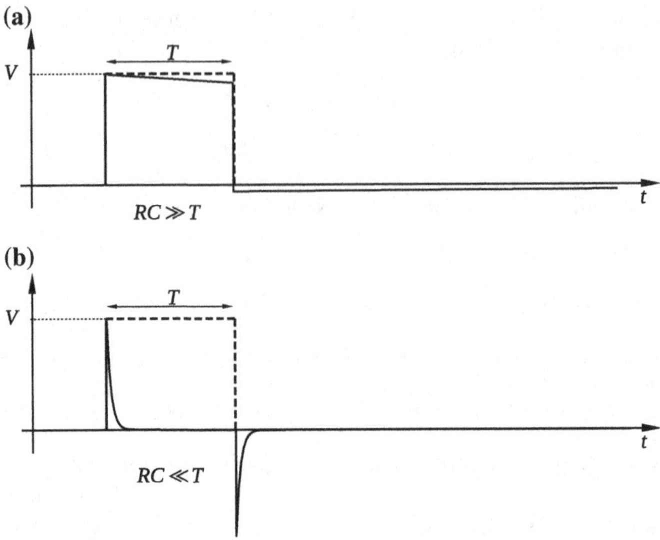

Fig. 9.3 Output signal shapes of a high-pass RC circuit: for $RC \gg T$ (*top plot*) the shape of output and input signals are almost equal, for $RC \ll T$ (*bottom plot*) the shape of output signal is significantly different from zero only in correspondence to the *steps* of the input signal (differentiating circuit). The *dashed line* represents the input signal

quite similar to the input, see Fig. 9.3a, while if $T \gg RC$ the output has spikes only in proximity of the rising and falling edges of the input pulse, see Fig. 9.3b. One can easily verify, using for example Eq. (9.8), that the total area under the output signal, for all values of the period T, is always *zero*.

9.3.3 Response to Rectangular Waveform

The rectangular waveform[7] is a periodic signal that in a fraction δ of its period T remains at a "high" constant value V_2 and in the rest of the period $T(1-\delta)$ remains at a "low" value V_1, see Fig. 9.4a. The steady state response of an RC high-pass circuit to a rectangular waveform is a superposition of signals described by expression (9.8) opportunely jointed. To find this output waveform, we consider a period of the input waveform $(0 - T)$, and we assume that at time $t = 0^-$ the output voltage is V_i, an unknown value to be determined. Taking into account the shape of the input signal, see Fig. 9.4a, the behavior of capacitors for impulsive signals and using the θ function, the output signal, in the time interval $0 - T$, can be written as

[7]This signal has already been described in Sect. 7.2.3.

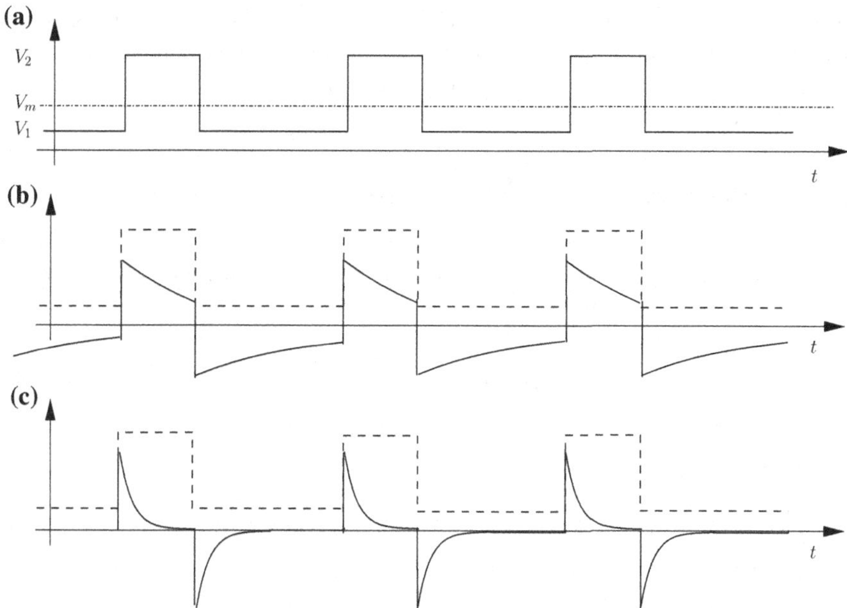

Fig. 9.4 Response of a RC high-pass circuit to a rectangular wave. Plot **a** input signal; plot **b** output signal (with $RC = T/2$); plot **c** output signal (with $RC = 0.2T$)

$$v_o(t) = (V_i + \Delta V)\theta(t)\exp\left(-\frac{t}{RC}\right) - \Delta V\theta(t - \delta T)\exp\left(-\frac{t - \delta T}{RC}\right) \quad (9.9)$$

where $\Delta V = V_2 - V_1$. By imposing the condition of periodicity of the output wave $v_o(T^-) = v_o(0^-)$, after some algebra one obtains $V_i = \Delta V[\exp(\delta T/RC) - 1]/[1 - \exp(T/RC)]$.

The response given by the expression (9.9) can also be obtained using the superposition principle. At any given time, the circuit response consists of the sum of three contributions: (i) the response to a continuous signal of amplitude V_1, (ii) the response to the infinite series of past transitions from V_1 to V_2, and (iii) the response to the infinite series of past transitions from V_2 to V_1.

For a high-pass RC circuit the first contribution is zero. Using relationship (9.7), the contribution of all the transitions in the past can be written as

$$(V_2 - V_1)\sum_{n=1}^{\infty} e^{-(t+nT)/RC}, \quad \text{and} \quad (V_1 - V_2)\sum_{n=1}^{\infty} e^{-(t+nT-\delta T)/RC}$$

After some simple algebra, we get

$$(V_2 - V_1)e^{-t/RC}\left(1 - e^{\delta T/RC}\right)\sum_{n=1}^{\infty} e^{-nT/RC} = (V_2 - V_1)e^{-t/RC}\frac{1 - e^{\delta T/RC}}{e^{T/RC} - 1}$$

where, in the last expression we used the formula for the sum of a geometric series of common ratio $e^{-T/RC}$ and first element $e^{-T/RC}$. To obtain the complete response we have to add to the previous expression the circuit response to the input signal in the $(0, T)$ time interval. In conclusion, the output signal is

$$v_o(t) = (V_2 - V_1)\left[\frac{e^{-t/RC}(1 - e^{\delta T/RC})}{e^{T/RC} - 1} + \theta(t)e^{-t/RC} - \theta(t - \delta T)e^{-(t-\delta T)/RC}\right]$$

that can be easily cast in the same form as Eq. (9.9). Let us show that this signal is periodic. At $t = 0^-$ we have

$$v_o(0^-) = (V_2 - V_1)\frac{1 - e^{\delta T/RC}}{e^{T/RC} - 1}$$

and at $t = T^-$

$$\begin{aligned} v_o(T^-) &= (V_2 - V_1)\left[\frac{e^{-T/RC}(1 - e^{-\delta T/RC})}{e^{T/RC} - 1} + e^{-T/RC} - e^{-(T-\delta T)/RC}\right] \\ &= (V_2 - V_1)\left[\frac{e^{-T/RC}(1 - e^{\delta T/RC}) + e^{-T/RC}(1 - e^{\delta T/RC})(e^{T/RC} - 1)}{e^{T/RC} - 1}\right] \\ &= (V_2 - V_1)\frac{1 - e^{\delta T/RC}}{e^{T/RC} - 1} = v_o(0^-) \end{aligned}$$

This expression shows that the output signal is periodic with the same period of input signal.

The response of a high-pass RC to a periodic signal of any shape is always an *alternating* signal, i.e., its mean value in one period is zero even when the input signal has a nonzero mean.[8] In fact, integrating in a period the two members of equation (9.3) one obtains

$$v_i(T) - v_i(0) = \frac{1}{RC}\int_0^T v_o(t)\,dt + v_o(T) - v_o(0) \tag{9.10}$$

Since we are looking for a *stationary state*, input and output signals are periodic: we have $v_i(T) = v_i(0)$ and $v_o(T) = v_o(0)$. Recalling the definition of mean value of a function, it follows from the previous equation that *the mean value of* $v_o(t)$ *in a period is zero*.

[8]The proof of this statement has already been given in a previous chapter in the *frequency domain* (see Sect. 7.3). The proof, by means of the Fourier series expansion, shows that the AC coupling, made through a high-pass RC filter, turns any periodic signal into an alternating signal, with zero average.

9.3.4 Response to Voltage Ramp

A voltage signal increasing linearly starting from a given time is called a *ramp signal* or a *voltage ramp*. Assuming $t = 0$ as starting time, the ramp signal has the expression: $v_i(t) = \alpha t$. The response of the *high-pass RC* circuit to the ramp is obtained solving Eq. (9.3), that in this case is written as:

$$\alpha = \frac{v_o}{RC} + \frac{dv_o(t)}{dt}. \tag{9.11}$$

The solution of (9.11) can be obtained either using the general solution formula of first-order differential equations (see footnote on p. 197) or, more directly, as the sum of the particular solution $v_o = \alpha RC$ and the solution $v_o = A\mathrm{e}^{-t/RC}$ (A being the integration constant) of the associated homogeneous equation. By imposing the initial condition $v_o(0) = 0$, we obtain

$$v_o(t) = \alpha RC(1 - \mathrm{e}^{-t/RC}) \tag{9.12}$$

It is a useful exercise to derive again the solution (9.12) using the convolution method expressed by Eq. (9.1). To apply it, we need to know the response $G(t)$ of the *high-pass RC* circuit to the unitary voltage step. This was already computed in Sect. 9.3.1 and is given by Eq. (9.7) with $V = 1$. Using this expression of $G(t)$, we obtain

$$v_o(t) = v_i(0)G(t) + \int_0^{\infty} \frac{dv_i(t')}{dt'} G(t - t')dt' = \int_0^{\infty} \frac{dv_i(t')}{dt'} \theta(t - t')\mathrm{e}^{-(t-t')/RC} dt'$$

where we used the relation $v_i(0) = 0$. Moreover $dv_i(t)/dt = \alpha$ and we can write

$$v_o(t) = \int_0^{\infty} \alpha\theta(t - t')\mathrm{e}^{-(t-t')/RC} dt' = \int_0^{t} \alpha\mathrm{e}^{-(t-t')/RC} dt' = \alpha\mathrm{e}^{-t/RC} \int_0^{t} \mathrm{e}^{t'/RC} dt'$$

Computing the integral, we obtain for the output signal expression (9.12) we found before solving the differential equation of the circuit.

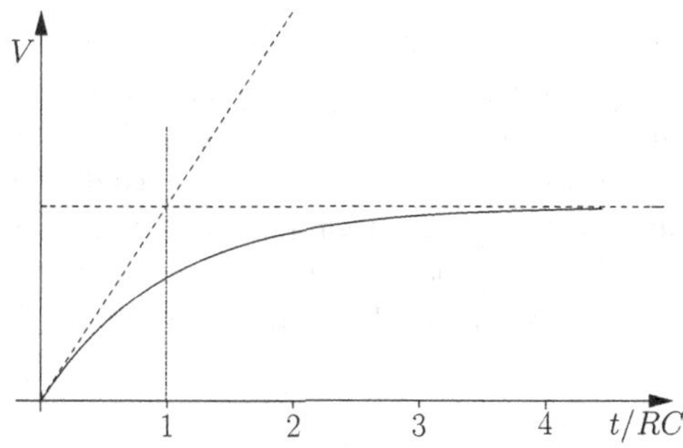

Fig. 9.5 Response of the *high-pass RC* circuit to a voltage ramp

Note also that for $t \gg RC$ the output tends to the constant value αRC, proportional to the time derivative of the input voltage, see Fig. 9.5. We will come back to this remark in the following Sect. 9.4 dealing with the differentiation properties of RC circuits.

9.3.5 Response to the Exponential Signal

A real step signal reaches its final value V in a finite time. A simple mathematical model, which takes into account the rise time of a real signal, is represented by the so-called exponential signal, whose expression, for $t > 0$, is

$$v_i(t) = V(1 - \mathrm{e}^{-t/\tau}) \tag{9.13}$$

where V is the final voltage of the signal and the parameter τ is related to the time needed to approach it. For an example of this waveform, see Fig. 9.6a.

The response of the *high-pass* RC circuit to the exponential signal can be obtained solving directly the differential Eq. (9.3) with $v_i(t)$ given by Eq. (9.13). However, also in this case it is more convenient to use the technique of the convolution with the unitary step response $G(t) = \theta(t)\mathrm{e}^{-t/RC}$ we derived in the previous section. Inserting it in Eq. (9.1) together with $v_i(0) = 0$ and $dv_i/dt = (V/\tau)\mathrm{e}^{-t/\tau}$, we can write

$$\begin{aligned} v_o(t) &= V\int_0^{\infty} \frac{\mathrm{e}^{-t'/\tau}}{\tau}\theta(t-t')\mathrm{e}^{-(t-t')/RC}dt' = V\frac{\mathrm{e}^{-t/RC}}{\tau}\int_0^t \mathrm{e}^{-t'(1/\tau-1/RC)}dt' \\ &= V\frac{\mathrm{e}^{-t/RC}}{\tau}\left[\frac{\mathrm{e}^{-(t-t')/RC}}{\frac{1}{RC}-\frac{1}{\tau}}\right]_0^t = V\frac{\mathrm{e}^{-t/RC}-\mathrm{e}^{-t/\tau}}{1-\frac{\tau}{RC}} \end{aligned} \tag{9.14}$$

Defining the variable $x = t/\tau$ and the parameter $\eta = RC/\tau$ (both dimensionless), Eq. (9.14) becomes

$$\frac{v_o(x)}{V} = \eta\frac{\mathrm{e}^{-x/\eta} - \mathrm{e}^{-x}}{\eta - 1} \tag{9.15}$$

For $\eta = 1$ this expression reduces to $x\mathrm{e}^{-x}$ as one obtains taking its limit for $\eta \to 1$. Figure 9.6 shows the output signal (9.15) for some values of the parameter η. Initially, the output signal follows the rise of the input signal, with characteristic time τ while its decay at longer times is linked to the time $\tau_C = RC$ needed to charge the capacitor. Note that when the rise time of the input signal is longer or equal to the characteristic time τ_C, the maximum value of the output is considerably lower than the corresponding value of the input signal.

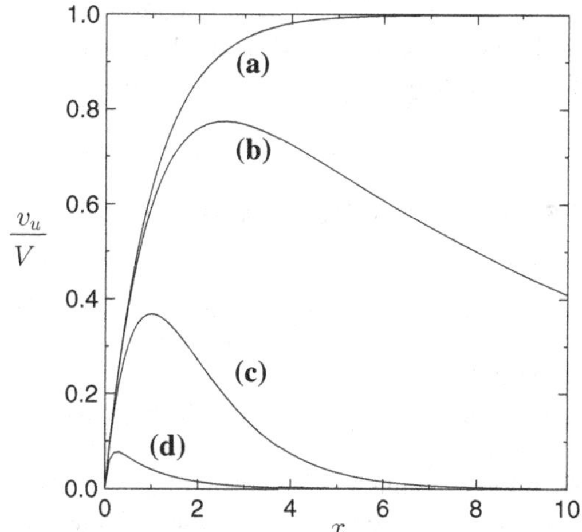

Fig. 9.6 Response of *high-pass RC* circuits with different time constant to an exponential signal. **a** input signal, **b** $\eta = 10$, **c** $\eta = 1.0$, **d** $\eta = 0.1$

9.4 High-Pass RC Circuit as Differentiator

In the *high-pass RC* circuit, the output signal is the voltage drop across the resistance R. When this voltage is much smaller than the voltage drop across the capacitance, the output signal is proportional to the time derivative of the input signal.[9] Indeed, if $v_R(t) \ll v_C(t)$, Eq. (9.2) can be approximated as $v_i(t) \simeq v_C(t) = \int i(t)dt/C$, so that $i(t) \simeq C dv_i(t)/dt$ and the output signal becomes

$$v_o(t) = Ri \simeq RC\frac{dv_i(t)}{dt} \tag{9.16}$$

Note that in the high-pass RC circuit, the condition $v_R \ll v_C$ is satisfied for $t \gg RC$, when the charge of the capacitor is nearly complete. In conclusion, if $v_R \ll v_C$ the output signal of a *high-pass RC* circuit is proportional to the time derivative of the input signal.

As an example, consider the response of the *high-pass RC* to a voltage ramp $v_i(t) = \alpha t$. Mathematically, the derivative of this signal is a constant function whose value is α. As we already noted commenting Eq. (9.12), the output signal for $t \gg RC$ becomes constant and proportional to α. Summarizing,

$$v_o(t) = \alpha RC(1 - \mathrm{e}^{-t/RC}) \underset{t \gg RC}{\sim} \alpha\, RC = RC\frac{dv_i}{dt}$$

Another example of the differentiation performed by the *high-pass RC* is provided by the plots in Fig. 9.3: when $RC \ll T$, see Fig. 9.3b, the output signal is significantly different from zero only in correspondence of the variations of the input signal while

[9]The same characteristic property was highlighted in the analysis of this circuit in the frequency domain. See Sect. 6.10.

it tends to zero when the input remains constant. On the contrary, when $RC \gg T$, see Fig. 9.3a, the voltage drop on the capacitance is small compared to the drop on the resistance and the output signal is only slightly different from the input.

9.5 Low-Pass RC Circuit

The circuit of Fig. 9.7 is a low-pass filter, as discussed in Sect. 6.7. This circuit has a particular importance because it represents, schematically, the input stage of most electronic devices.

In principle, the circuit of Fig. 9.7 is identical to that of Fig. 9.1; the only difference is that now the output signal is taken across the capacitor instead of the resistor. Therefore, the mathematical solution of the low-pass circuit can be obtained from equations obtained in Sect. 9.3 using the relation $v_o \equiv v_C = v_i - v_R$. However, given the importance of this circuit and the vastly different behavior of the two circuits, we will illustrate in detail the properties of the *low-pass RC* circuit too.

Differential equation for Low-Pass RC. Starting from Eq. (9.2) and setting $v_o(t) = (1/C)\int i(t)dt$, we derive $i = C dv_o(t)/dt$ and Kirchhoff's law of voltages yields the differential equation for the low-pass RC circuit:

$$v_i(t) = v_o(t) + RC\frac{dv_o(t)}{dt} \tag{9.17}$$

9.5.1 *Response to the Voltage Step*

The response of the *low-pass RC* circuit to the voltage step can be obtained directly from (9.5). At $t = 0$, when the input voltage jumps from zero to V, the output voltage remains at zero because the voltage drop across the capacitor cannot change instantaneously. Hence $V_i = 0$. Thereafter, the capacitance is charged through the resistor R and when its charge is completed the output voltage is $V_f = V$. Inserting these values in expression (9.5), we get

$$v_o(t) = V(1 - \mathrm{e}^{-t/RC}) \tag{9.18}$$

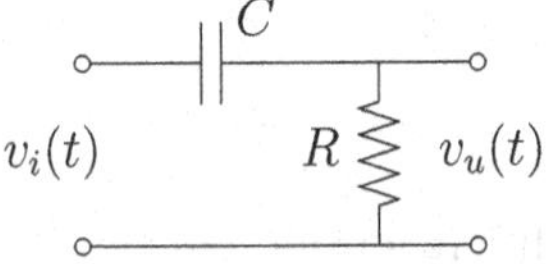

Fig. 9.7 The *low-pass RC* circuit

The shape of the output signal (9.18) is the same as the exponential signal defined by relation (9.13). It can be characterized with an experimental parameter called the pulse "rise time" and defined as the time taken by the signal to change from 10 to 90 % of its maximum value; this parameter gives an indication of how fast the circuit can react to input signals. We leave as a simple exercise to calculate that for the output signal of the low-pass RC (9.18) the rise time is

$$t_r = 2.2RC = \frac{0.35}{\nu_0}$$

where the parameter $\nu_0 = 1/(2\pi RC)$ is the circuit cutoff frequency.

9.5.2 Response to Voltage Pulse

The response of the *low-pass RC* circuit to a voltage pulse can be obtained using the superposition principle. We just need to add the circuit responses to two step signals: the first starting at $t = 0$ with an amplitude V, the second starting at $t = T$ (T pulse duration) with an amplitude $-V$. From Eq. (9.18) and using the θ function, we obtain

$$\begin{aligned} v_o(t) &= V\left[\theta(t)(1-\mathrm{e}^{-t/RC}) - \theta(t-T)(1-\mathrm{e}^{-(t-T)/RC})\right] \\ &= \begin{cases} V(1-\mathrm{e}^{-t/RC}) & 0 < t < T \\ V(1-\mathrm{e}^{-T/RC})\mathrm{e}^{-(t-T)/RC} & t > T \end{cases} \end{aligned}$$

Figure 9.8 shows a typical response of a *low-pass RC* circuit to a pulse signal.

9.5.3 Response to Rectangular Waveform

In this section, we assume as input signal a rectangular wave, already defined in Sect. 9.3.3 and shown in Fig. 9.9a. We start by showing that, for a periodic input voltage, in a low-pass RC circuit the average values of the input and the output signals are equal. We first notice that, when the input is periodic, in stationary conditions the output must be periodic too. Integrating Eq. (9.17) over a time period T, and taking

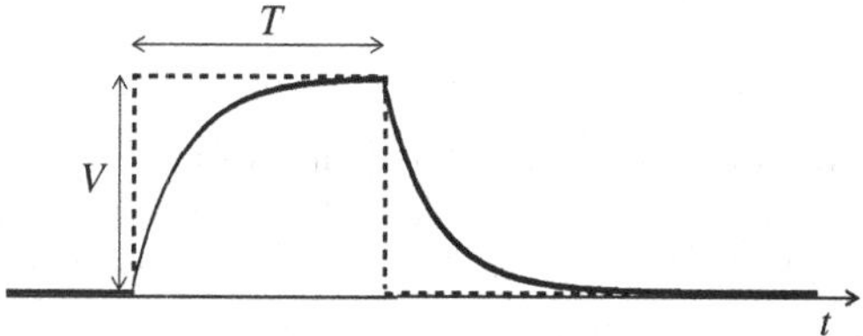

Fig. 9.8 Output signal (*continuous line*) of the *low-pass RC* to a pulse with duration T (*dashed line*). Here $RC = T/4$

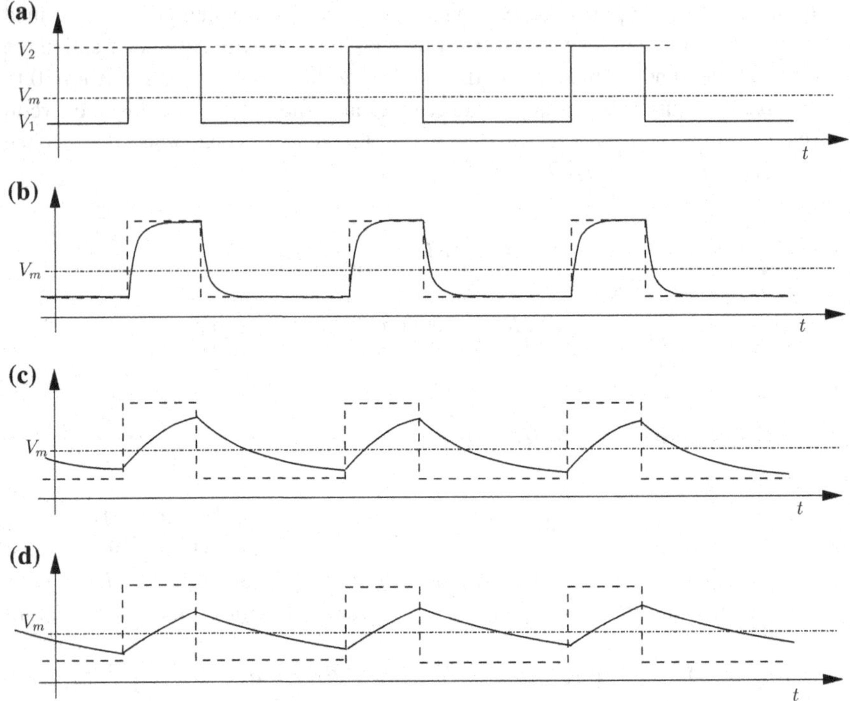

Fig. 9.9 Plot **a**: Rectangular wave of duty cycle $\delta = 0.3$. Plots **b**, **c** and **d**: responses of a *low-pass* *RC* circuit with characteristic time $RC = 0.03\,T$, $RC = 0.27\,T$ and $RC = 0.70\,T$ respectively, to the waveform shown in (**a**)

into account that $v_o(T) = v_o(0)$, we get

$$\frac{1}{T}\int_0^T v_i(t')dt' = \frac{1}{T}\int_0^T v_o(t')dt' + \frac{RC}{T}(v_o(T) - v_o(0)) = \frac{1}{T}\int_0^T v_o(t')dt'$$

Since the rectangular wave oscillates between two constant levels V_1 and V_2, we can apply Eq. (9.5) to find the output signal. The output signal is periodic and we limit the calculation to one period. Suppose, without loss of generality, that at $t = 0$, when the input voltage jumps from V_1 to V_2, the output voltage is V_l, an *unknown* value to be determined. In this case, we have $V_i = V_l$ and $V_f = V_2$, and Eq. (9.5) yields

$$v_{oa}(t) = V_2 + (V_l - V_2)\,\mathrm{e}^{-t/RC} \qquad 0 < t < \delta T$$

At $t = \delta T$ the input voltage jumps back from V_2 to V_1 while the output has reached a value V_h to be determined. In the time interval $\delta T < t < T$, we apply again the (9.5) with $V_f = V_1$ and with $V_i = V_h$ and we obtain

$$v_{ob}(t) = V_1 + (V_h - V_1)\,\mathrm{e}^{-(t-\delta T)/RC} \qquad \delta T < t < T$$

The values of V_l e V_h can be easily calculated imposing the continuity condition $v_{oa}(\delta T) = v_{ob}(\delta T)$ (the voltage across the output capacitor cannot change abruptly) and the periodicity condition of the stationary solution $v_{oa}(0) = v_{ob}(T)$. The details of this calculation are left to the reader as an exercise. The results are

$$V_l = V_1 + \frac{\Delta V\,(\mathrm{e}^{\delta T/RC} - 1)}{\mathrm{e}^{T/RC} - 1}; \quad V_h = V_2 - \frac{\Delta V\,(\mathrm{e}^{T(1-\delta)/RC} - 1)}{\mathrm{e}^{T/RC} - 1} \tag{9.19}$$

where we have used the position $\Delta V = V_2 - V_1$.

Another way to get the response to a rectangular wave of the low-pass circuit is to exploit the superposition principle following the same procedure previously used for the high-pass circuit.[10] The only difference consists in the response to the unitary voltage step given in this case by (9.18). The output signal in the time interval $0 < t < T$ is easily found as

$$v_o(t) = V_1 + \Delta V\left[\mathrm{e}^{-t/RC}\frac{\mathrm{e}^{\delta T/RC} - 1}{\mathrm{e}^{T/RC} - 1} + \theta(t)(1 - \mathrm{e}^{-t/RC}) - \theta(t - \delta T)((1 - \mathrm{e}^{-(t-\delta T)/RC})\right] \tag{9.20}$$

With a little algebra we obtain

$$v_o(0) = V_1 + \Delta V\frac{\mathrm{e}^{\delta T/RC} - 1}{\mathrm{e}^{T/RC} - 1}; \quad v_o(\delta T) = V_2 - \frac{\Delta V\,(\mathrm{e}^{T(1-\delta)/RC} - 1)}{\mathrm{e}^{T/RC} - 1}$$

respectively, equal to V_l and V_h we found in (9.19). Equation (9.20) makes it possible to show directly that the output signal is periodic with period T. Indeed $v_o(T)$ is given by

$$\begin{aligned} v_o(T) &= V_1 + \Delta V\left[\mathrm{e}^{T/RC}\frac{\mathrm{e}^{\delta T/RC} - 1}{\mathrm{e}^{T/RC} - 1} + (1 - \mathrm{e}^{t/RC}) - ((1 - \mathrm{e}^{-(T-\delta T)/RC})\right] \\ &= V_1 + \Delta V\frac{\mathrm{e}^{\delta T/RC} - 1}{\mathrm{e}^{T/RC} - 1} = v_o(0) \end{aligned}$$

This relation shows that the output voltage has the same periodicity of the input signal.

9.5.4 Response to Voltage Ramp

The output signal of *low-pass RC* circuit to a voltage ramp signal ($v_i(t) = \alpha t$) can be obtained exploiting the Kirchhoff's law of voltages applied to the circuit: $v_i(t) = v_R(t) + v_C(t)$. Using for $v_R(t)$ the expression (9.12) already found for the

---

[10] See p. 201 for details.

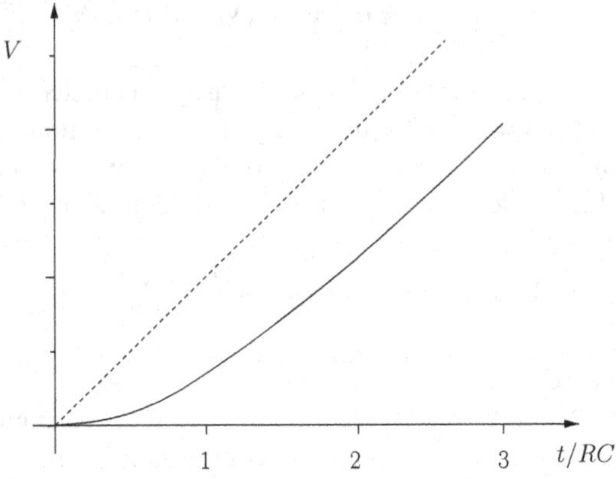

Fig. 9.10 Response of a low-pass *RC* circuit to a ramp signal

high-pass circuit, we obtain

$$v_o(t) = v_C(t) = \alpha(t - RC) + \alpha RC e^{-t/RC} \tag{9.21}$$

Output and input voltages for the ramp signal are shown in Fig. 9.10. Notice that, when time becomes large ($t \gg RC$), the output tends to follow the input signal with a delay equal to RC, showing a behavior complementary with respect to the high-pass case.

9.6 Low-Pass RC Circuit as Integrator

The output signal in a *low-pass RC* circuit is the voltage drop across the capacitance C. We show that, when this voltage is much smaller than the voltage drop across the resistance, the output is proportional to the time integral of the input signal.[11] In fact, if $v_R(t) \gg v_C(t)$ Eq. (9.2) reduces to $v_i(t) \simeq v_R(t)$ yielding $i(t) \simeq v_i(t)/R$. Therefore, the output signal can be expressed as

$$v_o(t) = \frac{1}{C}\int_0^t i(t')\,dt' \simeq \frac{1}{RC}\int_0^t v_i(t')\,dt' \tag{9.22}$$

Equation (9.22) shows that, if $v_R(t) \gg v_C(t)$ corresponding to $t \ll RC$, the output signal of a low-pass RC circuit is the integral of its input signal. This result justifies the name "integrator circuit" often used to refer to this circuit.

As an example, let us consider the response to a voltage step starting at time $t = 0$: $v_o(t) = V[1 - \exp(-t/RC)]$. For time values much lower than the characteristic time RC, the conditions for the integration ($v_R \gg v_c \equiv v_o$) are fulfilled and, expanding

[11] The same characteristic was highlighted in the analysis of this circuit in the frequency domain. See Sect. 6.10.

the exponential function in series of powers, we obtain

$$v_o(t) = V[1 - 1 + \frac{t}{RC} - \frac{1}{2}\left(\frac{t}{RC}\right)^2 + \cdots] = V\left\{\frac{t}{RC} + \mathcal{O}\left[\left(\frac{t}{RC}\right)^2\right]\right\}$$

Mathematically speaking, the integral of the voltage step is the product of the amplitude V and the time t. This shows that as long as $(t \ll RC)$ the first-order approximation is acceptable and the output voltage is proportional to the integral of the input signal.

9.7 RL Circuits

The *high-pass* and *low-pass RL* circuits, whose diagrams are shown in Fig. 9.11, can be solved in the time domain with the same approach adopted above for *RC* circuits. As already observed with the analysis in the frequency domain, see Sect. 6.9, we shall see that the solutions of *RL* circuits can be derived from those of the corresponding *RC* circuits, the only difference being the expression for the circuit characteristic time, $\tau = R/L$ in *RL* circuits (as we will show again below), instead of $\tau = RC$ in *RC* circuits.

As an example of solution of a *RL* circuit, consider the circuit in Fig. 9.11a. It is easily seen that it is the high-pass circuit since the output signal is taken across the inductance, whose impedance increases with the signal frequency. The Kirchhoff's law of voltages in this case yields $v_i(t) = v_R(t) + v_L(t) = Ri(t) + Ldi(t)/dt$. Differentiating it and taking into account that the output voltage expression is $v_o(t) = Ldi(t)/dt$, we get

$$\frac{dv_i(t)}{dt} = \frac{R}{L}v_o(t) + \frac{dv_o(t)}{dt}$$

As expected, this differential equation is similar to the one obtained for the *RC* circuits in Sect. 9.3, and, under the same assumptions adopted there, the solution is still given by Eq. (9.5) provided we use the appropriate expression for the characteristic time. However, before we can compute the constants V_i and V_f for RL circuits, we have to analyze the response of an inductance to pulsed signals, similar to what we did for a capacitance in Sect. 9.3.

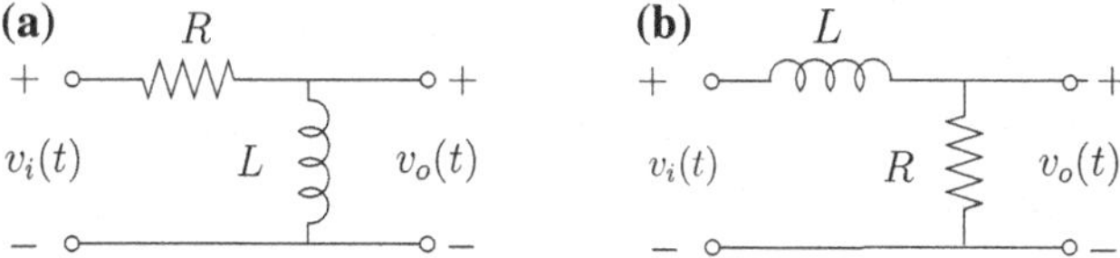

Fig. 9.11 The *RL* circuits: high-pass (**a**) and low-pass (**b**) configuration

The voltage drop across an inductance is given by $v(t) = L\,di/dt$, so the current in the inductor is $i = (1/L)\int v(t)dt + I_o$ (I_o, integration constant). For $t \to 0$ the integral, which gives the variation of the current, tends to zero, unless the value of the voltage $v(t)$ is infinite (not physically acceptable). One can therefore state the following important rule for the behavior of an inductance:
the current through the inductance cannot change instantaneously if the voltage drop across it remains finite

Using the results obtained above, we can show that the responses of ideal *RL* circuits can be deduced from those of the corresponding ideal *RC* circuits. However, although in principle, for example, a low-pass filter can be assembled with a resistor and a capacitor or with a resistor and an inductor, in practice the real behavior will be different because of differences in parasitic effects. In general, the influence of the parasitic parameters, as we have seen in the first chapter, is greater in inductors than in capacitors. In particular, at low frequency the ohmic resistance of the winding of the inductor puts a lower limit, greater than zero, to the inductor impedance while, at high frequency, the parasitic capacitance between the turns of the inductor will decrease its impedance. This can affect appreciably the circuit response both at short and long timescales.

9.8 RLC Parallel Circuit: Response to Voltage Step

Consider the circuit *RLC* parallel shown in Fig. 9.12 with a voltage step $v_i(t) = V\theta(t)$ as input signal. Applying Kirchhoff's law of currents to one of its nodes and taking the output $v_o(t)$ across the parallel *LC*, we can write

$$\frac{v_i - v_o}{R} - \frac{1}{L}\int v_o\,dt - C\frac{dv_o}{dt} = 0$$

Differentiating this expression with respect to time, we get a second-order differential equation with constant coefficients[12]:

$$C\frac{d^2v_o}{dt^2} + \frac{1}{R}\frac{dv_o}{dt} + \frac{1}{L}v_o = \frac{1}{R}\frac{dv_i}{dt} \tag{9.23}$$

Since the time derivative of the function θ is zero *almost everywhere*, the solution of (9.23) depends only on the roots of the characteristic equation

$$Cs^2 + \frac{1}{R}s + \frac{1}{L} = 0 \tag{9.24}$$

[12]Note that we already encountered this equation in studying the motion of the movable coil in the D'Arsonval ammeter; obviously, owing to the similarity of the two differential equations, the solution we obtain in this section are qualitatively similar to those obtained in Sect. 4.3.2 for the position of the moving coil.

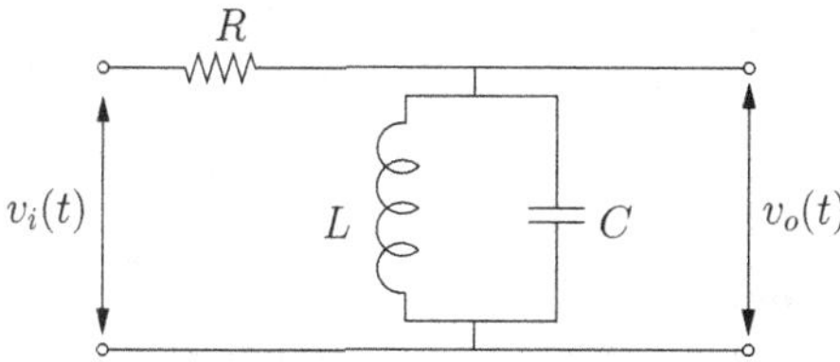

Fig. 9.12 RLC parallel circuit

Its solutions are:

$$s_{1,2} = -\frac{1}{2RC} \pm \sqrt{\left(\frac{1}{2RC}\right)^2 - \frac{1}{LC}} \tag{9.25}$$

Introducing the two parameters k, *attenuation constant*, and T_o, *period without attenuation*,[13] defined by

$$k = \frac{1}{2R}\sqrt{\frac{L}{C}} \qquad T_o = 2\pi\sqrt{LC} \tag{9.26}$$

the solutions of the characteristic Eq. (9.25) can be rewritten in a more compact way as

$$s_{1,2} = \frac{2\pi}{T_o}\left(-k \pm j\sqrt{1-k^2}\right) \tag{9.27}$$

The value of the parameter k determines the qualitative behavior of the output signal.

Critically damped response. If $k = 1$, Eq. (9.27) has two coincident real negative roots $s_{1,2} \equiv s = -2\pi/T_o$. The solution of differential equation is

$$v_o(t) = V_o\alpha\, te^{st} \tag{9.28}$$

where V_o and α are integration constants to be determined considering the capacitance and the inductance response to the impulsive signal. At the time $t = 0^-$ both the voltage across the capacitor and the current flowing through the inductor are zero. As we have already seen in the preceding sections, at time $t = 0^+$ both these quantities will remain zero. This implies that $v_o(0^+) = 0$ and that the current flowing in R is $i(0^+) = (v_i(0^+) - v_o(0^+))/R = V/R$. This current can only flow through the capacitance since the inductance blocks fast current changes. Using Eq. (9.28), the current in the capacitance C is $i(0^+) = Cdv_o/dt = V_oC\alpha$. Equating the currents through R and C, we get

$$\frac{V}{R} = V_oC\alpha \qquad \text{from which we get:} \qquad V_o\alpha = \frac{V}{RC}$$

[13]Note that k and T_o have already been introduced in the analysis of the resonant circuit RLC parallel (Chap. 6). In fact $k = 1/2Q_p$, where Q_p is the *quality factor* of the circuit and T_o is exactly the period corresponding to the resonant frequency of the circuit.

Using this value, we finally obtain

$$v_o(t) = \frac{V}{RC}\, t\, e^{-2\pi t/T_o} \tag{9.29}$$

Underdamped response. If $k < 1$, the solutions s_1 e s_2, given by expression (9.27) are complex with the real part negative. The output signal will be a dumped oscillation whose expression is given by

$$v_o(t) = \exp(-2\pi kt/T_o)[A\exp(j2\pi\sqrt{1-k^2}t/T_o) + B\exp(-j2\pi\sqrt{1-k^2}t/T_o)].$$

with A and B integration constants. Imposing $v_o(0) = 0$, we get $A = -B$. Moreover, since the current at $t = 0^+$ flows only through the capacitance, we have in addition $dv_o(0)/dt = V/CR$ yielding $A = -jT_o/4\pi C\sqrt{1-k^2}$. With a little algebra, we obtain

$$\frac{v_o}{V} = \frac{2k}{\sqrt{1-k^2}} e^{-2\pi kt/T_o} \sin\left(2\pi\sqrt{1-k^2}\frac{t}{T_o}\right) \tag{9.30}$$

As shown in Fig. 9.13, the maximum signal amplitude decreases for lower k values while the oscillation persistence increases.

Overdamped response. If $k > 1$, both solutions (9.27) are real and negative and the output signal is the sum of two decreasing exponentials. With the same procedure we used above we arrive at the expression

$$\frac{v_o(t)}{V} = \frac{2k}{\sqrt{k^2-1}} e^{-2\pi kt/T_o} \sinh\left(-\frac{2\pi\sqrt{k^2-1}}{T_o} t\right) \tag{9.31}$$

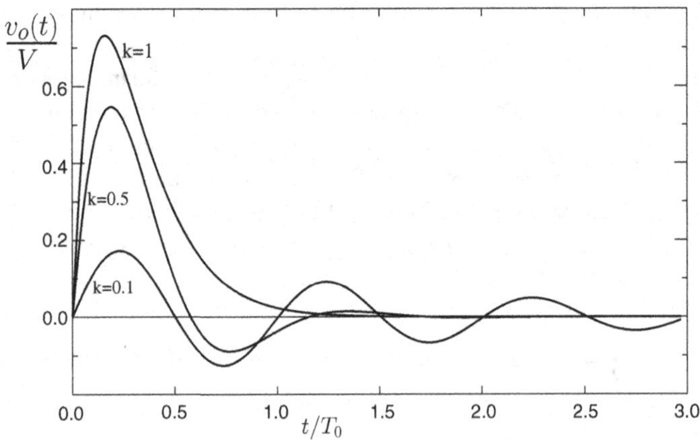

Fig. 9.13 *RLC* parallel output response signal to a voltage step as input for three different values of the k parameter ($k = 1$: critical damping, $k < 1$ underdamped signals)

When $k \gg 1$ it is possible to simplify formula (9.27) using

$$s_{1,2} = -\frac{2\pi k}{T_o} \pm \frac{2\pi k}{T_o}\sqrt{1 - \frac{1}{k^2}} \tag{9.32}$$

Expanding the square root in a power series, we get

$$\sqrt{1 - \frac{1}{k^2}} \simeq 1 - \frac{1}{2k^2} + \mathcal{O}\left[\frac{1}{k}\right]^3$$

so that, when $2k^2 \gg 1$, we have

$$s_1 = -\frac{\pi}{T_o k} \quad \text{e} \quad s_2 = -\frac{4\pi k}{T_o}$$

It is easy to show that $s_1 \ll s_2$ and, except for time values close to $t = 0$, the output voltage is well approximated by the following relation:

$$\frac{v_o}{V} \simeq \mathrm{e}^{-\pi t/kT_o} = \mathrm{e}^{-Rt/L}$$

Note that this expression is the output signal of a *RL* circuit: indeed, for given R and L, large value of k corresponds to small values of C meaning that the presence of the capacitor in the circuit can be neglected.

9.9 RLC Series Circuit: Response to Voltage Step

The *RCL series* response to voltage step is analogous to that of the *RCL parallel*. In fact the differential equation for the current $i(t)$ flowing through the unique loop of the circuit (see Fig. 9.14), as obtained by applying Kirchhoff's law of voltages, is given by

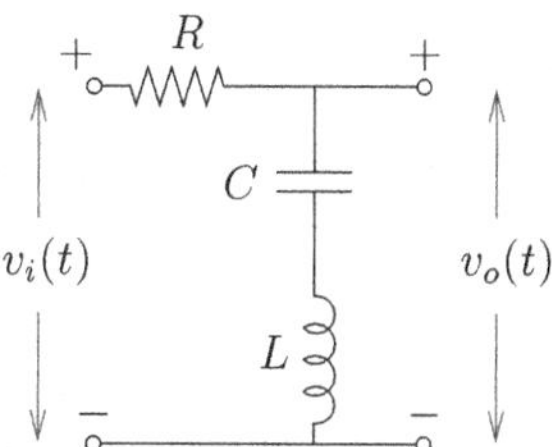

Fig. 9.14 The *RLC* series circuit

$$L\frac{d^2i(t)}{dt^2} + R\frac{di(t)}{dt} + \frac{1}{C}i(t) = \frac{dv_i(t)}{dt} \tag{9.33}$$

This equation is similar to Eq. (9.23) for the output voltage of the *RLC* parallel circuit. Using the same procedure as in the previous section, we can write for the current the expression: $i(t) = Ae^{s_1 t} + Be^{s_2 t}$, where s_1 e s_2, are the solutions of the characteristic equation

$$s_{1,2} = -R/2L \pm \sqrt{(R/2L)^2 - 1/LC}$$

For the initial conditions, we first note that, due to the inductance L, $i(0^+) = 0$, yielding $B = -A$. Furthermore, at $t = 0^+$ the generator voltage drops across the inductance because $v_R(0^+) = Ri(0^+) = 0$ and $v_C(0^+) = 0$ because the capacitance remains not charged. This gives $di/dt|_{t=0^+} = V/L$ from which we obtain $A = (V/L)/(s_1 - s_2)$, and the solution $i(t)$ of Eq. (9.33) is

$$i(t) = \frac{V/L}{s_1 - s_2}\left(e^{s_1 t} - e^{s_2 t}\right) \tag{9.34}$$

Note that defining the parameter $k_s = R/2\sqrt{C/L}$ and $T_0 = 2\pi\sqrt{LC}$, we can write the solutions $s_{1,2}$ as in formula (9.27) substituting k with k_s. Therefore, all the observations and classifications we made in Sect. 9.8 on the waveform of the output voltage signal of the *RLC* parallel circuit can be used for the current signal of the *RLC* series circuit.

9.9.1 Exponential Pulse

The study of the response of a *RLC* series circuit to an exponential pulse $v_i(t) = V(1 - e^{-t/\tau})$, besides yielding a result that can be compared with experimental data, is useful to describe an application presented in the next section.

This response is conveniently obtained exploiting the convolution method expressed by Eq. (9.1). Using the response to unitary voltage step given by Eq. (9.34), we can easily obtain

$$i(t) = \frac{V/L}{s_1 - s_2}\left[\frac{e^{s_1 t} - e^{-t/\tau}}{1 + \tau s_1} - \frac{e^{s_2 t} - e^{-t/\tau}}{1 + \tau s_2}\right]$$

With a little algebra, we can reformulate this expression as

$$i(t) = \frac{V\tau/L}{(1 + \tau s_1)(1 + \tau s_2)}\left[\frac{1 + \tau s_2}{\tau(s_1 - s_2)}e^{s_1 t} - \frac{1 + \tau s_1}{\tau(s_1 - s_2)}e^{s_2 t} + e^{-t/\tau}\right]$$

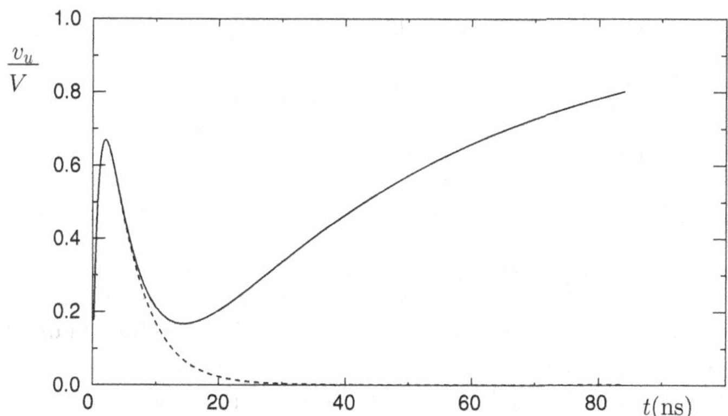

Fig. 9.15 The *solid line* represents the response of a *RLC* series to an exponential pulse. The values of the circuit components are $R = 50\,\Omega,\ L = 0.25\,\mu\text{H},\ C = 1\,\text{nF}$ and the exponential pulse characteristic time is $\tau = 1\,\text{ns}$. *Dashed line* represents the response of the circuit when $C \to \infty$. In this latter case, the circuit behaves like a *RL* high pass filter

When the voltage output is taken across the series of C and L (see Fig. 9.14), we have

$$\begin{aligned} v_o(t) &= v_i(t) - Ri(t) \\ &= V(1 - \mathrm{e}^{-t/\tau}) - \frac{V\tau R/L}{(1+\tau s_1)(1+\tau s_2)}\left[\frac{1+\tau s_2}{\tau(s_1 - s_2)}\mathrm{e}^{s_1 t} - \frac{1+\tau s_1}{\tau(s_1 - s_2)}\mathrm{e}^{s_2 t} + \mathrm{e}^{-t/\tau}\right] \end{aligned} \tag{9.35}$$

This signal $v_o(t)$ has a rather complex expression formed by the sum of three exponential functions with three different characteristic times, namely the rise time τ of the input signal, $1/s_1$ and $1/s_2$ both determined by the solution of the characteristic equation associated to the differential Eq. (9.33). Introducing the characteristic times of the circuit: $\tau_L = R/L$ and $\tau_C = RC$[14] we can write s_1 and s_2 as

$$s_{1,2} = -\frac{1}{2\tau_L}\left(1 \mp \sqrt{1 - 4\frac{\tau_L}{\tau_C}}\right) \quad \text{yielding:} \quad s_1 - s_2 = \frac{1}{\tau_L}\sqrt{1 - 4\frac{\tau_L}{\tau_C}}$$

[14]The time $\tau_L = R/L$ is the characteristic times of the circuit shown in Fig. 9.14 for $C \to \infty$ (this means that this circuit behaves like a *high-pass RL*), and similarly the time $\tau_C = RC$ is the characteristic times of the circuit for $L \to 0$ (this means that this circuit behaves like a *low-pass RC*).

and the expression (9.35), after some algebra, becomes

$$v_o(t) = V(1 - \mathrm{e}^{-t/\tau}) - V\frac{\tau\tau_C}{\tau_C\tau_L - \tau\tau_C + \tau^2}$$
$$\times\left\{\mathrm{e}^{-t/\tau} - \frac{1}{2}\left[1 - \frac{2\tau_L/\tau - 1}{\sqrt{1 - 4\tau_L/\tau_C}}\right]\mathrm{e}^{s_1 t} - \frac{1}{2}\left[1 + \frac{2\tau_L/\tau - 1}{\sqrt{1 - 4\tau_L/\tau_C}}\right]\mathrm{e}^{s_2 t}\right\} \quad (9.36)$$

In Fig. 9.15, we plot the exponential pulse response for the series RLC circuit given by Eq. (9.36) and the response of the RL filter obtained by short-circuiting the capacitor. As can be seen from the comparison of the two waveforms shown in the figure, the short timescale response of the circuit (the first peak) is mainly determined by the impedance of the inductance, the capacitor behaving as a short circuit on timescales $t \ll \tau_C$. On the contrary, the second part of the response can be explained by the impedance of the capacitance, the inductance behaving as a short circuit on timescales $t \gg \tau_L$.

Measurement of the parasitic inductance of a capacitor. As shown in Fig. 9.15 the inductance determines the short time response of the circuit. On timescales $t \ll \tau_C$, the circuit can be considered a *high-pass RL* circuit, whose response function is obtained either as limit of the expression (9.36) for $\tau_C \to \infty$ or directly from (9.15) taking into account that for a *high-pass RL* the η parameter becomes $\eta = \tau_L/\tau = L/R\tau$. We can write the response function as

$$v_o(t) = V\frac{\eta}{\eta - 1}\left(\mathrm{e}^{-x/\eta} - \mathrm{e}^{-x}\right) \quad \text{with} \quad x = \frac{t}{\eta} \quad \text{and} \quad \eta = \frac{\tau_L}{\tau}$$

It easy to show that the maximum of $v_o(t)$ is $v_{\max} = V\eta^{1/(1-\eta)}$ and if $\eta \ll 1$, then

$$v_{\max} = V\eta^{1/(1-\eta)} \simeq V\eta = \frac{VL}{R\tau}$$

This expression shows that in this limit the first peak amplitude of the output voltage is proportional to the value L of the inductance and, when V, R, and τ are known, its value can be used to estimate L.

This relationship suggests a method for the measurement of the stray inductance of a capacitor, applying an exponential pulse to the capacitor under test. Figure 9.16 shows the voltage measured across a 10 nF real capacitor for exponential pulse with rise time of 10 ns and amplitude of 5 V supplied by a voltage generator with 50 Ω output impedance. From the sharp peak in picture we obtain $L \simeq 0.2\,\mu$H.

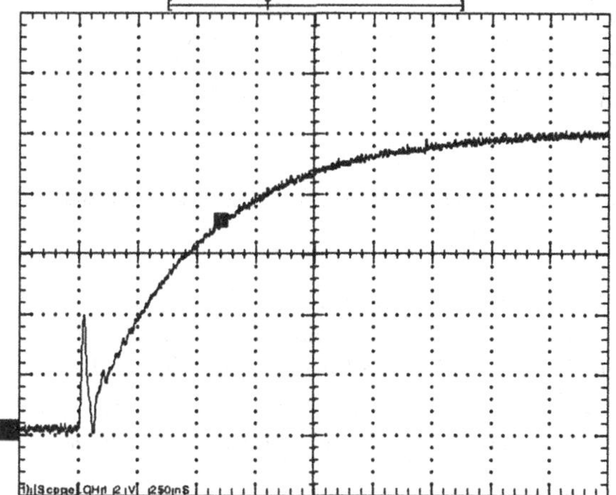

Fig. 9.16 The oscillogram shows the voltage across a 10 nF capacitor as a function of time for an input given by an exponential pulse with rise time $\tau_r = 10$ ns and amplitude 5 V obtained with a 50 Ω voltage generator. The initial peak is due to the parasitic inductance that can be estimated to be 0.2 μH. The oscilloscope settings: Vertical 1 V/div., Horizontal 250 ns/div. The bandwidth is 100 MHz

9.10 Compensated Voltage Divider

The last section of this chapter is devoted to the detailed illustration of a simple but important circuit, known as the *compensated voltage divider*. It is used to attenuate (i.e., reduce the amplitude without deformation) electrical signals. In principle the attenuation of a voltage signal can be obtained using the simple resistive divider shown in Fig. 9.17a whose transfer function (attenuation) is $A = R_2/(R_1 + R_2)$, ideally independent of frequency.

However, in practice, stray capacitances can significantly affect the behavior of the circuit. Consider, for example, the case of a stray capacitance in parallel to R_2 (see Fig. 9.17b).[15] The transfer function of the circuit in Fig. (9.17b) can be found exploiting the Thévenin equivalent circuit as "seen" from the resistance R_2, where we pick up the attenuated signal. This circuit, shown in Fig. 9.17c, is a *low-pass RC* filter that attenuates high frequency more than low frequency components and integrates the input signal. Its transfer function is given by expression (6.22) with $\omega_o = 1/R_{eq}C$ and $R_{eq} = R_1 \parallel R_2$.

As a numerical example, we consider $R_1 = 9\,\text{M}\Omega$, $R_2 = 1\,\text{M}\Omega$, $C = 20\,\text{pF}$ (these are the typical values of a "×10" oscilloscope probe with an attenuation ratio of a factor 10); we get $\omega_o = 1/R_{eq}C = 55 \times 10^3\,\text{s}^{-1}$ corresponding to a frequency $\nu_o = 8.8\,\text{kHz}$, an unacceptably low value even for oscilloscopes of average quality. The rise time of the response of this circuit to a voltage step is $t_r = 40\,\mu\text{s}$ and the signal shape is heavily distorted, see Fig. 9.8.

[15]For example, this is the case when R_2 represents the input resistance of a measuring instrument. In fact any real device has a finite value of the input capacitance. The input capacitance of oscilloscopes varies in the range 15–50 pF. Furthermore, the connection cables between the attenuator and the measuring instrument add a capacitance of the order of 100 pF/m and an inductance of the order of 0.1 μH/m.

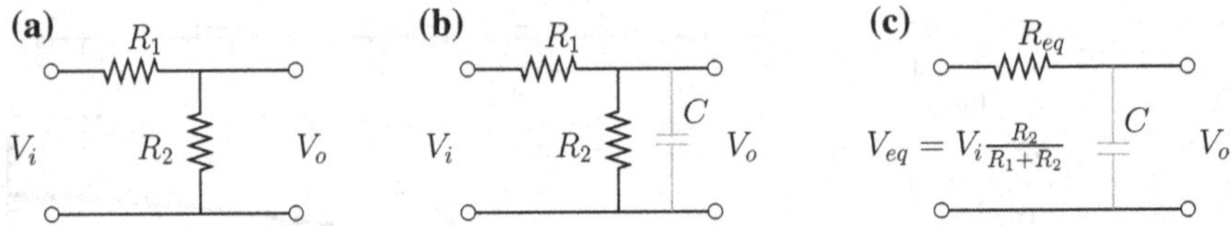

Fig. 9.17 **a** Resistive attenuator. **b** Resistive attenuator with a stray capacitance C. **c** Thévenin equivalent of the circuit (**b**)

We can obtain an attenuator with a frequency independent transfer function shunting both R_1 and R_2 with capacitances larger than the strays as shown in Fig. 9.18. In the next section, we analyze this circuit to find the conditions that must be satisfied by its components to build a *compensated attenuator*.

9.10.1 Compensated Attenuator—Analysis in Frequency Domain

The transfer function $H(\omega)$ of the circuit shown in Fig. 9.18 is easily computed:

$$H(\omega) = \frac{V_o}{V_i} = \frac{Z_2}{Z_1 + Z_2} = \frac{1}{1 + \dfrac{Z_1}{Z_2}} = \frac{1}{1 + \dfrac{R_1}{R_2} \dfrac{1 + j\omega R_2 C_2}{1 + j\omega R_1 C_1}} \tag{9.37}$$

where $Z_1 = R_1/(1 + j\omega R_1 C_1)$ is the impedance of the parallel of R_1 and C_1 and $Z_2 = R_2/(1 + j\omega R_2 C_2)$ is the impedance of the parallel of R_2 and C_2. In general, H, given by expression (9.37), depends on frequency except when

$$R_1 C_1 = R_2 C_2 \tag{9.38}$$

If this condition is verified, the transfer function becomes

$$H(\omega) = \frac{R_2}{R_1 + R_2} = \frac{C_1}{C_1 + C_2} \tag{9.39}$$

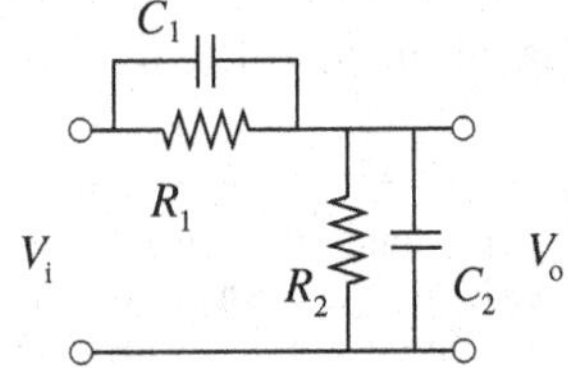

Fig. 9.18 Schematic of the compensated attenuator

In this case, we say that the *attenuator is compensated*.[16]

In the general case ($R_1C_1 \neq R_2C_2$), the $H(\omega)$ modulus and phase, omitting the simple but rather long mathematical derivation, are given by

$$|H(\omega)| = \frac{R_2}{R_1 + R_2}\sqrt{\frac{1 + \omega^2 R_1^2 C_1^2}{1 + \omega^2\left(\frac{R_1 R_2}{R_1 + R_2}\right)^2 (C_1 + C_2)^2}} \tag{9.40}$$

$$\tan\phi = \frac{\omega R_1(C_1 R_1 - C_2 R_2)}{R_1 + R_2 + \omega^2 C_1(C_1 + C_2)R_1^2 R_2} \tag{9.41}$$

The asymptotic behavior of the attenuation $H(\omega)$ for $\omega \to 0$ and for $\omega \to \infty$ can be obtained by Eq. (9.37) or (9.40):

$$H(\omega) \underset{\omega\to 0}{\sim} R_2/(R_1 + R_2) \equiv A_R \qquad H(\omega) \underset{\omega\to\infty}{\sim} C_1/(C_1 + C_2) \equiv A_C \tag{9.42}$$

As expected, at sufficiently low frequency the impedance of the capacitances is very high, and the attenuation is determined only by the resistors (A_R). The circuit behaves as a purely resistive voltage divider. On the contrary, at sufficiently high frequency the impedance of the capacitors becomes small enough to allow neglecting the resistances and the attenuation is given only by the capacitances (A_C). The circuit behaves as a purely capacitive voltage divider. Using definition (9.42) of A_R and A_C, and with the position $\omega_o = 1/R_1 C_1$, Eqs. (9.40) and (9.41) can be written in a more understandable way:

$$|H(\omega)| = A_R\sqrt{\frac{1 + \left(\frac{\omega}{\omega_o}\right)^2}{1 + \left(\frac{\omega}{\omega_o}\right)^2\left(\frac{A_R}{A_C}\right)^2}} \tag{9.43}$$

$$\tan\phi = \frac{\omega}{\omega_o}\frac{A_C - A_R}{A_C + A_R\left(\frac{\omega}{\omega_o}\right)^2} \tag{9.44}$$

These relations show that the compensation of our voltage divider is obtained when the resistive attenuation A_R is equal to the capacitive attenuation A_C. It is easy to show that $A_R = A_C$ is equivalent to relation (9.38).

---

[16]Oscilloscope probes consist of a compensated voltage divider were R_2 is the input resistance of the instrument and C_2 its stray input capacitance. Resistance R_1 and capacitance C_1 depend upon the attenuation factor of the probe. Capacitor C_1 is variable and is adjusted to compensate the probe response.

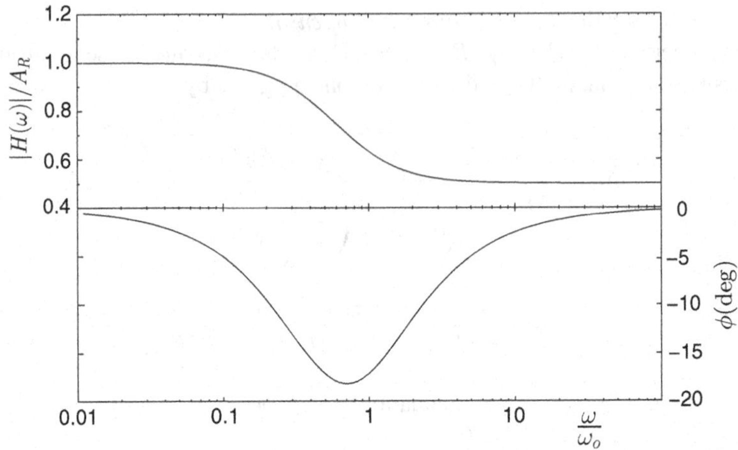

Fig. 9.19 Transfer function of an undercompensated attenuator as function of the variable ω/ω_o where $\omega_o = 1/R_1C_1$. *Upper plot* shows the modulus of $H(\omega)$ divided by A_R, *lower plot* shows the phase of $H(\omega)$ in the same frequency interval. In these plots $A_C = A_R/2$. The meaning of symbols is given in the text

When the attenuator is not compensated, there are two possibilities:

- $A_R > A_C$, or $R_1C_1 < R_2C_2$. The attenuator is *undercompensated.* High frequencies are attenuated more than low frequencies and the circuit tends to "integrate" the input signal.
- $A_R < A_C$, or $R_1C_1 > R_2C_2$. The attenuator is *overcompensated.* Low frequencies are attenuated more than high frequencies and the circuit tends to "differentiate" the input signal.

Figure 9.19 shows an example of an undercompensated attenuator with a ratio between the two asymptotic attenuations given by $A_R/A_C = 1/2$.

9.10.2 Compensated Attenuator—Analysis in Time Domain

In this section, we study the attenuator of Fig. 9.18 in the time domain and we derive its response to a voltage step.

We denote with $v_i(t)$ and $v_o(t)$, respectively, the input and the output signals of the circuit. Equating the current in the parallel of R_1 and C_1 to the current in the parallel of R_2 and C_2, we can write

$$\frac{v_i - v_o}{R_1} + C_1\frac{d(v_i - v_o)}{dt} = \frac{v_o}{R_2} + C_2\frac{dv_o}{dt}$$

or equivalently

$$(C_1 + C_2)\frac{dv_o}{dt} + \left(\frac{1}{R_1} + \frac{1}{R_2}\right) v_o = \frac{v_i}{R_1} + C_1\frac{dv_i}{dt} \tag{9.45}$$

Introducing the parameters $C_p = C_1 + C_2$ and $R_p = R_1 R_2/(R_1 + R_2)$, with $v_i(t) = V\theta(t)$ Eq. (9.45) becomes

$$C_p\frac{dv_o}{dt} + \frac{1}{R_p}v_o = \frac{V}{R_1}$$

where we have taken into account that the time derivative of input signal is zero *almost everywhere*. The solution of this first-order differential equation is

$$v_o(t) = V_C e^{-t/R_pC_p} + V\frac{R_2}{R_1 + R_2}$$

where the first term is the general solution of the homogeneous equation and the second the particular solution $v_o(t) = const$. The integration constant V_C can be determined in first instance assuming that, at $t = 0$ both capacitors charge *instantaneously*[17] so that the capacitive attenuation is immediately obtained:

$$v_o(0) = V\frac{C_1}{C_1 + C_2} \quad \text{yielding} \quad V_C = V\left(\frac{C_1}{C_1 + C_2} - \frac{R_2}{R_1 + R_2}\right)$$

Finally, we obtain for the output signal as

$$v_o(t) = V\left[\left(\frac{C_1}{C_1 + C_2} - \frac{R_2}{R_1 + R_2}\right) e^{-t/R_pC_p} + \frac{R_2}{R_1 + R_2}\right]$$

Because of this equation, we can state

- in general if $R_1C_1 \neq R_2C_2$, for "short times" ($t \ll R_pC_p$) the output voltage is determined by the values of the capacities, in particular at $t = 0$ we have $v_o = VC_1/(C_1 + C_2)$, while for "large times" ($t \gg R_pC_p$) the output voltage is determined by the values of the resistances, in particular for $t \to \infty$ we have $v_o = VR_2/(R_1 + R_2)$; see Fig. 9.20.
- if the compensation condition $R_1C_1 = R_2C_2$ is satisfied, the attenuation at "short times" and at "large times" are equal and the shape of the signal remains unchanged, confirming the result obtained in frequency domain.

Circuit with a real generator. The derivation in previous paragraph is unsatisfactory as it admits, not realistically, that at time $t = 0$ an infinite current flows in the capacities. It is well known that real voltage generators have a finite internal resistance

[17]Note that, from a mathematical point of view, at $t = 0$ also the current has a discontinuity and through the capacitors flow an "infinite" current that loads them instantaneously. More realistic physical considerations will be made in the following.

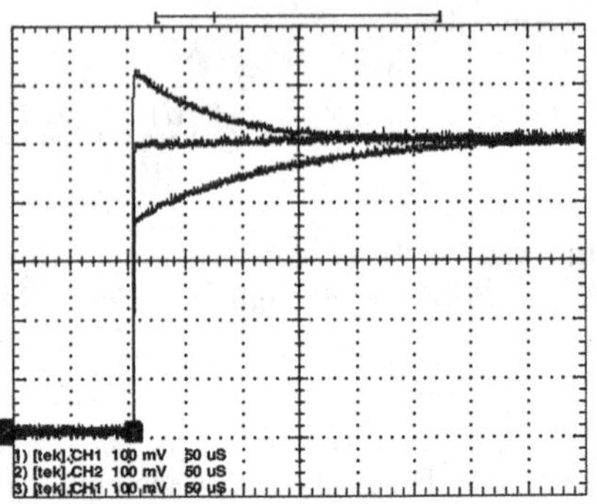

Fig. 9.20 The figure shows the response to voltage step of a compensated (*middle track*), overcompensated (*upper track*), and undercompensated (*lower track*) voltage divider

R_g limiting the maximum current they can supply. A more realistic description of the behavior of a compensated attenuator must include the resistance R_g in the circuit schematic as shown in Fig. 9.21. If v_1 denotes the voltage drop on the parallel of R_1 and C_1, Kirchhoff's law of voltages yields

$$v_g - v_1 - v_o = i R_g \tag{9.46}$$

where v_o is the output voltage and v_g and i are voltage and current output of the generator. This current flows through the two parallel of resistance and capacitance and can be expressed as

$$i = \frac{v_1}{R_1} + C_1 \frac{dv_1}{dt} \quad \text{and} \quad i = \frac{v_o}{R_2} + C_2 \frac{dv_o}{dt} \tag{9.47}$$

Using Eq. (9.46) to eliminate v_1 and assuming a voltage step function as input signal, $v_g(t) = V\theta(t)$, we have[18]:

$$v_1 = V - v_o - i R_g \quad \text{and} \quad \frac{dv_1}{dt} = -\frac{dv_o}{dt} - R_g \frac{di}{dt} \tag{9.48}$$

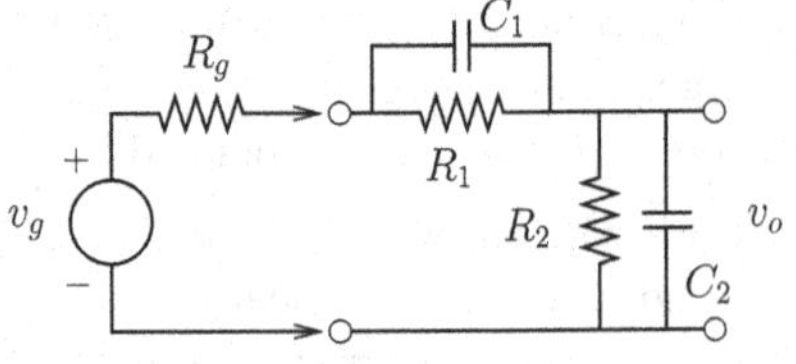

Fig. 9.21 Compensated attenuator

[18]We remind that the derivative of $v_g(t)$ proportional to the $\theta(t)$ function is zero "almost everywhere".

Equating the two expressions of the current given in relations (9.47) we get

$$\frac{V - v_o - iR_g}{R_1} - C_1\left(\frac{dv_o}{dt} + R_g\frac{di}{dt}\right) = \frac{v_o}{R_2} + C_2\frac{dv_o}{dt} \tag{9.49}$$

Isolating the known term:

$$\frac{V}{R_1} = v_o\left(\frac{1}{R_1} + \frac{1}{R_2}\right) + \frac{dv_o}{dt}(C_1 + C_2) + R_g\left(\frac{i}{R_1} + C_1\frac{di}{dt}\right)$$

Using again the second Eq. (9.47), this expression becomes

$$\frac{V}{R_1} = v_o\left(\frac{R_1 + R_2}{R_1 R_2}\right) + \frac{dv_o}{dt}(C_1+C_2) + R_g\left(\frac{v_o}{R_1 R_2} + \frac{C_2}{R_1}\frac{dv_o}{dt} + \frac{C_1}{R_2}\frac{dv_o}{dt} + C_1 C_2\frac{d^2 v_o}{dt^2}\right)$$

Dividing by $R_g C_1 C_2$, the differential equation for $v_o(t)$ becomes

$$\frac{d^2 v_o}{dt^2} + \left[\left(\frac{1}{C_1} + \frac{1}{C_2}\right)\frac{1}{R_g} + \frac{1}{R_1 C_1} + \frac{1}{R_2 C_2}\right]\frac{dv_o}{dt} + \left[\frac{1}{R_g C_2 C_1}\frac{R_1 + R_2}{R_1 R_2} + \frac{1}{R_1 R_2 C_1 C_2}\right]v_o$$
$$= \frac{V}{R_g R_1 C_1 C_2}$$

This equation gives the circuit response to voltage step when the generator internal resistance is not negligible. However, in most cases of practical interest, R_g is small with respect to the two resistors R_1 and R_2. In these cases, we can simplify it neglecting the terms without R_g in the denominator obtaining

$$\frac{d^2 v_o}{dt^2} + \frac{1}{R_g C_s}\frac{dv_o}{dt} + \frac{v_o}{R_g C_s R_p C_p} = \frac{V}{R_g R_1 C_s C_p} \tag{9.50}$$

where $C_s = C_1 C_2/(C_1 + C_2)$ and $C_p = C_1 + C_2$ are, respectively, the capacitance of the series and of the parallel of C_1 and C_2 and $R_p = R_1 R_2/(R_1 + R_2)$ is the parallel of R_1 and R_2. The characteristic equation of (9.50) has the roots

$$m_{1,2} = -\frac{1}{2R_g C_s}\left(1 \pm \sqrt{1 - 4\frac{R_g C_s}{R_p C_p}}\right)$$

When $R_g \ll R_p C_p / C_s$, expanding the square root to the first order, we obtain

$$m_{1,2} = -\frac{1}{2R_g C_s}\left[1 \pm \left(1 - 2\frac{R_g C_s}{R_p C_p}\right)\right] \quad \text{or explicitly}$$

$$\begin{cases} m_1 = -\dfrac{1}{R_g C_s} + \dfrac{1}{R_p C_p} \simeq -\dfrac{1}{R_g C_s} \\ m_2 = -\dfrac{1}{R_p C_p} \end{cases}$$

As expected, this second-order circuit has two characteristic times: a "short" time $\tau_f = R_g C_s$ that describes, as will be shown in the following, the charge of C_s (series of C_1 and C_2) and a "long" time $\tau_s = R_p C_p$ that determines the dynamics of charge partitioning through the resistances and the time needed to reach the resistive value of attenuation.

To obtain the general solution of Eq. (9.50) we need a particular solution that can be easily obtained looking for a time-independent solution:

$$v_o = V\frac{R_p}{R_1} = V\frac{R_2}{R_1 + R_2}$$

Finally, we get:

$$v_o(t) = V_f \mathrm{e}^{-t/\tau_f} + V_s \mathrm{e}^{-t/\tau_s} + V\frac{R_2}{R_1 + R_2}$$

The two integration constants V_f and V_s can be determined imposing the initial condition at $t = 0$. Since initially C_2 is uncharged, we have $v_o(0) = 0$, so that

$$V_f + V_s = -V\frac{R_2}{R_1 + R_2} \tag{9.51}$$

Also C_1 is initially uncharged and we have $v_1(0) = 0$. Applying Eq. (9.46), we get $i(0) = V/R_g$. The second equation in (9.47) gives

$$\left.\frac{dv_o}{dt}\right|_{t=0} = \frac{V}{R_g C_2}$$

and we finally obtain

$$\frac{V_f}{\tau_f} + \frac{V_s}{\tau_S} = -\frac{V}{R_g C_2} \tag{9.52}$$

Equations (9.51) and (9.52) form a system whose solution is

$$V_f = -V\frac{C_1}{C_1 + C_2} \quad \text{and} \quad V_s = \left(\frac{C_1}{C_1 + C_2} - \frac{R_2}{R_1 + R_2}\right) V$$

Finally, the attenuation as a function if the time is

$$\frac{v_o(t)}{V} = \left(\frac{C_1}{C_1 + C_2} - \frac{R_2}{R_1 + R_2}\right) \mathrm{e}^{-t/\tau_s} - \frac{C_1}{C_1 + C_2}\mathrm{e}^{-t/\tau_f} + \frac{R_2}{R_1 + R_2}$$

The "long times" asymptotic value is reached when $t \gg \tau_f$ and $t \gg \tau_s$. As expected, it is equal to the resistive partition ratio $R_2/(R_1 + R_2)$. When the two times $\tau_f = R_g C_s$ and $\tau_s = R_p C_p$ are significantly different with $\tau_f \ll \tau_s$, as it happens often in real devices, when $\tau_f \ll t \ll \tau_s$ we get:

$$\frac{v_o(t)}{V} \simeq \frac{C_1}{C_1 + C_2}$$

This is the time interval where the attenuation is given by the capacitive partition ratio. In these conditions τ_f is the signal rise time while τ_s is the timescale needed for the signal to reaches its asymptotic value $(R_2/(R_1 + R_2))$.

It is interesting to note that in a compensated divider, the two timescales remain different but the amplitude of the slow component of the signal V_s is null. This component is entirely due to the mismatch of the resistive and capacitive attenuations and, therefore, does not exist in a compensated attenuator.[19]

9.10.3 Frequency Limitation of Compensated Attenuators

A well-compensated attenuator, as we have shown in previous sections, has a transfer function that is independent of the signal frequency and its output is a reduced but nondeformed copy of the input signal. However, the results of the previous section suggest that this is only true for signal duration longer that the short timescale τ_f given by the output resistance of the signal generator and the series of the capacitances in the divider.

A similar conclusion can be reached also analyzing the circuit in the frequency domain. When the frequency is high enough we can neglect the two resistances and the attenuator is seen by the generator as the series of the two capacities C_1 and C_2. Taking into account the output resistance R_g of the generator, the circuit behaves like the *low-pass RC* shown in Fig. 9.7 with a characteristic time is given by $\tau = R_g C_1 C_2/(C_1 + C_2) \equiv \tau_f$. Therefore, signals with angular frequency greater then $\omega_o = 1/\tau_f = (C_1 + C_2)/R_g C_1 C_2$ will be attenuated. However, for many practical applications this frequency is quite high. If we take for example $R_g = 50\Omega$ and $C_2 = 200\,\text{pF}$, typical values of input resistance and capacitance of a commercial oscilloscope connected with a matched cable, and we assume an attenuation $\times 10$, so that $C_1 \simeq C_2/10$, we get a cutoff frequency of 160 MHz, adequate for most purposes.

Problems

Problem 1 **A**: Obtain the differential equation for the output voltage $v_o(t)$ across the inductance L of the *RL* circuit shown in Fig. 9.11a when the input voltage is $v_i(t)$. **B**: Determine under what conditions the voltage $v_o(t)$ is proportional to the derivative of $v_i(t)$ in the case the input signal is a pulse of duration τ. **C**: Solve the differential equation found in **A** in the case $v_i(t)$ is a voltage step function of amplitude V

---

[19] Oscilloscope probes are compensated minimizing the slow component of their response to a square wave calibration pulse supplied by the instrument.

using appropriate initial conditions. [A. A: $dv_i(t)/dt = (R/L)v_L(t) + dv_L(t)/dt$; B: $(L/R) \ll \tau$; C: $v_L(t) = V\exp(-Rt/L)$.]

Problem 2 **A**: Obtain the differential equation for the output voltage $v_o(t)$ across the resistance R of the RL circuit shown in Fig. 9.11b when the input voltage is $v_i(t)$ **B**: Determine under what conditions the voltage $v_R(t)$ is proportional to the integral of $v_i(t)$ in the case the input signal is a pulse of duration τ. **C**: Solve the differential equation found in **A** in the case $v_i(t)$ is a voltage step function of amplitude V using appropriate initial conditions. [A. A: $v_i(t) = (L/R)v_R(t) + dv_R(t)/dt$; B: $(L/R) \ll \tau$; C: $v_R(t) = V(1 - \exp(-Rt/L))$.]

Problem 3 In the circuit shown in the figure the switch T is open at time $t = 0$. Determine the voltage across the capacitance C as a function of time. [A. $v_C(t) = V_1 + (V_2 - V_1)\exp(-t/RC)$.]

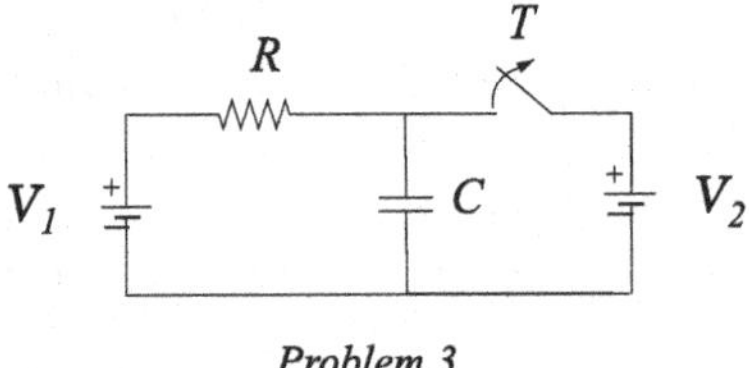

Problem 3

Problem 4 Obtain the differential equation for the voltage $v(t)$ across the capacitance C of the circuit shown in the figure. Solve the differential equation when v_i is a voltage step function of amplitude V. [A. $v(t) = VR_2/(R_1 + R_2)\exp(-t(R_1 + R_2)/(R_1R_2C))$.]

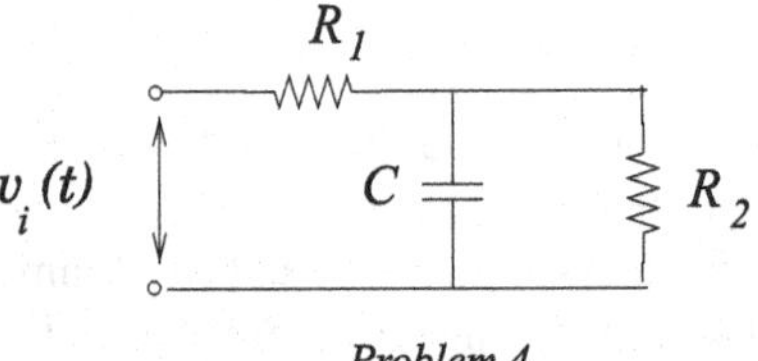

Problem 4

Problem 5 Solve the previous problem using the Thévenin's theorem.

Problem 6 In the circuit shown in the figure the switch T is closed at time $t = 0$. Determine the voltage across the capacitance C as a function of time. [A. $v(t) = VR_3/(R_1 + R_3)\exp(-t(R_3 + R_1)/(R_3C(R_1 + R_2 + R_1R_2/R_3)))$.]

Problem 7 Solve the previous problem using the Thévenin's theorem

Problem 8 Obtain the differential equation for the current $i_L(t)$ through the inductance L in the circuit shown in the figure, when $v_i(t)$ is the input voltage. [A. $v_i(t)R_3(R_1R_3+R_1R_2+R_2R_3) = i_L+L(R_1+R_3)/((R_1R_3+R_1R_2+R_2R_3)di_L/dt$.]

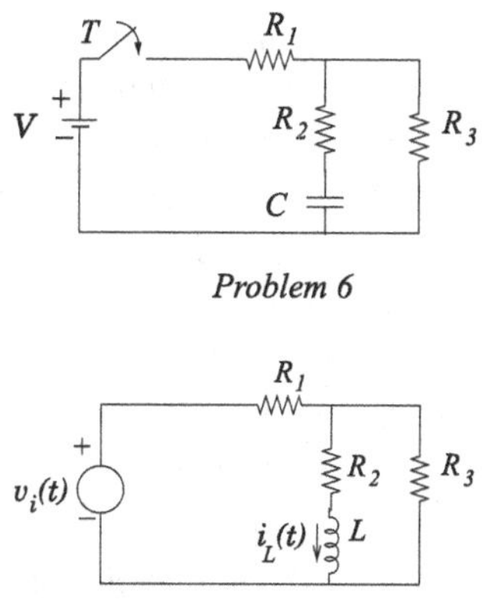

Problem 6

Problem 8

Problem 9 Solve the previous problem using the Norton's theorem

Problem 10 Obtain the differential equation for the voltage $v_C(t)$ across the capacitance C in a RLC series circuit when $v_i(t)$ is the input voltage. Solve the differential equation, using the appropriate boundary conditions, when v_i is a voltage step function of amplitude V. [A. $v_i(t) = v_C(t) + RCdv_C(t)/dt + LCd^2v_C(t)/dt^2$; $v_C(0) = 0, dv_C/dt(0) = 0$.]

Problem 11 As an alternative to the exponential pulse, a real voltage step can be described as a linear ramp of time duration T up to the constant value V. Using the convolution method, find the response of a *high-pass RC* circuit to this input waveform. Comment the solution obtained in terms of the behavior of the capacitor C. [A. $v_o(t) = 0$ for $t < 0$, $v_o(t) = \alpha RC[1 - \exp(-t/RC)]$ for $0 < t < T, v_o(t) = \alpha RC[1 - \exp(-T/RC)]\exp[-(t - T)/RC]$ for $t > T$.]

Problem 12 The circuit shown in the figure represents a real *RL* filter where the inductor L has a stray resistance R_L. Obtain the differential equation for the current $i_L(t)$ through the inductance L in response of an input voltage $v_i(t)$. Solve the differential equation when v_i is a voltage step function of amplitude V. Finally obtain the output voltage $v_o(t)$ across the real inductor. [A. $v_o(t) = VR_L/(R+R_L)+VR/(R+R_L)exp(-t/\tau)$ with $\tau = L/(R + R_L)$.]

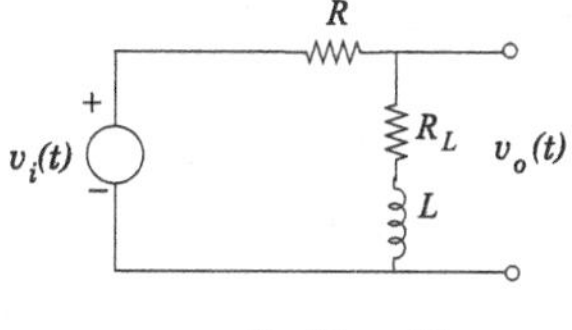

Problem 12

Problem 13 In the circuit shown in the figure the switch T is closed at time $t = 0$. Obtain the differential equation for the current $i_L(t)$ in the inductance L as a function of time and determine the appropriate boundary conditions for the solution. [A. $V = R_1 i_L(t) + (L + R_1 R_2 C) di_L(t)/dt + R_1 LC d^2 i_L(t)/dt^2$ where $i_L(0) = V/(R_1 + R_2)$, $di_L/dt(0) = V/L(R_1 + R_2)$.]

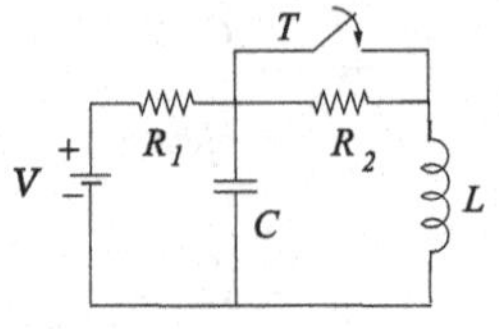

Problem 13

Problem 14 The circuit in the figure is a diagram of the cable connection of a signal to an oscilloscope. Derive in time domain the differential equation yielding the signal $v_o(t)$ across the resistance R_2 when the input signal is $v_i(t)$. Find the appropriate boundary conditions for the solution when the input is a voltage step function. [A. $v_i(t) = (1 + R_1/R_2)v_o(t) + (R_1 C + L/R_2) dv_o(t)/dt + LC d^2 v_o(t)/dt^2$; $v_o(0) = 0$, $dv_o/dt(0) = 0$.]

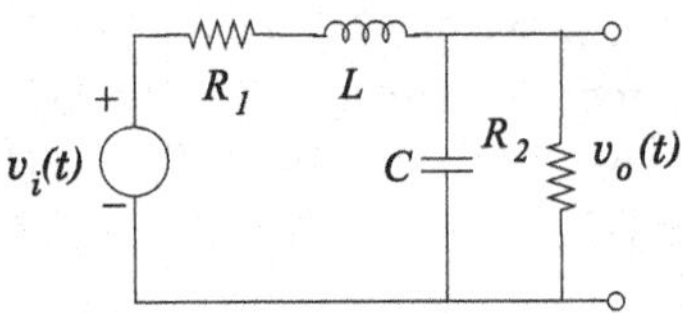

Problem 14

Problem 15 The circuit in the figure represents a filter LR realized with a real inductor whose series resistance is R_L and whose parasitic capacitance is C. Obtain the differential equation that describes the response $v_o(t)$ to a signal $v_i(t)$. Find the appropriate boundary conditions in case the input signal is a step function of amplitude V. [A. $v_i(t) + RC dv_i(t)/dt + LC d^2 v_i(t)/dt^2 = (1 + R_L/R)v_o(t) + (RC + L/R) dv_o(t)/dt + LC d^2 v_o(t)/dt^2$; $v_o(0) = V$, $dv_o/dt(0) = -V/RC$.]

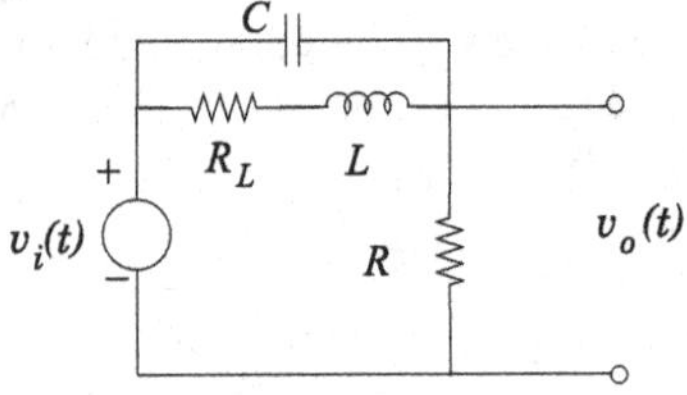

Problem 15

Chapter 10
The Transmission Line

10.1 Introduction

The methods adopted in the previous chapters and the results obtained are only valid when the size of components and circuits are small enough to allow neglecting variations of the electromagnetic (e.m.) field inside them. In this case, we say that we are dealing with *lumped elements* circuits. However, when the frequency of electrical signals increases, their wavelength λ decreases, and if λ becomes comparable with the physical dimensions of components, the lumped elements approximation is no more verified.

As an example, consider an e.m. sinusoidal signal of frequency $\nu = 300\,\text{MHz}$; assuming that this signal propagates in a real component at a speed comparable with light speed c, its wavelength is $\lambda = c/\nu \sim 1\,\text{m}$ and $\lambda/4 \sim 25\,\text{cm}$ is the space separation between the maximum and the minimum wave amplitude. If the component linear dimensions are comparable to $\lambda/4$, the electromagnetic field changes across it and the methods used for the solution of lumped circuits cannot be applied. In these conditions, the circuit is said to have *distributed constants* and its solution cannot be obtained from Kirchhoff's laws but requires in principle to use directly Maxwell's equations.

In various circumstances, it is necessary to transmit electrical signals between circuits far apart (from a few meters up to many kilometers of distance). A common example of such a connection is the home television equipment, which consists in an *antenna*,[1] the signal generator, and in the television set, a circuit that receives, amplifies, and decodes the signal. Typical distance between the antenna and television set is a few dozen meters and their connection is made via a third circuit, commonly known as the TV antenna cable, which transmits *without distortion* the

[1]More accurately, the antenna selects a signal transmitted in the air and transfers it to its output terminals. The large majority of television signals are broadcast in the "Ultra High Frequency" range (UHF), an international acronym to indicate the frequency range of 0.3–3 GHz, ($\lambda = 1$–$0.1\,\text{m}$).

R. Bartiromo and M. De Vincenzi, *Electrical Measurements in the Laboratory Practice*, Undergraduate Lecture Notes in Physics,
DOI 10.1007/978-3-319-31102-9_10

signal captured by the antenna to the TV set. Comparing the order of magnitude of its length ($\simeq$10 m) with the wavelength of broadcast television signals ($\simeq$1 m), it is clear that within the cable the electromagnetic field is not constant.

In this chapter, we will illustrate a model that helps to understand how the signal propagates along the cable and the methods used to solve circuits with distributed constants. We first give in the next Sect. 10.2 a precise definition of a transmission line and then introduce a model in terms of infinitesimal discrete components. The analysis of this model will lead us to the telegrapher equation and will uncover the role of two important parameters determining the line behavior, namely the propagation speed and the line characteristic impedance. The general solution of the telegrapher equation is the subject of Sect. 10.3 that also covers the particular case of a lossless line. Next, in Sect. 10.4 we discuss the properties of the coaxial cable while the problem of line termination is addressed in Sect. 10.5 where we describe the reflection of electromagnetic signals at the line ends. In Sect. 10.7, we consider signal attenuation observed in lossy lines and finally in Sect. 10.8 we compute the line impedance in the frequency domain.

10.2 Transmission Lines

The circuits used to carry electrical signals over long distances are called *transmission lines*. They can be described as a set of two conductors, electrically isolated, with one of the spatial dimensions much greater than the others and with the cross-section, normal to the long dimension, constant in shape and area. Ideally, the two conductors must be in conditions of complete electrostatic induction and no magnetic flux dispersion is allowed.

Examples of transmission lines are given in Fig. 10.1. They are two wires that run parallel to each other at a constant distance; a wire placed on an infinite ground plane; a conducting strip separated from a ground plane by a dielectric; two coaxial cylindrical conductors (coaxial cable).

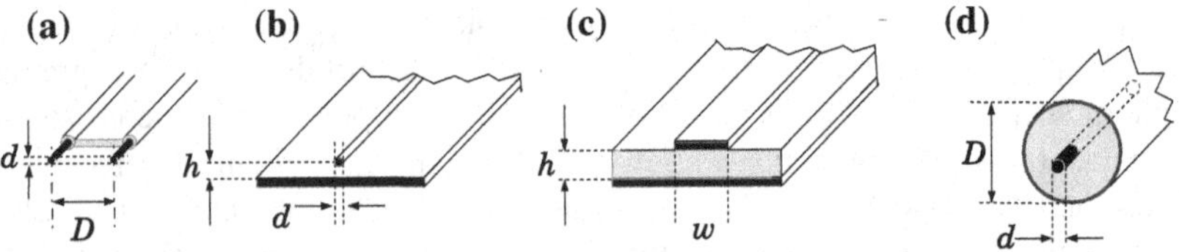

Fig. 10.1 Some types of transmission lines and the value of their capacity C and inductance L, per unit of length: **a** two-wire line, $C = \pi\varepsilon/\cosh^{-1}(D/d)$, $L = \mu/\pi \cosh^{-1}(D/d)$; **b** wire on a (infinite) ground plane, $C = \pi\varepsilon/\cosh^{-1}(2h/d)$, $L = \mu/\pi \cosh^{-1}(2h/d)$; **c** strip on a (infinite) ground plane $C \simeq \varepsilon w/h$, $L \simeq \mu h/w$; **d** coaxial cable, $C = 2\pi\varepsilon/\ln(D/d)$, $L = \mu/2\pi \ln(D/d)$

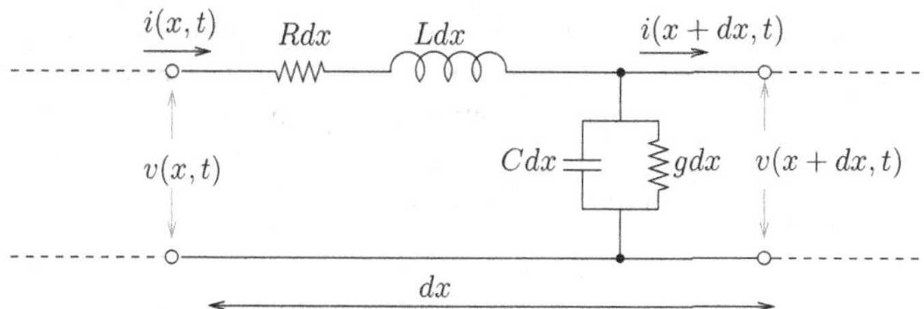

Fig. 10.2 Schematization of a transmission line element

10.2.1 Model of the Transmission Line

To analyze quantitatively the behavior of a transmission line we consider an *infinitesimal* element of the line of length dx where it is still possible to use the approximation of lumped components. The whole line is then modeled as a succession of such elements, as shown in Fig. 10.2. In this schematization, the transmission line is a passive circuit formed by components (resistors, inductors and capacitors) *distributed* along the entire length of the line. The main features of the line are described by the following four parameters:

1. the resistance per unit length R, which represents the ohmic resistance of its conductors (units: $\Omega \mathrm{m}^{-1}$)
2. the inductance per unit length L, which depends (except in the special case of a magnetic dielectric) only by the geometry of the line (units: Hm^{-1})
3. the capacity (known as shunt capacity) per unit length C, which depends on the geometry of the line and on the dielectric constant of the insulator that separates the two conductors forming the line (units: Fm^{-1})
4. The conductance per unit length g, which takes into account of the losses of the dielectric (units: $\Omega^{-1}\mathrm{m}^{-1}$).

The application of the Kirchhoff's laws to the line element shown in Fig. 10.2 brings to the following equations:

$$v = Rdx\, i + Ldx \frac{\partial i}{\partial t} + \left(v + \frac{\partial v}{\partial x} dx \right)$$

$$i = \left(i + \frac{\partial i}{\partial x} dx \right) + \left(v + \frac{\partial v}{\partial x} dx \right) gdx + Cdx \frac{\partial}{\partial t} \left(v + \frac{\partial v}{\partial x} dx \right)$$

Simplifying and disregarding the second-order terms, we get

$$\begin{cases} -\dfrac{\partial v}{\partial x} & = Ri + L\dfrac{\partial i}{\partial t} \\ -\dfrac{\partial i}{\partial x} & = gv + C\dfrac{\partial v}{\partial t} \end{cases} \tag{10.1}$$

The solution of this system of partial differential equations, together with the boundary conditions, allows computing the voltage and the current as a function of the time t and of the x coordinate along the line. To obtain an equation for the voltage $v(x, t)$, we take the derivative of the first of equations (10.1) with respect to x and of the second with respect to t. Removing the term of mixed derivative $\partial^2 i/\partial t\partial x$, we obtain

$$LC\frac{\partial^2 v}{\partial t^2} - \frac{\partial^2 v}{\partial x^2} + (gL + RC)\frac{\partial v}{\partial t} + gRv = 0 \qquad (10.2)$$

If we take the derivative of the first of equations (10.1) with respect to t and of the second with respect to x, we obtain an equation similar to (10.2) for the current $i(x, t)$:

$$LC\frac{\partial^2 i}{\partial t^2} - \frac{\partial^2 i}{\partial x^2} + (gL + RC)\frac{\partial i}{\partial t} + gRi = 0 \qquad (10.3)$$

This equation, either for the voltage or for the current, is known as the *telegrapher equation*, because it was used to describe the signal transmission along cables connecting far apart telegraph stations. We can easily recognize that Eqs. (10.2) and (10.3) describe a propagating signal by solving a simplified version in the time domain.

Lossless transmission line. When the energy dissipation in the line can be neglected, we can consider $R = g = 0$ in the Eq. (10.2) for the voltage (or in the Eq. (10.3) for the current), so that for a "lossless" line we get:

$$LC\frac{\partial^2 v}{\partial t^2} - \frac{\partial^2 v}{\partial x^2} = 0 \qquad (10.4)$$

This is the well-known equation describing wave propagation in one dimension with speed given by

$$u = \frac{1}{\sqrt{LC}}$$

The general solution of Eq. (10.4) is

$$v(x, t) = v_f(x - ut) + v_b(x + ut) \qquad (10.5)$$

where v_f and v_b are any two functions that fulfill the boundary conditions of the specific problem and that are differentiable up to second order.

It is easy to see that $v_f(x - ut)$ and $v_b(x + ut)$ represent signals that propagate respectively in positive and the negative direction of x (the subscript f and b stand respectively for *forward* and *backward*). Note that the signals described by the functions $v_f(x - ut)$ and $v_b(x + ut)$ propagate along the lossless line without undergoing deformations.

The expression for the current in the line is obtained substituting the voltage (10.5) in the second equation of system (10.1) with $g = 0$ and, noting that $\partial v_{f,b}/\partial t = \mp u\partial v_{f,b}/\partial x$, we get

$$-\frac{\partial i(x,t)}{\partial x} = -Cu\frac{\partial v_f}{\partial x} + Cu\frac{\partial v_b}{\partial x} \tag{10.6}$$

After integration over x, we obtain the following expression for the current:

$$i(x,t) = \frac{v_f(x-ut) - v_b(x+ut)}{Z_c} \tag{10.7}$$

where we introduce the constant

$$Z_c = \sqrt{\frac{L}{C}} \tag{10.8}$$

Z_c is the *characteristic impedance* of the line, an important parameter to describe the line behavior, as we will see soon.

10.3 Solution of Telegrapher Equation

The system (10.1) consists of two first order differential equations with constant coefficients and can be more easily resolved using the Fourier transform method that, as already seen for the solution of AC circuits, is equivalent to solve the circuit equations when voltages and currents have sinusoidal waveform. Writing $v(x,t) = V(x)\exp(j\omega t)$, the Eq. (10.1) become

$$\begin{cases} -\dfrac{dV}{dx} = (R + j\omega L)I = Z_o I \\ -\dfrac{dI}{dx} = (g + j\omega C)V = Y_o V \end{cases} \tag{10.9}$$

where $V = V(x)$ and $I = I(x)$ are respectively the complex tension and complex current. In Eq. (10.9) we introduced the quantities Z_o and Y_0, called respectively the *line complex impedance* and the *line complex admittance* per unit's length. Note that, in this case, $Z_o \neq Y_o^{-1}$.

Differentiating with respect to x the Eq. (10.9) we get, after simple algebra,

$$\frac{d^2V}{dx^2} = Z_o Y_o V \tag{10.10}$$

$$\frac{d^2I}{dx^2} = Z_o Y_o I \tag{10.11}$$

The Eqs. (10.10) and (10.11) are another form in which we can write the telegrapher equation when we search for sinusoidal function solutions. The relationships (10.10) and (10.11) show that complex voltage and current in a transmission line must satisfy exactly the same equation.

The general solution of differential equation (10.10) is

$$V(x) = V_f\, e^{-\gamma x} + V_b\, e^{\gamma x} \tag{10.12}$$

where the constant γ is the *complex propagation constant* that in terms of the line parameters, is given by

$$\gamma = \sqrt{Z_o Y_o} = \sqrt{(R + j\omega L)(g + j\omega C)}, \tag{10.13}$$

The constant γ has the physical dimension of the inverse of length while V_f and V_b are the (complex) integration constants with physical dimension of voltage.

The expression for the current can be derived from the Eq. (10.9)

$$I = -\frac{1}{Z_o}\frac{dV}{dx} = \sqrt{\frac{Y_o}{Z_o}}\,(V_f e^{-\gamma x} - V_b e^{\gamma x})$$

Generalizing relation (10.8), we can define the *characteristic impedance* of the transmission line as

$$Z_c = \sqrt{\frac{Z_o}{Y_o}} = \sqrt{\frac{R + j\omega L}{g + j\omega C}} \tag{10.14}$$

that in general is a complex quantity with the physical dimension of impedance. The current I takes the following expression:

$$I = \frac{V_f}{Z_c} e^{-\gamma x} - \frac{V_b}{Z_c} e^{\gamma x} \tag{10.15}$$

The expressions of the space and time evolution of voltage and current along the line are given by

$$v(x,t) = V_f\, e^{-\alpha x} e^{j(\omega t - \beta x)} + V_b\, e^{\alpha x} e^{j(\omega t + \beta x)} \tag{10.16}$$

$$i(x,t) = \frac{V_f}{|Z_c|}\, e^{-\alpha x} e^{j(\omega t - \beta x + \phi)} - \frac{V_b}{|Z_c|}\, e^{\alpha x} e^{j(\omega t + \beta x + \phi)} \tag{10.17}$$

where ϕ is the phase difference between voltage and current (for both forward and backward signals) and the complex constant γ has been separated in its real (α) and imaginary (β) part: $\gamma = \alpha + j\beta$.

In both expressions of the voltage and current, it is easy to recognize

- a progressive wave $e^{j(\omega t - \beta x)}$, i.e., traveling forward, in the positive x direction with speed $u = \omega/\beta$, whose amplitude V_f is modulated by a decreasing exponential $e^{-\alpha x}$
- a regressive wave $e^{j(\omega t + \beta x)}$, i.e., traveling backward, in the negative x direction with the same speed u and whose amplitude V_f is modulated by increasing exponential $e^{\alpha x}$.

Note that, whatever is the propagation direction of the wave, the modulus of the wave amplitude decreases with time. The current in the line, as function of coordinate and time, is given in (10.17); it is worth to note that the proportionality between current and voltage is valid only for each separate mode of propagation, but not for the superposition of the two modes.

10.3.1 Lossless Line

As stated above, for a lossless or ideal transmission line, all the dissipative elements (R and g) can be neglected. In this approximation, the *characteristic impedance* Z_c, defined in (10.14), becomes

$$Z_c = \sqrt{\frac{Z_o}{Y_o}} = \sqrt{\frac{R + j\omega L}{g + j\omega C}} \rightarrow \sqrt{\frac{L}{C}} \tag{10.18}$$

and the propagation constant γ, defined in (10.13) becomes

$$\gamma = \sqrt{Z_o Y_o} \rightarrow j\omega\sqrt{LC} \equiv j\beta \tag{10.19}$$

Before describing some important properties of the ideal line, we observe that for sufficiently high frequencies a real line, with R and g different from zero, tend to behave as an ideal line. In fact, if $\omega \gg R/L$ and $\omega \gg g/C$ the expression of Z_c and γ are simplified and (10.18) and (10.19) can be considered valid in any transmission line for signals of sufficiently high frequency.

The expressions of Z_c and γ characterizing the ideal line, (10.18) and (10.19), allow us to say

- Z_c becomes real, and then the phase shift between the current signal and voltage is zero, as happens with the resistors.
- The propagation constant γ is purely imaginary and therefore the amplitude of the signal remains constant during propagation.
- Z_c and u do not depend on ω and therefore all signals (even non-sinusoidal) propagate along the line without deformation (neglecting dispersion in the line dielectric).

Using the relation (10.19), we obtain the expression for the propagation velocity of the signals along the ideal line: q

$$u = \omega/\beta = \frac{1}{\sqrt{LC}} \tag{10.20}$$

that is the result already obtained in the time domain in Sect. 10.2.1. Both L and C depend on the geometry of the line, so that one could deduce that the propagation velocity of signals along the line depends on its geometry. However, it is possible to show that, when the cross section of the line is constant and the dielectric is homogeneous,[2] the product LC is independent of the geometry and is equal to $\mu\varepsilon$. This means that the signal velocity in the line depends only on the dielectric properties of the insulator used in the line.

$$u = \frac{1}{\sqrt{LC}} = \frac{1}{\sqrt{\mu\varepsilon}} \simeq \frac{c}{\sqrt{\varepsilon}} \tag{10.21}$$

where c is light speed in vacuum and where we accounted for the observation that materials used in real lines are not ferromagnetic ($\mu \simeq \mu_o$). A general proof of (10.21) can be found in Ref. [1] (Sect. 1 of Chap. 10).

10.4 The Coaxial Cable

The most widespread type of transmission lines is the coaxial cable whose sketch is shown in Fig. 10.1d. In the following, we compute its main characteristic parameters.

Capacity per unit length. Consider a length l of a coaxial cable, the capacity of the two conductors (internal and external cylinder) is given by the ratio $Q/\Delta V$ with Q charge on the two conductors and ΔV tension between the two conductors. The flux of electric field through the cylindrical surface of radius r and length l is

$$\Phi(E) = \int \mathbf{E} \cdot d\mathbf{s} = 2\pi r\, l E(r)$$

and applying the Gauss's law we obtain $E(r) = 1/(2\pi\varepsilon\, l)\, Q/r$, where ε is the dielectric constant of the insulator between the two conductors. The voltage difference is obtained integrating the electric field: $\Delta V = \int_{d/2}^{D/2} E(r)\, dr = Q/(2\pi\varepsilon\, l) \ln D/d$ and finally, the capacity per unit length of a coaxial cable is

$$C = \varepsilon \frac{2\pi}{\ln D/d} \tag{10.22}$$

Inductance per unit length. Consider a rectangular surface with one side of length l along the cable direction and the other side along the radial direction from $d/2$ to $D/2$, see Fig. 10.1d. If i is the current in the inner conductor, the magnetic field $\mathbf{B}$, by Biot–Savart law, is normal to the considered rectangular surface and its modulus is $B(r) = \mu/(2\pi)(i/r)$, where μ is the magnetic permeability of the insulator material. The magnetic field flux through the considered rectangular surface is

[2]These properties can be satisfied by the line types shown in Fig. 10.1b, d, but not of the type Fig. 10.1a, c, the reason is that, for the latter, the dielectric is not homogeneous and the relation (10.21) is only approximately verified.

Table 10.1 Common coaxial cables parameters

Coax	R_c (Ω)	C (pF/m)	L (nH/m)	D (mm)	d (mm)	R(Ω/km)	ε_r	u/c
RG58C/U	50	101	253	3.55	0.90	34	2.3	0.66
RG174/U	50	100	253	1.55	0.48	140	2.3	0.66
RG59/U	75	65	380	4.5	0.65	161	2.3	0.66

$$\Phi(B) = \int \mathbf{B} \cdot d\mathbf{A} = l \int_{d/2}^{D/2} B(r)\, dr = \mu i l/(2\pi) \ln(D/d)$$

Recalling the definition of inductance, given by the relationship $\Phi(B) = Li$, the inductance per unit length of a coaxial cable is given by:

$$L = \mu \frac{\ln D/d}{2\pi} = \mu\varepsilon \frac{1}{C} \tag{10.23}$$

From the explicit expression for C and L, we note that in the product LC the geometrical properties of the line cancel out, as we stated in general way at the end of previous section. This computation confirms that the propagation velocity of signals in coaxial cables depends only on electromagnetic properties (ε and μ) of the material used as insulator. From relations (10.22) and (10.23), we obtain the explicit expression of the characteristic impedance of a coaxial line:

$$Z_c = \sqrt{\frac{L}{C}} = \frac{1}{2\pi}\sqrt{\frac{\mu}{\varepsilon}} \ln \frac{D}{d}$$

Note that the logarithmic dependence of Z_c on the geometrical parameters D and d limits the range of the values of characteristic impedance practically achievable to the interval (10–200) Ω. In Table 10.1 we report the values of some parameters of coaxial cables of common use in the laboratory.

10.5 Reflections on Transmission Lines

In this section, we will study the behavior in the time domain of an ideal lossless transmission line terminated on a resistive load (see Fig. 10.3). We limit our study to a simple resistive load in order to focus the reader's attention to the specific phenomenon of signal reflection in lines avoiding complications due to the change in signal shape brought by the load reactance.

The general solution for voltage and current signals in an ideal transmission line, as indicated by the Eqs. (10.5) and (10.7), is formed by the sum of two voltage and current signals: a progressive one propagating in the positive x direction and a regressive one propagating in the negative x direction. Consider a line, of length l

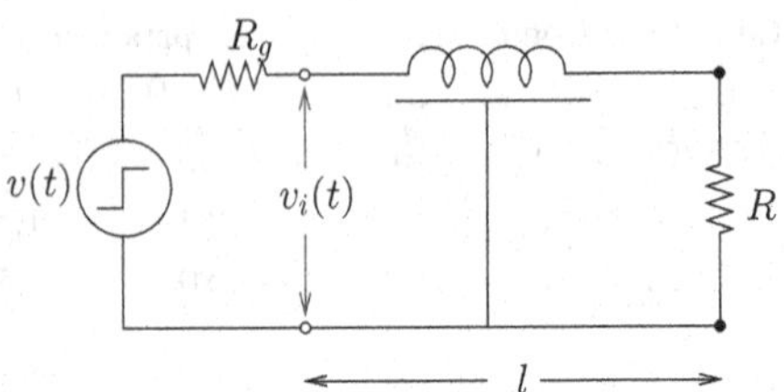

Fig. 10.3 Transmission line connected to a voltage generator of internal impedance R_g and terminated with a load R

and characteristic impedance R_c, connected at one end to a voltage step generator, of amplitude V_o and internal impedance R_g, and terminated at the other end on the load R (see Fig. 10.3). At time $t = 0$ the ideal line behaves like a resistance of value R_c and at its input terminal a voltage $V_f = V_o R_c/(R_c + R_g)$ and a current $I_f = V_f/R_c$ appear. Using these as boundary conditions, the solution for the voltage and current forward signals propagating along the line, are:

$$v(x,t) = V_f \theta(x - ut) \quad i(x,t) = I_f \theta(x - ut) \tag{10.24}$$

The relations (10.24) show that the voltage and current steps propagate in the line with a speed $u = 1/\sqrt{LC}$ so that after a time interval l/u they reach the line termination. The load R represents a discontinuity and the solution (10.24) is no longer valid. In order to satisfy the boundary conditions imposed by the load R, we must return to the general solution (10.5) and (10.7) to find a combination of progressive and regressive waves that satisfy the Ohm's law on the load R. This requires that the current in the resistor $\frac{V_f+V_b}{R}$ is equal to the current in the line $\frac{V_f-V_b}{R_c}$. We easily obtain

$$R = \frac{V_f + V_b}{(V_f - V_b)/R_c} \tag{10.25}$$

Introducing the *reflection coefficient* ρ, the amplitude of the *reflected wave* becomes $V_b = \rho V_f$. Substituting this expression in Eq. (10.25), we find the expression of reflection coefficient as function of the line characteristic impedance R_c and of the load resistance R:

$$\rho = \frac{R - R_c}{R + R_c} \tag{10.26}$$

The following cases are noteworthy:

- $R = \infty$: *Open Line*. In this case $\rho = 1$, the progressive signal is reflected with *the same sign and amplitude.*
- $R = 0$: *Shorted Line*. In this case $\rho = -1$, the progressive signal is reflected with *opposite sign and same amplitude.*
- $R = R_c$: *Matched Line*. In this case $\rho = 0$, no reflection takes place and the line is equivalent to a line of infinite length.

10.5.1 Shorted and Open Line

Consider an ideal transmission line of characteristic impedance R_c, connected to a voltage step generator $v(t) = V_o\theta(t)$ of internal impedance $R_g = R_c$. Suppose that the line, of length l, is terminated with a short circuit (see Fig. 10.3 where $R = 0$). We now illustrate the time evolution of the voltage signal at the line input. At $t = 0$ the step signal starts to propagate along the line with amplitude $V_f = V_oR_c/(R_g+R_c) = V_o/2$. Defining the line time delay as $t_d = l/u$, when $t = t_d$ the signal reaches the short-circuited end of the line ($R = 0$) and it is reflected with inverted amplitude ($\rho = -1$). For $t > l/u$, in the line there are two signals: the forward signal of amplitude $+V_o/2$ and the reflected one of amplitude $-V_o/2$. The total signal is sum of the two, and the reflected signal cancel out the forward one while propagating toward the generator. At $t = 2l/u$ the reflected signal reaches the line input. Since at this end the line is matched ($R_g = R_c$, i.e., $\rho = 0$), the steady state is reached and the total signal at the input is reduced to zero. The mathematical expression of the voltage signal at the line input, for a short-circuited line, is the following:

$$v_i(t) = \frac{V_o}{2}[\theta(t) - \theta(t - 2l/u)]$$

If the line end is terminated with an open-circuit ($R = \infty$), the reflection coefficient is $\rho = +1$ and reasoning as in the previous case, we deduce that for $t > 2l/u$ at the line input the total signal is doubled with respect to the signal at $t = 0$. The mathematical expression of the voltage signal at the line input, for an open line, is the following:

$$v_i(t) = \frac{V_o}{2}[\theta(t) + \theta(t - 2l/u)]$$

The voltage signals as function of time are shown in Fig. 10.4 for the two different cases discussed above. It worth to note that before the time $t = 2l/u$, the signal at the line input is the same independently of the line termination. It is also worth noting that in both cases at steady state the line behaves as a lumped parameter component of negligible resistance.

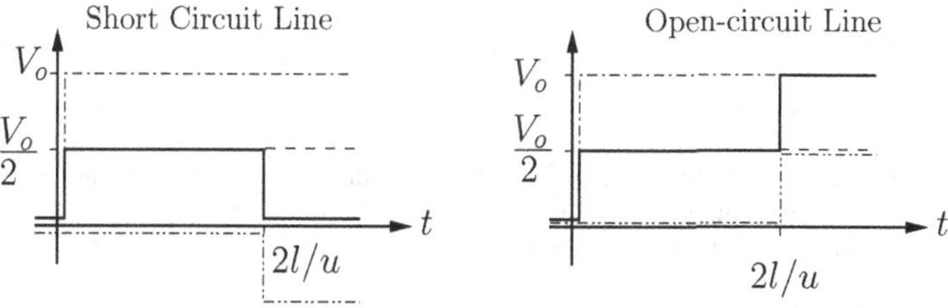

Fig. 10.4 Signals at the input of a transmission line of length l (*continuous line*) when terminated with short circuit (*left*) or with an open-circuit (*right*). Generator signal is a voltage step of amplitude V_o. The generator signal is drawn as *dot-dash line*, the forward signal is drawn as *dashed line* and the reflected signal id drawn as *two dots* and *dash line*

10.5.2 Multiple Reflections

When a line is not matched at both ends, the phenomenon of *multiple reflections* takes place. Consider again the circuit of Fig. 10.3 and assume that the internal impedance of the generator R_g is different from the characteristic impedance of the line $R_g \neq R_c$ and that the line is closed on a load with $R \neq R_c$. When a progressive wave reaches the load R at the line end, it will be reflected with a reflection coefficient $\rho_f = (R-R_c)/(R+R_c)$, while regressive waves will be reflected with a reflection coefficient $\rho_b = (R_g - R_c)/(R_c + R_g)$ when they reach the line input. With a voltage step generator $v(t) = V_o\theta(t)$ connected to the line, at $t = 0$ the amplitude of the forward signal propagating in the line is $V_1 = V_oR_c/(R_g + R_c)$. When this signal reaches the line termination, it is reflected with a reflection coefficient ρ_f; the amplitude of the reflected signal (propagating backward) is $V_2 = \rho_f V_1$, and, once it reaches the line input, this signal is again reflected with a reflection coefficient ρ_b and a new forward signal is generated with amplitude $V_3 = \rho_b\rho_f V_1 = \rho_b V_2$. For ideal lines (without energy dissipation), the signal will bounce indefinitely back and forth between the two terminations. The amplitude of the signal at a given point and at a given time is the sum of all signals present at that time in the location considered; the Fig. 10.5 shows a diagram useful for calculating this amplitude.

Figure 10.6 shows an example of multiple reflections in a line. Since the absolute value of ρ is always smaller than unity, the reflections occur at both ends of the line with progressively smaller wave amplitude.

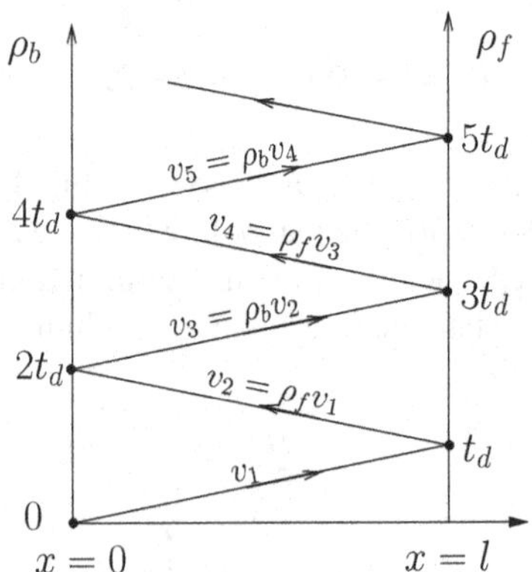

Fig. 10.5 Multiple reflection chart for an ideal line that is not matched at either end. ρ_b and ρ_f are respectively the reflection coefficients for waves traveling backward and forward. The *vertical axis* shows the time and on *inclined lines*, we show the amplitude of the traveling wave. Figure 10.6 shows the waveforms at $x = 0$ (line input) and at $x = l$ (line end) for particular values of ρ_b and ρ_f

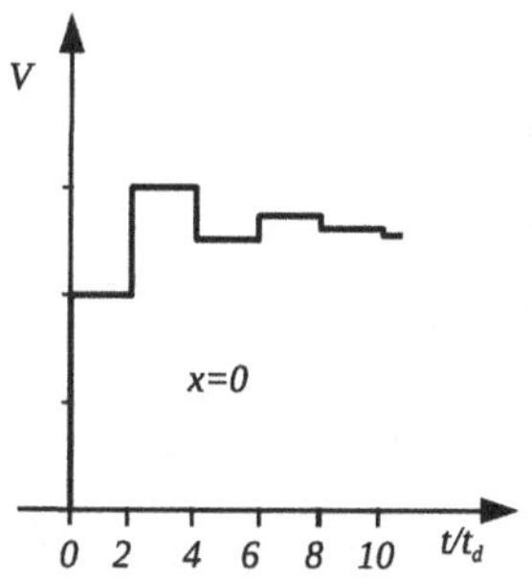

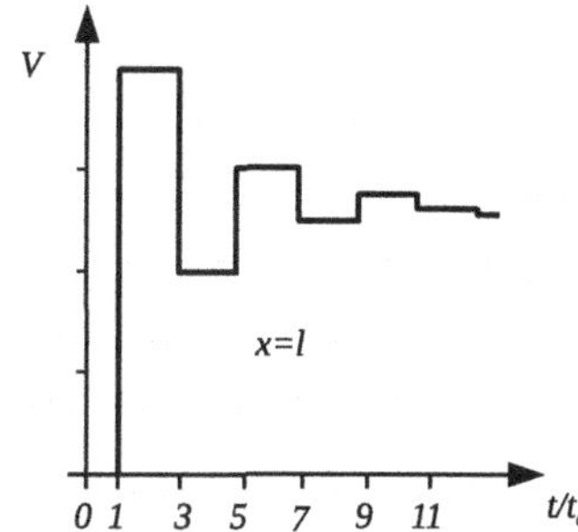

Fig. 10.6 Waveforms of signals at the ends of a not matched line when the input is a step function. At line input the resistance is $R_c/3$ with reflection coefficient $\rho_b = -1/2$ and at the line end termination is an open circuit with $\rho_f = 1$

10.6 Line Discharge

Consider an ideal transmission line uniformly charged at constant voltage V as shown in Fig. 10.7 with the switch S closed. In this condition, the current flowing along the line is $i = V/R_2$. If at $t = 0$ the switch is open, the line starts to discharge and when $t = 0^+$ at terminal 1 both voltage and current change to new values V' and i'. This results in a step discontinuity that propagates along the line with voltage and current amplitude respectively equal to $V_f = V' - V$ and $i_f = i' - i$.

When $t = 0^+$, in R_1 flows a current equal to V'/R_1. This current comes from the line where it flows in the opposite of the propagation direction. Therefore, $i' = -V'/R_1$ yielding $i_f = -V'/R_1 - V/R_2$. In the line we have $i_f = V_f/R_c$, so that

$$i_f = \frac{V_f}{R_c} = \frac{V' - V}{R_c} = -\frac{V'}{R_1} - \frac{V}{R_2}$$

Solving for V', one obtains

$$V' = V\frac{R_1}{R_2} \cdot \frac{R_2 - R_c}{R_1 + R_c} \quad \text{and} \quad V_f = -V\frac{R_c}{R_2} \cdot \frac{R_1 + R_2}{R_1 + R_c} \tag{10.27}$$

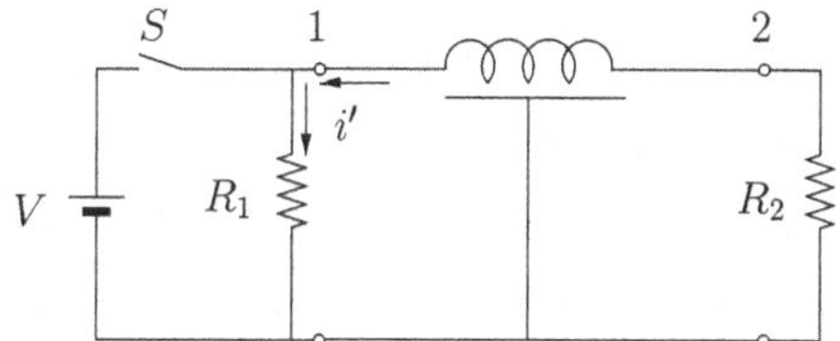

Fig. 10.7 Discharge of a transmission line. The figure shows the current flowing through terminal 1 immediately after the switch S is open

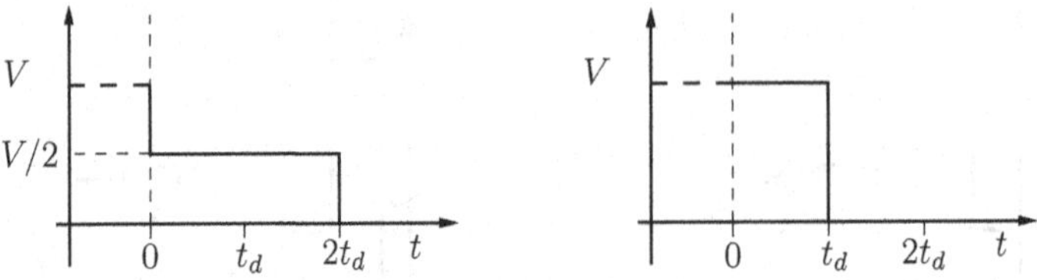

Fig. 10.8 Waveforms at both ends of a discharging transmission line

The V_f signal is reflected at the end of the line with a reflection coefficient $\rho = (R_2 - R_c)/(R_2 + R_c)$ and after a time $2t_d$ (t_d is the line delay time) the voltage at the line input is

$$V'(2t_d) = V + V_f + \rho V_f = V\left(1 - \frac{2R_c(R_1 + R_2)}{(R_1 + R_c)(R_2 + R_c)}\right) \tag{10.28}$$

When the line is matched in output ($R_2 = R_c$), Eq. (10.27) shows that $V' = 0$. This means that each point along the line is grounded when reached by the step propagating in the forward direction. In this case the reflected step vanishes. When the line is matched in input ($R_1 = R_c$), Eq. (10.28) gives $V'(2t_d) = 0$. This means that each point along the line is grounded only when reached by the reflected step. When the line is not matched at one of its terminals, its discharge requires a number of multiple reflections.[3]

As an example, assuming $R_1 = R_c$ and $R_2 = \infty$, we have $V' = -V_f = V/2$ and $\rho = 1$. The resulting voltage waveforms at the two line ends are shown in Fig. 10.8. At terminal 1, the voltage is immediately halved as the switch is open and vanishes when the reflected negative step reaches its location. On the contrary, terminal 2 is grounded when the forward negative step arrives and is reflected at the open line end.

10.7 Lossy Transmission Line

The real part of the propagation constant γ, which vanishes for a lossless line, becomes finite when the dissipative parameters R and g cannot be neglected. Recalling relations (10.16) and (10.17) we recognize that in a lossy line the signal attenuates while propagating along the line. The expression (10.13) of the propagation constant γ can be rearranged as

$$\gamma = \sqrt{(R + j\omega L)(g + j\omega C)} = j\omega\sqrt{LC}\sqrt{\left(1 + \frac{R}{j\omega L}\right)\left(1 + \frac{g}{j\omega C}\right)}$$

[3]The interested reader can find valuable information on this subject in Ref. [2].

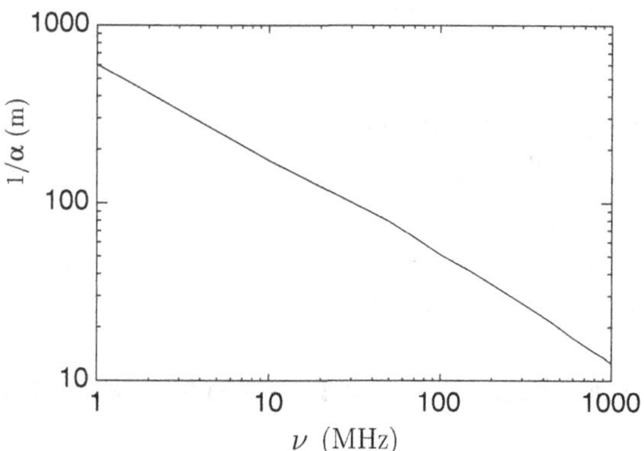

Fig. 10.9 Attenuation length of the coaxial cable *RG58C/U* as a function of signal frequency

where $j\omega\sqrt{LC}$ is the value of γ for lossless lines. In the high frequency limit, if $\omega \gg R/L$ and $\omega \gg g/C$ we can expand the second term of the γ expression in the variable $1/\omega$ neglecting the second-order term:

$$\gamma \simeq j\omega\sqrt{LC}\left[1 - j\frac{1}{2}\left(\frac{R}{\omega L} + \frac{g}{\omega C}\right)\right] = j\omega\sqrt{LC} + \frac{1}{2}\left(\frac{R}{R_c} + gR_c\right) \tag{10.29}$$

where $R_c = \sqrt{L/C}$ is the characteristic impedance of the line. The expression (10.29) seems to imply that at high frequency, i.e., for short pulses, the behavior of a real line deviates from the ideal one for the constant attenuation coefficient $\alpha = (R/R_c + gR_c)/2$. However, this is not the case in practice because in real lines the attenuation coefficient α is frequency dependent. This is due to the skin effect that progressively reduces the effective section of the conductors, and increases the value of the parameter R for increasing frequency. In addition, dielectric losses, and the related parameter g, typically also increase with the frequency. The Fig. 10.9 shows the dependence of the attenuation length[4] $1/\alpha$, as a function of the frequency for a cable type widely used in laboratories.

Line Attenuation Measurement. In this paragraph, we show how to measure the attenuation coefficient α of a real transmission line. Consider a line of characteristic impedance R_c and length l terminated with a short circuit. The line is fed with a voltage step generator of internal impedance $R_g = R_c$ (see Fig. 10.3 with $R_g = R_c$ and $R = 0$). If V_o is the step amplitude, at $t = 0^+$, as already shown before, the amplitude of the forward signal is $V_o/2$. After a time interval $2l/u$ the reflected signal reaches the line input and is added to the generator signal. The reflected signal, reversed in sign ($\rho = -1$), travels in the real line for a total length $2l$, so its amplitude is attenuated by a factor $e^{-2l\alpha}$. If ε is the voltage measured at the line input at the time $t = 2l/u$, then

[4]The attenuation length is the length of the line corresponding to a signal amplitude reduction by a factor e, the Neper number.

$$\varepsilon = \frac{V_o}{2}(1 - e^{-2\alpha l}) \simeq V_o \alpha l \quad \left(\text{for } t = \frac{2l}{u}\right)$$

where we used a power series expansion of the exponential limited to the first order in l. From the previous equation, we can easily obtain α:

$$\alpha = \frac{\varepsilon}{V_o l}$$

10.8 Transmission Lines in Sinusoidal Steady State

Consider a transmission line of length l connected to a sinusoidal voltage generator of internal impedance R_g equal to the characteristic impedance of the line. If the line is terminated with a load Z_L, the reflection coefficient, for forward traveling signals is $(Z_L - R_C)/(Z_L + R_C)$ and is zero for backward signals. In steady state,[5] the input (complex) impedance of the line is given by the ratio between the (complex) voltage present at the input and the input (complex) current.

The voltage at the line input is given by the sum of direct and reflected wave as follows:

$$v_{in}(x = 0, t) = V_f e^{j\omega t} + V_b e^{j\omega t - \gamma 2l} = V_f \left(e^{j\omega t} + \rho e^{j\omega t - \gamma 2l}\right) \tag{10.30}$$

Using Eq. (10.17), we get the input current as

$$i_{in}(0, t) = [v_f(e^{j\omega t} - \rho\, e^{j\omega t - \gamma 2l})]/R_c \tag{10.31}$$

and finally, we get the expression of the (complex) input impedance of the line as:

$$Z_{in} = R_c \frac{1 + \rho\, e^{-\gamma 2l}}{1 - \rho\, e^{-\gamma 2l}} \tag{10.32}$$

For a lossless line, $\gamma = j\beta = j\omega\sqrt{LC}$ and the (10.32) becomes[6]:

$$Z_{in} = R_c \frac{Z_L + jR_c \tan \beta l}{R_c + jZ_L \tan \beta l} \tag{10.33}$$

The previous expression is further simplified assuming that the load of the line is a short circuit ($Z_L = 0$); in this condition, the input impedance is

$$Z_{in} = jR_c \tan \beta l \tag{10.34}$$

[5] Steady state means that all the voltage and current transients are over.

[6] To obtain relation (10.33) it is useful to multiply numerator and denominator of (10.32) by $e^{+\gamma l}$.

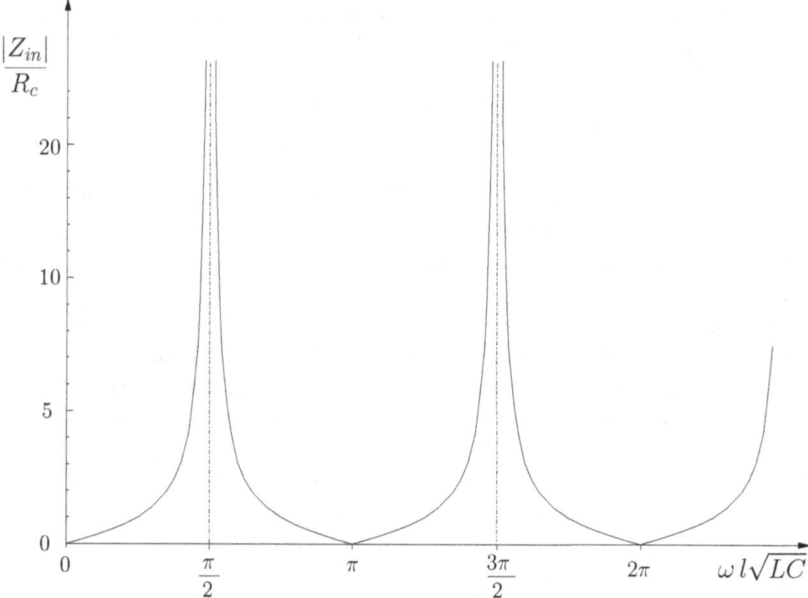

Fig. 10.10 Ratio of the amplitude of the input impedance of a shorted ideal line to its characteristic impedance, as a function of the dimensionless parameter $\omega l\sqrt{(LC)}$

The ratio Z_{in}/R_c as a function of $\omega\, l\sqrt{LC}$ (a sort of "reduced angular frequency") is shown in Fig. 10.10. Note that, in ideal conditions, this impedance varies between 0 and ∞. This behavior can be easily verified in the laboratory by measuring the input voltage amplitude as a function of the signal frequency. In particular, it is easy to verify that the impedance has a minimum for βl equal to integer multiple values of $\pi/2$. This behavior can be understood in terms of interference between the direct and the reflected signals at the line input. In fact, at this point the two signals have a phase shift equal to the value of $2\beta l$. The Eq. (10.31), with $\rho = -1$, shows that the input current vanishes for βl equal to an odd multiple of $\pi/2$ while the tension remains at a finite value, as seen from the Eq. (10.30). In this (ideal) condition, the impedance seen by the generator diverges. Conversely for βl equal to an integer multiple of $\pi/2$ and with $\rho = -1$, it is now easily shown, using again (10.30) and (10.31), that the input voltage is zero while the current remains at a finite value. In this (ideal) condition, the impedance seen by the generator is zero. Finally, it is easy to show that in case of real lines, the input impedance remains always finite and different from zero.

If the line termination is an open circuit ($Z_L \to \infty$), the input line impedance is $Z_{in} = R_c/j\tan\beta l$ and the role of zeros and divergence points is exchanged with respect to previous case when the line termination was a short circuit.

"Short line". Another interesting point on the input impedance of a line regards the case when $\beta l \ll 1$, i.e., when we can consider the line as a short line. In this limit, which can be applied to the majority of the cables length used in the laboratory, it is possible to make the approximation tan $\beta l \simeq \beta l$, so the Eq. (10.33) becomes

$$Z_{in} \simeq R_c \frac{Z_L + j\omega R_c \beta l}{R_c + jZ_L \beta l} = \frac{Z_L + j\omega Ll}{1 + j\omega Z_L Cl}$$

where we used the definitions of β and R_c. If the line termination is an open circuit ($Z_L = \infty$) its total input impedance becomes $Z_{in} = 1/j\omega Cl$, that is the impedance of the capacity of a cable of length l (remember that C is the capacitance per unit length of the line). Conversely, if the line termination is a short circuit ($Z_L = 0$) its total input impedance becomes $Z_{in} = j\omega Ll$, i.e., the impedance of the inductance of a cable of length l. Only in the case where $Z_L = R_C = \sqrt{L/C}$ the cable is "seen" as a resistance value of R_C, as it happens for matched lines. For values of $Z_L < R_C$ the line impedance can be interpreted as "inductor like" while if $Z_L > R_C$ the line impedance can be interpreted as "capacity like".

These results show that, when it is not possible to match the impedances of cables connecting different parts of a measurement apparatus, it will be necessary, especially if one works at high frequency, to maintain their length to a minimum to reduce the side effects caused by parasitic inductances or capacitances.

Problems

Problem 1 Compute the delay of a 100 MHz signal when is transmitted by a 10 m length ideal transmission line. It is known that the wavelength of this signal in the line is 1 m. [R $\times 10^{-8}$ s.]

Problem 2 The setup shown in figure is used to avoid reflections in splitting signals that propagate in transmission lines. Supposing the characteristic impedance of the line is 75 Ω, compute the value of R that avoid reflections. [A. $R = 25\ \Omega$.]

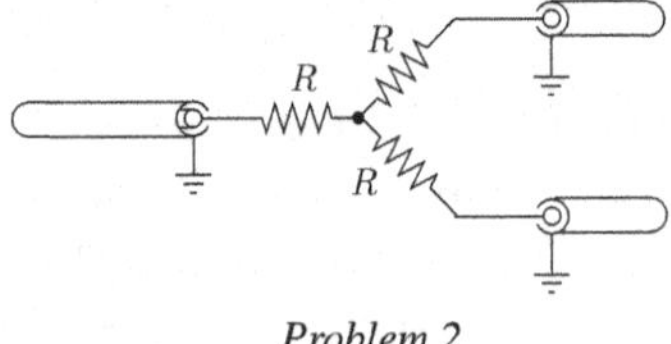

Problem 2

Problem 3 The setup in the figure can be used instead of that of the previous problem to avoid reflections in signals split. Compute the value of R assuming that the characteristic impedance of the lines is 75 Ω. [A. $R = 75\ \Omega$.]

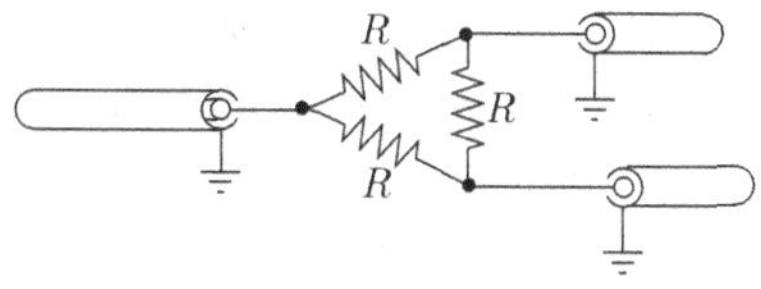

Problem 3

Problem 4 An ideal transmission line of length l and with characteristic impedance R_c is connected to another ideal line of length $l/2$ and with characteristic impedance $R_c/3$. The latter line ends with an open circuit. A voltage step generator $v(t) = V_o\theta(t)$ is connected to the former as shown in the figure. Compute the voltage waveform at the input of the first line, supposing that the propagation velocity in the two lines is the same. Assume that all lines are lossless.

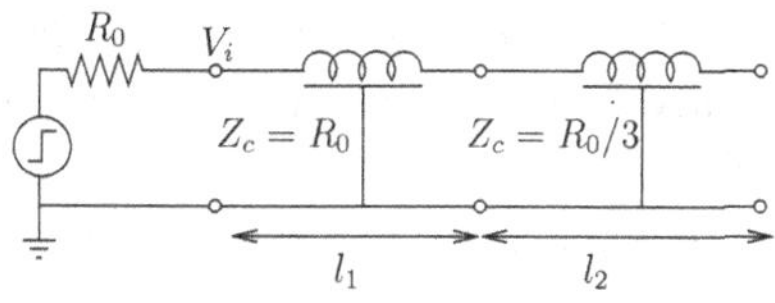

Problem 4

Problem 5 The "Blumlein line" is a device used to generate short high voltage pulses. The time duration can be few nanoseconds and the voltage may reach hundreds kilovolts. One particular realization of this device is shown in the figure. It consists of two parallel strips making a line of characteristic impedance R_c. The line is matched, i.e., is terminated with its characteristic impedance R_c. At the other end, a third strip is inserted, half way between the first two strips, for a fraction of the total length of the line. In this way, we obtain two lines of impedance $R_c/2$ and of the same length of the third strip. The central strip is charged at potential V and then is short-circuited, by means of a spark gap (SG in figure) that works like a switch. Determine the duration and the amplitude of the pulse on the resistance at the termination. [*Hint: the structure in the figure is equivalent to two lines of characteristic impedance $R_c/2$ connected in series through a resistor R_c*]

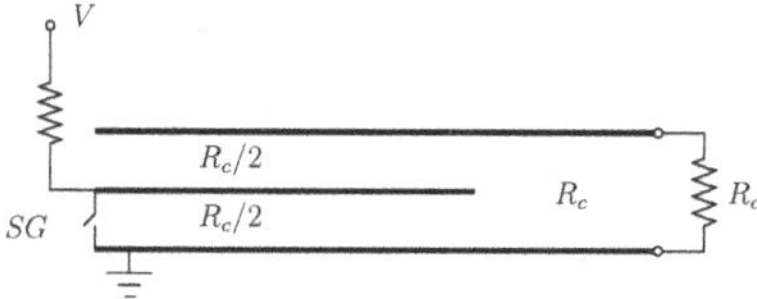

Problem 5

Problem 6 A pulse generator using a relay with contacts wetted in mercury was commonly used, until the 70's of the last century, to obtain voltage pulses with rise times of the order of one nanosecond. The schematic of the circuit is shown in the figure. To obtain a pulse, the relay contact is closed after charging the line on left at voltage V_g. Prove that the signal on the load R_c (equal to the characteristic impedance of the line) is a pulse of amplitude V_g and time width τl, τ being the specific delay of the line and l the length of the line charged at V_g. Suppose that $R_g \gg R_c$ and assume that the *relay with contacts wetted in mercury* is an ideal switch.

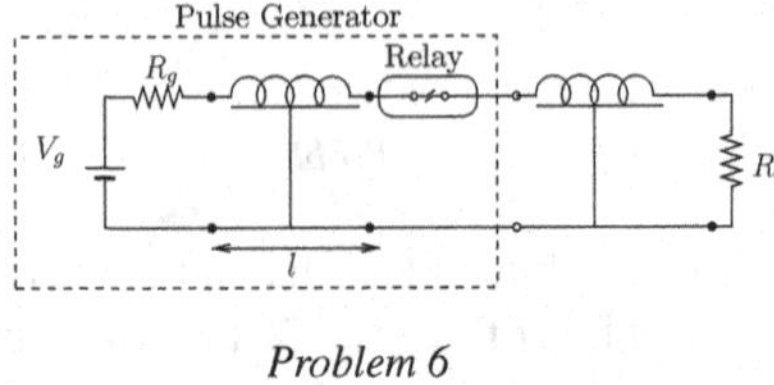

Problem 6

Problem 7 A voltage step generator is connected to the parallel of two cables of identical impedance R_L and specific delay τ. The first cable, shown on the left of the figure, is short circuited and its length is equal to l. The second cable, of arbitrary length, connects to a matched load R_L. Show that at this load we obtain a voltage pulse of duration $2\tau l$.

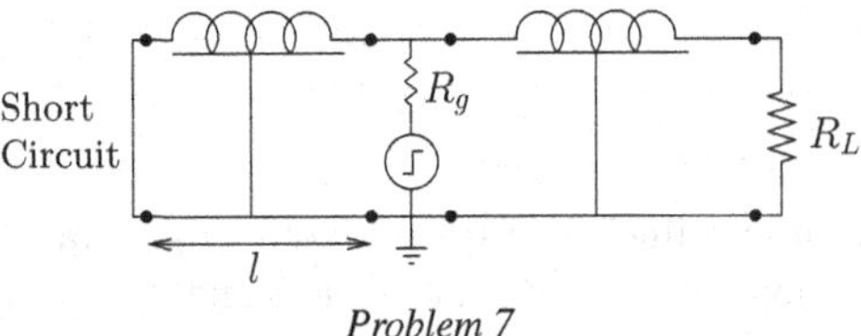

Problem 7

References

1. S.J. Orfanidiis, *Electromagnetic Waves and Antennas* (Rutgers University, New Jersey, 2002)
2. A.F. Peterson, G.D. Durgin, *Transient Signals on Transmission Lines* (Morgan & Claypool, San Rafael, 2009)

Appendix A
Laboratory Experiments

A.1 Introduction

In this appendix, we illustrate nine experiments that we have used extensively in our laboratory classes. They are designed as the necessary complement to the matters dealt with in the main text and they allow for the practical implementation of the many concepts that we deem students should internalize when attending the lecture classes.

The sequence and the content of these experiments have been designed keeping in mind a series of considerations. First, we required that in each laboratory session students should produce quantitative information about the relevant physical quantities, a fundamental requirement for a course aimed at teaching the principles of the measurement science.

Second, we devised a learning path whereby students begin by using simple instruments and by performing basic data analysis and conclude their laboratory experience exploiting advanced instrumentation and devising rather complex procedures of data handling. In this perspective, we started by using simple analog devices in order to introduce later in the sequence the use of the most modern digital instrumentation and to allow data transfer to a computer for further specialized elaboration.

The third consideration is perhaps the most important. We thought that it is important that students understand that it is never easy to obtain accurate results in experimental science. We wanted to make them aware that the measurement process can have by itself an important impact on the measured quantity, and the analog instrumentation, when still available, can be extremely useful to exemplify this concept in practice. Similarly, we wanted to make them able to minimize the impact of parasitic elements related to cables connecting the measured circuit to the measuring instrumentation. Finally, we wanted that students learn that there are always variables of influence that can have an important effect on the quantity of interest and that critical thinking is the most effective approach to gain control of this important aspect of experimental science.

R. Bartiromo and M. De Vincenzi, *Electrical Measurements in the Laboratory Practice*, Undergraduate Lecture Notes in Physics,
DOI 10.1007/978-3-319-31102-9

In the following sections, we illustrate the aim of each of these experiments, followed by a list of the necessary equipment. Then we give a plan of action that students should adopt to fulfill the requirement of each experimental session. Students should be required to write a report on each experiment adopting the proper style of a scientific publication. Finally, we give for each experiment a note that can be useful to tutors in the preparation of the experiment and in its illustration to students. We remark that these notes are written for experienced teachers and, therefore, they can turn out to be of little use for most undergraduate students without appropriate help from their tutor.

A.2 Experiment: Laboratory Instrumentation and the Measure of Resistance

Aim of the experiment: to gain confidence with electrical instrumentation and to learn how to evaluate correctly the effects of averaging on uncertainties.

Material available for the experiment:

- Analog multimeter
- Digital multimeter
- 100 resistors with identical nominal value
- Cables and connectors

Plan of the Experiment

Measure the resistance of 100 resistors with identical nominal value. For each resistor, measure the resistance value with both the analog and the digital ohmmeter. For the analog instrument, it is required to interpolate its reading between the divisions of its ruler. Collect the values and their uncertainty in a spreadsheet. Use the data to perform the following tasks.

1. Compare the two averages obtained from data collected with the digital instrument and data collected with the analog meter.
2. Verify the compatibility of these two values taking into account the uncertainties provided by the user manuals of the instruments.
3. Build a histogram of the resistance values measured with the digital instrument. Check compatibility with the nominal value of resistors taking into account its uncertainty as stated by the color code.
4. Build a histogram of the difference between each analog value and the corresponding digital measurement. Calculate the estimated standard deviation of the distribution. Discuss the origin of the dispersion.
5. In this experimental session, your colleagues are measuring the same resistors with different instruments. Collect the average values obtained by them and use these data to

- obtain a more accurate estimate of the average value of the 100 measured resistances and
- verify the compatibility of your analog result with the class of the instrument to assess if a new calibration is needed.

Notes to the Tutor

In this first experimental session, students must gain confidence with instrumentation and learn how to evaluate the effects of averaging on uncertainties. The tutor will introduce them to the electrical laboratory instrumentation: resistors, and the color code to read their resistance value, solder-less breadboards to mount circuits and cables to connect them to power supplies and measuring instruments.

Then he will illustrate the use of measuring instruments, namely an analog and a digital multimeter. For the analog instrument, he will illustrate the use of the mirror for compensation of parallax error and will discuss the need to interpolate the reading between the divisions of the graduated scale. For the digital instrument, he will discuss the two contributions to the uncertainty, namely the calibration factor and the quantization error, and the different nature of their correlations. The tutor should explain that in digital instruments in general, uncertainties of type B due to calibration are different from those due to quantization. These two contributions are not correlated with each other and should be added in quadrature.

Students must learn to consult the user manual of each instrument to evaluate features and capabilities, identifying formulas and parameters needed to assess the uncertainty of measurement. They will configure them for use as ohmmeter and control that the instrumental zero is properly set by measuring a short circuit.

In the preparation of this experiment, the tutor must choose a nominal value of the resistance such that the nonlinear scale of the analog instrument is used in the low resistance end so that the distance of its divisions is sufficiently wide to allow for a visive interpolation between them. For the execution of the measures, resistors can be mounted in groups of ten on breadboards with ten resistors each. Rotating them among students, one can obtain multiple measurements of the same resistances with different ohmmeters. Each student will measure with both the analog and the digital instrument.

By making sure that each student measures at least a hundred resistors, after data collection each student should perform the following analysis:

- Calculate and compare the two mean values using digital or analog data. Evaluate the uncertainty on these averages taking into account the correlation between the different components of the uncertainties of the individual measures. After completing this task, students should have understood that, since the experimental uncertainty is strongly correlated, in first approximation the relative uncertainty of the average is equal to the relative uncertainty of the single measurements.
- Verify the compatibility of the average values obtained with the two instruments, taking into account the uncertainties supplied by the manufacturers of the instruments (type B uncertainties).

- Build a histogram of the values of the resistances measured with the digital instrument, which usually present a smaller uncertainty. Assess their compatibility with the accuracy of the nominal value of the resistance provided by their manufacturer, typically 5 %. Discussing the shape of the histogram obtained, explain that the resistance of a resistor depends on the setting of the machine that produced it. If care is taken to avoid choosing all resistors from the same batch, the histogram has more than one peak.
- Build a histogram of the difference between digital and analog measurement of each resistor and calculate the mean and standard deviation of the estimate. Discuss the shape of the histogram (if well done, it will be a bell curve that resembles a Gaussian) and the origin of the dispersion (reading error manly of the analog instrument, which are random and not correlated, and generally better half of the difference between the divisions on the graduated scale). Students should realize that by visual inspection they could interpolate much better than half the division spacing. In absolute terms, a skilled eye can distinguish a thickness with an uncertainty better than 0.1 mm.
- Compare the average values obtained by different students to verify that none of the instruments used requires a calibration check. Usually digital instruments maintain calibration over time and the distribution of the observed values falls within the manufacturer's specifications. Therefore, from these measurements one can get a more accurate estimate of the average resistance using all available values since now they are not correlated. An analysis of the difference in analog measurements with this value is now an accurate test to identify any instrument out of specification.

A.3 The Voltmeter–Ammeter Method and Ohm's Law

Aim of the experiment: to learn how to implement the voltmeter-ammeter method for measuring resistance values and to validate Ohm's law for two different types of conductors.

Material Available:

- DC voltage generator ($V = 0 \div 30$ V).
- Digital multimeter and/or analog multimeter (two instruments) for measuring voltage, current and DC resistance.
- Solder-less breadboard for assembling the circuit. Resistors of different values.
- Light bulb (rated at 5 V).
- Wires for connecting the components of the circuit.

Plan of the Experiment

1. Choose a resistor based on the maximum allowed power dissipation and the maximum voltage planned for the experiment. Use a large safety limit and check that the resistor remains cold when the maximum voltage is applied.

2. Choose the instrument to use as voltmeter and explain the motivation of your choice.
3. Assemble on the breadboard the circuit with the ammeter "upstream" of the voltmeter and perform a series of measurements of the voltage drop V across the resistor and the current in the circuit for a series of predetermined values of the generator output voltage V_0.
4. Calculate initially the resistance value with its uncertainty from a single pair of measurements and assess the extent of possible systematic effects.
5. Make a plot of the values of the voltage drop V as a function of the current I flowing in the resistor (after making the correction for systematic effects, if needed).
6. Apply the weighted least squares method to fit a straight line $y = mx + q$ to the data obtained. Choose whether current or voltage should be represented with x and justify your choice. Calculate the parameters, and their uncertainties, that best fit the results. Compare the value and the uncertainty of the resistance measured in this way with those obtained in Step 4 and comment on the results.
7. Comment on the value of the offset q. Is a value of q different from zero acceptable for an ohmic conductor? Is your result compatible with the hypothesis $q = 0$?
8. Replace the resistance with the light bulb and repeat the measurements of voltage and current. Carry out the measurements by increasing the values of V_0 until the bulb becomes incandescent, paying attention to avoiding burning it. For each experimental point, wait for the measurement to stabilize (the bulb must reach thermal equilibrium with the environment). Draw a graph of the values of the voltage as a function of current and comment on the result.

Notes to the Tutor

The experiment consists in determining the voltage-current characteristic of a resistor and an incandescent lamp.

Preliminarily, the students must choose the configuration of instruments. In general, they should perform the measurement with an ammeter in series with the resistor and the voltmeter in parallel to it. Discuss the reasons for this choice keeping in mind that the resistance of the ammeter can change depending on the used range and that digital voltmeter presents quite high internal resistance. Explain to the students that, because of internal resistance, the voltage meter of the power supply does not measure the voltage drop across the resistor.

Determine the maximum power that can be dissipated by the available resistors and measure the maximum voltage that can be supplied by the DC voltage generator. The value of the resistance must allow that measures extend up to the full scale of the analog ammeter where the relative uncertainty is smaller. If P_{max} is the maximum allowable power dissipation with appropriate safety factor (as it will be evident at the end of the experiment, we must avoid heating the component) and V_{max} is the maximum voltage available, the full scale of the ammeter I_{fs} should be chosen so that $I_{fs} \cdot V_{max} < P_{max}$. At this point, the optimal value of the resistor for the validation of Ohm's law is equal to V_{max}/I_{fs}.

Students must perform the measurements by varying the applied voltage up to the maximum possible: discuss the choice of the number of measurement points and their separation, paying attention to the need to determine with good accuracy the value of the intercept with the currents axis for Ohm's law validation.

After evaluating the uncertainties of measured values of currents and voltages, students must fit a generic straight line to their data points using the least squares method without weighting for errors. If data points have very different uncertainties, discuss the need for taking into account the errors on the ordinate. In this case, discuss the criteria for choosing the quantity to use as ordinate. Evaluate slope and intercept with related uncertainties. Use the slope for the evaluation of the resistance value and its uncertainty. Discuss possible corrections due to the impedance of the voltmeter. Compare the value found with the nominal value and the value measured with digital multimeter.

Discuss the significance of the intercept by comparing it with its uncertainty. If the measurement was carried out with care, in general the value of the intercept is well compatible with the zero. If it does not, probably low voltage data points were either too few or too inaccurate. It is also possible that not enough attention was paid to the need to avoid heating the resistor.

In the second part of the experiment, students will exchange the resistor for the light bulb and increase gradually the applied voltage so it becomes incandescent, being careful not to burn it. After deciding the number of measurement points, they will measure the voltage-current characteristic. They should be brought to identify the nonlinearity and discuss the cause. Once the students realize that it is extrinsic nonlinearity caused by the temperature change, they will perform measurement again with a number of data points adequate to document the nonlinear behavior and allowing for time to reach thermal equilibrium at each change of voltage to obtain a reproducible result.

Possibly, require students to derive an estimate of the temperature of the filament from the resistance measurements carried out and the temperature coefficient of tungsten resistivity. Seize the opportunity to introduce Wien's displacement law for blackbody radiation.

A.4 Experiment: Resistivity Measurements

Aim of the experiment: to measure the resistance of different samples of graphite mixtures as a function of their length to derive their resistivity, with its uncertainty, from the second Ohm's law.

Material available:

- Digital Multimeter
- Cables and terminals
- Caliper
- Samples of graphite mixtures (pencil leads)

Plan of the Experiment

1. Connect the ohmmeter to the sample through a fixed and a sliding contact.
2. For each of the available samples, measure the value of resistance R as a function of the distance l between contacts along the axis of the cylindrical pencil lead.
3. Use the digital multimeter for the resistance measurement and the caliper for the measurement of l. For the interpretation of the experiment remember that the resistance of a conductor is directly proportional to its length l, and inversely proportional to its cross sectional area S.
4. Construct the plot of R as a function of the length; evaluate the resistivity from the slope of the best fitting straight line and the diameter of the sample as measured by the caliper.
5. Comment on the results obtained comparing resistivity and hardness of samples.
6. Discuss the values obtained for the intercept, its significance, and its possible origin.

Notes to the Tutor

Aim of the experiment is to learn how to measure the coefficient of electrical resistivity using a simple setup. After completing this work, students should have realized how they avoided important systematic effects obtaining the required quantity through a difference measurement.

The experiment can be done with samples obtained using leads for pencil of varying hardness. Pencil leads are made of a mixture of graphite and clay in which this last component, the more resistive, increases in percentage with the hardness. The instruments to use are a digital ohmmeter to measure the resistance and a caliper for measuring the length and diameter of cylindrical pencil leads (2 mm diameter is a good choice).

For contact between the sample and the ohmmeter, the brass contacts of a cable joiner strip can be used.

Students must first carry out measurements on the same pencil lead at different lengths. They should realize the presence of systematic effects comparing the value of the resistivity obtained from different lengths of sample. Next, they should realize that using the difference between two measurements with different lengths, reliable results can be obtained. The source of systematic effect should be identified in the contact with the sample: extended contacts lead to poor definition of the sample length and possibly intrinsic contact resistance.

At this point students should be encouraged to measure pencil leads of different hardness. They will perform a linear fit of the results to evaluate the slope and the intercept with related uncertainty. They will obtain the resistivity of each sample, with its uncertainty, using the slope of the fit and the value of the sample diameter as measured by a caliper. The relation between resistivity and hardness should become apparent.

Finally, students should concentrate on the analysis of intercepts of the fitting lines. They should realize that they are significantly different from zero and that their value correlates with the sample resistivity. This observation lends support to the hypothesis that their origin is due to a poor definition of the contact localization producing an offset in the length measurements. However, the tutor should make students aware of the existence of non-ohmic contact resistance that possibly contributes to the intercept value.

In the discussion of these experiments, the tutor should explain to students how the impact of the contact localization and contact resistance on the measurement can be made irrelevant. This requires using 2 power contacts to inject current in the sample and 2 sensing knife contacts, well defined and placed inside the power contacts, to measure the voltage drop with a high impedance voltmeter. This configuration would allow the use of a fixed length of sample.

A.5 Experiment: Measurement of the Partition Ratio Along a Chain of Resistors

Aim of the experiment: to assess the disturbance induced by a voltmeter on the measured voltage.

Material Available:

- DC voltage generator ($V_0 = 1 \div 30\,\text{V}$).
- Analog voltmeter.
- Digital multimeter for measuring voltage and resistance.
- A resistor chain of 10 elements.

Plan of the Experiment

1. Measure the resistance value of individual resistors
2. Power the resistor chain with a voltage of 10 V and measure the voltage along the chain using the analog voltmeter with a full-scale range of 10 V
3. Power the resistor chain with a voltage of 2 V and measure the voltage along the chain using the analog voltmeter with a full-scale range of 2 V
4. Use the results obtained to calculate the partition ratio as a function of the resistor number and compare with their unperturbed values.
5. Use the analog voltmeter impedance to correct the systematic effect observed. To this purpose, assume that all resistances in the chain are equal to the average of their measured values with uncertainty equal to the standard deviation of their measured distribution.

Notes to the Tutor

With this experiment, students have the opportunity to assess an example of the disturbance induced by measuring instruments on measured parameters.

The experiment consists in measuring the partition ratio of the electrical voltage along a chain of N resistors of equal resistance as a function of the order number n that distinguishes the single resistor. The expected value, neglecting the small fluctuation of the value of individual resistances, is equal to the ratio n/N.

In preparation of the experiment, choose resistors such that the resistance of the chain is comparable to internal resistance of the analog voltmeter ($100\,\mathrm{k}\Omega$) but small compared to that of a digital voltmeter ($10\,\mathrm{M}\Omega$).

During the experiment, the students will measure the partition ratio for two different supply voltages in order to use the analog voltmeter with two different full-scale ranges, and hence with two different values of the internal resistance. Having found that the two measures of the partition ratio do not coincide with each other and that none of them coincides with the expected outcome, students can repeat the measurements with a digital voltmeter and obtain a measure consistent with expectations.

Once the cause of the discrepancy is identified in the finite resistance of the analog voltmeter, students will proceed with the necessary analysis to take into account its effect. For this purpose, they can assume that all resistances in the chain are equal to the average of their measured values with uncertainty equal to the standard deviation of their measured distribution. Then they can apply Thevenin's theorem and obtain the correct expression for the expected voltage in the presence of the connection to the voltmeter, see Chap. 4, Problem 14.

At this point, they can use this result in two different ways:

- Recover the voltmeter internal resistance R_v from the user manual and correct the measured voltages to estimate the unperturbed values.
- Use the difference between the unperturbed and the measured partition ratios to derive N−1 estimates of the resistance R_v with the relative uncertainty. Note that, as it is intuitive, this measure is less uncertain when the perturbation is larger and, consequently, take the weighted average of the $N-1$ estimates to obtain the most accurate value for R_v with its uncertainty.

A.6 Experiment: Characterization of RC Filters

Aim of the experiment: to measure the transfer function, amplitude and phase, of a low-pass and a high-pass RC filter.

Material Available:

- Digital oscilloscope with voltage probes
- Waveform generator
- Resistors and capacitors
- Connecting cables
- Solder-less breadboard

Plan of the Experiment

1. Design an RC filter with cutoff frequency of the order of 1 kHz, choosing its components in such a way as to minimize the impact of the output impedance of the waveform generator and of the input impedance of the oscilloscope, or the probe used to connect to it, during the measurements of the transfer function.

2. Measure the value of resistance and capacitance of the two selected components.
3. Assemble the low pass filter and measure attenuation and phase of its transfer function as a function of the frequency of the input sinusoidal signal. Choose the value of the input voltage taking into account the presence of ambient noise and the full-scale range of the oscilloscope. Choose the number of data points and their spacing to optimize the information needed to accomplish the task in the next step. Evaluate uncertainties for all attenuation and phase data points.
4. Determine the value of the cutoff frequency, and its uncertainty, from both attenuation and phase data. Compare with the value obtained from measured values of resistance and capacity.
5. Repeat function transfer characterization for the high pass filter and compare with theoretical expectation computed using the data on frequency cutoff obtained in the previous step.

Notes to the Tutor

In this experimental session, students must first become acquainted with the use of an oscilloscope and a waveform generator. The tutor should first explain how to perform simple operations with these two instruments.

The tutor should discuss the systematic effect of the output impedance of the voltage generator, of the oscilloscope input impedance, and of the stray capacitance of cables used to connect to it. As an alternative to such cables, the use of a compensated probe can be illustrated.

The tutor should help students to choose among the different voltage amplitude measurements provided by a digital oscilloscope. Peak-to-peak amplitude can be used when the signal is much higher than ambient noise. Otherwise, effective amplitude should be preferred.

The tutor should also show how to use a digital oscilloscope to measure the phase delay between two sinusoidal signals, see Sect. 8.4 in Chap. 8.

The tutor should discuss the uncertainty of the measurements of amplitude and time obtained via the oscilloscope and their propagation on the measurement of the attenuation and phase.

For the design of the filter, its resistance R must be sufficiently higher than the output resistance of the waveform generator, usually equal to 50 Ω, and sufficiently lower than the input resistance of the oscilloscope, usually of the order of 1 MΩ. A value of R in the range 1–10 kΩ is therefore adequate.

For the characterization of the filter transfer function, students must understand that they need to plan the number of data points taking into account that a Bode plot should be drawn in logarithmic scale. They will be led to think in decades and to space data point accordingly.

The cutoff frequency of the filter corresponds to an attenuation $1/\sqrt{2}$ or to a phase delay of 45 degrees. The student should be encouraged to find these value interpolating between suitable data points. For optimal results, the acquisition of new data points may be required. In the evaluation of the cutoff frequency by interpolation, students must take into account that the uncertainties of the attenuation measurements are correlated while those of phase measurements are uncorrelated.

Students should compare these two determinations of the cutoff frequency with the value given by the product of the resistance and capacity values. A best estimate of this quantity should then be obtained as the weighted average of the three available results. This value will be used to compute the transfer function of the high pass filter built with the same capacitor and resistor. A comparison of this function with experimental data will be used as a validation check of the work done before.

A.7 Experiment: Characterization of an RLC Series Resonant Circuit

Aim of the experiment: to measure the transfer function, amplitude and phase, of a passband RLC filter.

Material Available:

- Digital oscilloscope with voltage probes
- Waveform generator
- Two resistors of nominal resistance 470 and 4.7 Ω
- A capacitor of nominal capacity 10 nF
- An inductor of nominal inductance 10 mH
- Digital multimeter
- Connecting cables
- Solder-less breadboard

Plan of the Experiment

1. Measure of components parameters, possibly with a vectorial bridge
2. Compute the expected value of the resonance frequency
3. Compute the quality factor of the circuit for the two values of the available resistors
4. Using computed values, choose an adequate number of frequencies for the measurement of the transfer function
5. Measure amplitude and phase of the transfer function for the two values of the available resistors
6. Measure an accurate value of the resonant frequency from the phase of the transfer function
7. Measure the value of the quality factor for the two resistors
8. Compare measurements with expected values and comment on the results

Notes to the Tutor

In this experiment, very accurate measurements can highlight a number of small parasitic effects. Therefore, the tutor will advise students to use an ×10 probe and will show how to compensate it. Moreover, he will suggest using a single channel with external trigger for the best accuracy of phase measurements.

Students will measure first the value of resistances, capacitance and inductance and evaluate their uncertainties. The tutor should make sure that the capacitors used have a low thermal coefficient, to avoid drifts of the resonance frequency when students touch components.

For the measurement of the transfer function, the notions learned in the previous session must be used.

The resonant frequency is best measured as the point corresponding to the null phase, obtained by means of an interpolation procedure between two appropriate measured values.

The resonant frequency measured in this way is usually compatible with the value obtained from components parameters. However, the values measured with the two different resistors may not turn out compatible among them if an accurate measurement is performed and if the inductor has a ferromagnetic core. However, changing the input voltage to make the current in the coil at the resonance equal in the two cases, the nonlinear response of this core can be made irrelevant and a good matching of the two frequencies recovered.

After measuring the resonant frequency, students should check that it corresponds to the maximum of the transfer function amplitude.

At this point, they can measure the quality factor identifying, on the two sides of the resonance, the frequency value corresponding to a reduction of this amplitude of a factor $\sqrt{2}$.

Phase measurements can be used to obtain quality factor from the two frequencies corresponding to a phase shift of $+45$ and -45 degrees. The phase and amplitude determination of the quality factor are in general compatible but they can be different from the value expected from component parameters, at least when the lower resistance is used, unless the series resistance of the inductor and the capacitor are taken into account.

Students should compare the value of the attenuation at the resonance obtained with the two resistors and come up with an explanation for their difference. They should link this observation with the findings on the quality factors.

If a vectorial bridge is available, the tutor will encourage students to characterize the components with both real and imaginary parts of their impedance. The frequency dependence of these values should be remarked, assuming the available instrumentation allows for it.

For an advanced version of this experiment, see Ref. [1].

A.8 Experiment: Study of Voltage Dividers

Aim of the experiment: to measure the transfer function (amplitude and phase) and the response to a voltage step of different kinds of voltage dividers.

Material available:

- Digital oscilloscope with voltage probes
- Waveform generator
- Resistors and capacitors

- Connecting cables
- Solder-less breadboard

Plan of the Experiment

1. Build a voltage divider with an attenuation of 20 dB, using for the grounded impedance a 470 Ω resistor in parallel to a 33 nF capacitor.
2. Choose for the remaining impedance a resistor in parallel to a capacitor to obtain respectively a compensated, over-compensated, and under-compensated divider. For the three cases, measure the attenuation and the phase delay as a function of frequency for sinusoidal signals.
3. Using a rectangular pulse of suitable duration, document the step function response for the three cases and comment on the results.

Notes to the Tutor

When we take into account stray capacitances, a voltage divider is characterized by the low frequency attenuation given by the resistive partition ratio, and the high frequency attenuation given by the capacitive partition ratio. These two values are in general different. In a compensated divider, the high frequency attenuation is made equal to the low frequency value. This leads to the well-known relation among components value, see main text, Chap. 9.

In this laboratory session, the students will work with three different dividers with different frequency response. They will characterize them in the frequency domain and compare with theoretical predictions as given in Chap. 9 of the book.

They will then move to the time domain to observe the divider's response to a step function. The tutor can use the experimental findings to introduce them to the time-frequency duality. He will point out the two phenomena of overshooting, when the circuit response is higher for high frequency components of the input signal, and of undershooting, when the opposite is true.

In these experiments, it is possible to identify the sharp pulse due to the stray inductance in series with the capacitors as described in Sect. 9.10 of the main text.

A.9 Experiment: Study of RC Circuits in the Time Domain

Aim of the experiment: to measure the time constant of an RC circuit from its response to a step function and to demonstrate the use of the same circuit as an integrator or, after the inversion of its components, as a differentiator.

Material Available:

- Digital oscilloscope with voltage probes
- Waveform generator
- Resistors and capacitors
- Digital multimeter
- Connecting cables

- Solder-less breadboard

Plan of the Experiment

1. Build an RC circuit in the low-pass configuration with a characteristic time of 1 ms, choosing its components in such a way so as to minimize the impact of the output impedance of the waveform generator and of the input impedance of the oscilloscope, or of the probe used to connect to it.
2. Select from the waveform generator a unipolar pulse of duration suitable to study both the charge and the discharge of the capacitor.
3. Measure the input and output signals with the digital oscilloscope over a time span adequate to the determination of the time constant of the circuit. Read the data from the oscilloscope with a personal computer.
4. Linearize the time response of the circuit and recover the time constant from a linear fit from both the charge phase and the discharge phase of the capacitor. Compare the two values obtained in this way.
5. Use the best determination of the time constant and evaluate theoretically the maximum duration of a rectangular input pulse to obtain in output its integral with an error lower than 3 %. Verify your finding with an appropriate measurement.
6. Change the input pulse from rectangular to sawtooth and document the circuit response. Comment on the result.
7. Modify the circuit in a high-pass filter using the same components. Use a triangular input signal in a range of parameters where it works as a differentiator, and document that:

    - for the constant duration, the output signal amplitude is proportional to input amplitude;
    - for the constant input amplitude, the output signal amplitude is inversely proportional to the duration of the input.

Notes to the Tutor

In this experimental session, the students will build and study an RC integrator circuit with assigned characteristic time. They must select suitably the resistance R, taking into account the output impedance of the signal generator and the input impedance of the oscilloscope used to measure the voltages, and consequently the capacitor C.

Students will measure the response function of the circuit to a unipolar pulse of sufficiently long duration to achieve complete charge and discharge of the capacitor. The data measured for input and output voltage will be imported from the oscilloscope to the computer for data analysis.

The theoretical expression for the output voltage needs to be linearized through an appropriate logarithmic transformation to obtain the characteristic time by a linear fit possibly weighing data with their uncertainties. Note that the uncertainty of transformed data can become very large toward the end of the capacitor charge (or discharge). A criterion should be worked out to exclude them from the fitting range.

In this experimental session, it is important to discuss the existence of an offset in the response of the analog-digital converter. It can be corrected through the acquisition of the signal when the circuit input is left open (this works in the presence of a little noise, otherwise the offset measurement should be made with a zero-mean noise generator or by measuring a small periodic signal over an integer number of periods). With a good determination of the full charge voltage, the values obtained for the characteristic time for the charge and the discharge of the capacitor will be compatible between them.

Note that this experiment lends itself to a detailed discussion of the uncertainties in the measurement of voltages with an analog-digital converter. In particular, it will be possible to distinguish the contribution of digitization from that of the overall calibration and from that of the differential nonlinearity of the converter and the nonlinearity of the oscilloscope input amplifiers. The study of the residues of the fit, insensitive to the calibration integral, can be used to obtain an evaluation of the importance of the nonlinearity with respect to digitization if the ambient noise is made negligible.[1]

In the second part of the session, students need to compute the maximum useful pulse width for which the RC circuit provides in output the integral of the input with a relative error lower than an assigned value. This requires the solution of a transcendental equation that can be solved numerically in various ways (for example by the method of Newton).

The result will be tested experimentally by integrating a square wave of the required width. Finally, the circuit will be used with the same pulse duration for integrating a ramp obtaining a parabola at the output.

In the last part of the experiment, the two components of the filter will be inverted to observe the response of a differentiator. Triangular input pulses with duration longer than the circuit characteristic time must be selected to obtain a rectangular output. The student will document that amplitude of the output is proportional to the input derivative.

A.10 Experiment: The Toroidal Transformer

Aim of the experiment: to measure the transfer function, amplitude and phase, of a real transformer and to document the hysteresis loop of its magnetic core.

Material Available:

- A ferrite toroidal transformer with given number of turns for its windings
- Digital oscilloscope with voltage probes
- Waveform generator
- A resistor of nominal resistance 10 Ω
- Digital multimeter
- Connecting cables
- Solder-less breadboard

[1]This is an advanced topic.

Plan of the Experiment

1. Use a sinusoidal waveform of variable frequency to measure the transfer function, amplitude and phase, of the transformer leaving its secondary winding open and feeding its primary directly with the function generator.
2. Explore the range of frequencies from a few kilohertz to a few hertz and plot the transformation ratio as a function of the frequency for two different amplitudes of the input voltage (first 100 mV and subsequently 10 V). If necessary, use the oscilloscope in the averaging mode to increase the signal-to-noise ratio.
3. Document carefully the waveforms, with special attention to the case of low frequency and high input voltage. Describe and comment your results.
4. Subsequently, connect in series to the primary a 10 Ω resistance and, instead of the applied voltage, measure the signal at its terminals with the oscilloscope to obtain a measurement of the current flowing in the primary.
5. Compute the H field amplitude in the toroidal solenoid from the circulating current and use the voltage at the terminals of the open secondary winding to obtain the B field amplitude. For this purpose, use the digital data transferred to a computer from the memory of the oscilloscope to integrate numerically the voltage signal of the secondary winding. Plot the hysteresis loop of the transformer core.

Notes to the Tutor

For this experimental session, it is necessary to prepare toroidal transformers with ferrite core. In the design phase of these components, you need to know the value of the magnetic field H required to saturate the magnetic material. For the toroidal core, it is advisable to use a ceramic ferrite with the widest hysteresis loop available.

The number of turns, equal for both windings, must be chosen taking into account the output impedance and voltage of the function generator, in order to obtain the saturation of the magnetic core at low frequency with the available maximum input voltage.

In the first part of the session, students will measure amplitude and phase on the secondary as a function of the frequency taking care to use a low voltage on the primary to avoid the nonlinearity of the ferrite. A constant transformation ratio should then be observed, decreasing toward the low frequency range where the resistance of the primary is no longer negligible with respect to its reactance. In correspondence, the onset of a phase shift between primary and secondary voltage should be detected. At sufficiently high frequency, a reduction of the transformation ratio could be observed due to the effects of inter-turn capacity.

In the second part of the experiment students will increase the voltage on the primary winding to a value that makes visible the nonlinearity of the ferrite core. They will document and describe the distortions observed.

In the last part of the experiment, students will change the circuit on the primary winding by inserting a 10 Ω resistance to obtain a signal proportional to the current flowing in the primary, and therefore the H-field in the ferrite. This signal will be digitized by the oscilloscope simultaneously to the voltage signal induced on the secondary, proportional to the derivative of the B field in the ferrite.

Transferring data from the oscilloscope to a digital computer, students can evaluate B by numerical integration of the signal on the secondary and obtain a graph of the hysteresis loop by plotting B as a function of H.

Special attention must be paid to cancel the offset of the analog-digital converter. This can be done via hardware, subtracting to the secondary voltage a measurement of white noise, if enough is available. Alternatively, it can be done via software, adding to the secondary voltage an increasing fraction of the last significant bit prior to integration until the hysteresis loop closes upon itself.

The symmetry of the hysteresis loop must be exploited to find the initial value of the B field in the numerical integration.

Reference

1. R. Bartiromo, M. De Vincenzi, AJP **82**, 1067–1076 (2014)

Appendix B
Skin Effect

To calculate the distribution of AC current inside a conductor, it is necessary to abandon the simplification that, considering the circuit components dimensionless, led to the formulation of the two Kirchhoff's laws. Instead, we need to use directly Maxwell's equations to solve the problem.

Consider a current flowing in a homogeneous conductor of resistivity ρ. If the current oscillation frequency is low enough to neglect displacement current, Ampère's law allows writing

$$\nabla \times \mathbf{B} = \mu_0 \mathbf{J} \tag{B.1}$$

where $\mathbf{J}(t, r)$ describes the space distribution of the current density. In addition, from the law of Faraday-Lenz we obtain

$$\frac{\partial \mathbf{B}}{\partial t} = -\nabla \times \mathbf{E} = -\nabla \times (\rho \mathbf{J}) \tag{B.2}$$

where in the last expression we made use of Ohm's in the formulation $\mathbf{E} = \rho \mathbf{J}$. Assuming that resistivity ρ is uniform and taking the rotor of this expression, we obtain

$$\frac{\partial}{\partial t}(\nabla \times \mathbf{B}) = -\nabla \times (\nabla \times \mathbf{J}) = \rho \nabla^2 \mathbf{J} \tag{B.3}$$

where we used the vector identity

$$\nabla \times (\nabla \times \mathbf{J}) = \nabla(\nabla \cdot \mathbf{J}) - \nabla^2 \mathbf{J}$$

and the continuity equation for electrical current $\nabla \cdot \mathbf{J} = 0$. Taking the time derivative of Ampère's law, we get the relation

R. Bartiromo and M. De Vincenzi, *Electrical Measurements in the Laboratory Practice*, Undergraduate Lecture Notes in Physics,
DOI 10.1007/978-3-319-31102-9

$$\frac{\partial \mathbf{J}}{\partial t} = \frac{\rho}{\mu_0} \nabla^2 \mathbf{J} \tag{B.4}$$

This *diffusion equation* describes the space-time evolution of many physical phenomena.[2]

The solution of Eq. (B.4) depends upon the geometry of the problem and its boundary conditions. In general, it is obtained with a complex procedure, which is beyond the scope of these notes. On the contrary, it is rather easy to show that the time and space scales characterizing its solutions are related. Indeed, given that with the assigned boundary conditions the solution of Eq. (B.4) has a time evolution described by an angular frequency ω, we can define a new dimensionless time variable $\tau = \omega t$. The partial derivative with respect to t can be expressed as

$$\frac{\partial \mathbf{J}(t, x)}{\partial t} = \omega \frac{\partial \mathbf{J}(\tau, x)}{\partial \tau}$$

and the original Eq. (B.4) becomes

$$\frac{\partial \mathbf{J}(\tau, \mathbf{r})}{\partial \tau} = \frac{\rho}{\mu_0 \omega} \nabla^2 \mathbf{J}(\tau, \mathbf{r})$$

The quantity $\Delta = \sqrt{\rho / \mu_o \omega}$ has the physical dimension of a length and yields the space scale of the solution. Indeed, defining a new dimensionless space variable $\hat{\mathbf{r}} = \mathbf{r}/\Delta$, it is possible to reduce the equation to a form that is independent on both ρ and ω and on the unity of measure of the independent variables

$$\frac{\partial \mathbf{J}(t, \hat{\mathbf{r}})}{\partial \tau} = \nabla^2 \mathbf{J}(\tau, \hat{\mathbf{r}}) \tag{B.5}$$

Once solved with the appropriate boundary conditions, the solution describes any system with the same geometry and the same ratio of physical dimensions to the quantity Δ.

We now consider again the simple case of the infinite conducting plane we discussed in Sect. 1.3.4. We choose a reference system with the x-axis normal to the plane and the y- and z-axes running along the plane surface. In these conditions the current density is bound by symmetry to change only along the direction x perpendicular to the plane. We make use of the symbolic method (see Chap. 5) to describe sinusoidal time dependence. In addition, we adopt the dimensionless variables defined above. The z component of the current density can be expressed as

$$\hat{J}_z(\tau, \hat{x}) = \hat{J}_z(\hat{x}) e^{j\tau}$$

[2]For example the diffusion equation describes how the temperature varies in time along a rod heated from one extreme, how a drop of milk spreads in coffee or how charge carriers move in a semiconductor. In general, the diffusion equation describes all those phenomena induced by a random walk at the microscopic level.

and, upon insertion in Eq. (B.5), we get

$$\frac{\partial^2 \hat{J}_z(\hat{x})}{\partial \hat{x}^2} = j\hat{J}_z(\hat{x})$$

The characteristic of this differential equation has solutions given by $\pm\sqrt{j} = \pm(1+j)/\sqrt{2}$. We get easily

$$\hat{J}_z(\hat{x}) = J_+ \mathrm{e}^{\frac{1+j}{\sqrt{2}}\hat{x}} + J_- \mathrm{e}^{-\frac{1+j}{\sqrt{2}}\hat{x}}$$

The current density must remain bounded when moving inside the conducting plane. Therefore $J_+ = 0$. In the original physical variables the solution becomes

$$J_z(x) = J_z(0)\mathrm{e}^{-\frac{x}{\Delta\sqrt{2}}}\mathrm{e}^{-j\frac{x}{\Delta\sqrt{2}}}$$

A similar expression holds for the component in the other direction in the plane y. Using the continuity equation for the electric charge $\nabla \cdot \mathbf{J} = 0$, we can show that in the direction normal to the plane the component of the current density is null.[3] Therefore, we obtain

$$\mathbf{J}(t, x) = \mathbf{J}(0)\mathrm{e}^{-\frac{x}{\Delta\sqrt{2}}}\mathrm{e}^{-j\left(\omega t - \frac{x}{\Delta\sqrt{2}}\right)}$$

In conclusion, the module of current density decreases exponentially moving inside the plane with a decay length $\delta = \sqrt{2\rho/\mu_o\omega}$ as stated at the end of the Chap. 1 in the main text.

[3]Since for all values of x we have $\partial J_z/\partial z = 0$ and $\partial J_y/\partial y = 0$, we also obtain $\partial J_x/\partial x = 0$ everywhere. Using this result in Eq. (B.4) we get $\partial J_x/\partial t = 0$. Therefore J_x is constant in time and space. Since it is null at the beginning of the experiment it remains identically null everywhere.

Appendix C
Fourier Analysis

Fourier analysis allows representing a large class of mathematical functions as linear superposition of sinusoidal functions. This appendix summarizes fundamental formulas of this analysis and shows some examples of Fourier series and Fourier integral representation of functions of particular interest for circuit analysis. Proofs of theorems and mathematical details of Fourier analysis are beyond the scope of this appendix and the reader is referred to the numerous and valuable textbooks available on this topic.

C.1 Fourier Series

Any periodic function $s(t)$ with period T can be expanded as a Fourier series, i.e., as an infinite sum of sinusoidal functions.[4] The Fourier series can be written in three equivalent formulations, the choice among them being a matter of convenience:

$$s(t) = \frac{a_0}{2} + \sum_{n=1}^{+\infty} (a_n \cos n\omega_1 t + b_n \sin n\omega_1 t) \tag{C.1}$$

$$s(t) = \frac{a_0}{2} + \sum_{n=1}^{+\infty} A_n \cos(n\omega_1 t - \phi_n) \tag{C.2}$$

$$s(t) = \sum_{n=-\infty}^{+\infty} c_n \mathrm{e}^{j\omega_1 nt} \tag{C.3}$$

[4]More rigorously, a mathematical function can be be expanded as a Fourier series only if it meets the conditions known as the Dirichlet conditions. The periodic functions, even discontinuous, used as models of physical signals always meet these conditions.

R. Bartiromo and M. De Vincenzi, *Electrical Measurements in the Laboratory Practice*, Undergraduate Lecture Notes in Physics,
DOI 10.1007/978-3-319-31102-9

where the parameter $\omega_1 = 2\pi/T$ is known as *fundamental angular frequency* and $\nu_1 = \omega_1/2\pi$ is the *fundamental frequency*. It can be shown that the coefficients a_n and b_n in previous expression (C.1) are given by

$$\begin{cases} a_0 = \frac{2}{T}\int_{-\frac{T}{2}}^{\frac{T}{2}} s(t)\,dt \\ a_n = \frac{2}{T}\int_{-\frac{T}{2}}^{\frac{T}{2}} s(t)\cos n\omega_1 t\,dt \\ b_n = \frac{2}{T}\int_{-\frac{T}{2}}^{\frac{T}{2}} s(t)\sin n\omega_1 t\,dt \end{cases} \tag{C.4}$$

With simple algebra and using trigonometric identities, we obtain the following relationships among the parameters in the Eqs. (C.1)–(C.3):

$$A_n = \sqrt{a_n^2 + b_n^2} \qquad \phi_n = \arctan\frac{b_n}{a_n}$$

$$c_n = \frac{a_n - jb_n}{2}, \qquad c_{-n} = c_n^* = \frac{a_n + jb_n}{2}, \qquad c_0 = a_0 \tag{C.5}$$

The Fourier series of a periodic function is the sum of a time-independent constant $a_0/2$, which is the average value of the function over the period T, and an infinite number of *harmonic functions*[5] each with a frequency multiple of the fundamental one: $\omega_n = n\omega_1 (n = 1, 2, \ldots)$.

The formulation of the Fourier series given in (C.1) is useful when the periodic function has a definite parity: for even functions ($s(t) = s(-t)$), all the sine coefficients are zero ($b_n = 0, n = 1, \ldots, \infty$), whereas for odd functions ($s(t) = -s(-t)$), all the cosine coefficients are zero ($a_n = 0, n = 1, \ldots, \infty$).

The formulation of the Fourier series given in (C.2) shows explicitly the amplitude A_n of each individual harmonic in the signal.

The compact expression given in (C.3) is derived with the use of Euler's formula; this formulation of Fourier series is the starting point to obtain the expansion of a non periodic function in terms of sinusoidal functions (the *Fourier integral* or *Fourier transform*) as it will be shown in Sect. C.1.2.

C.1.1 The Spectral Diagram. Examples

The *spectral diagram* consist in a plot of amplitudes A_n and phases ϕ_n, as defined in (C.5) as a function of the harmonic number n. The spectral diagram of amplitudes gives at a glance the "weight" of each harmonic contained in the time dependent signal.

[5]Here, harmonic function means sine or cosine function.

Example 1: Rectangular Waveform. As discussed in Sect. 7.2, the rectangular waveform is defined by its period T, the *duty factor* δ and the amplitude V_o. To find its spectral diagram, it is convenient to choose the origin of time ($t = 0$) in such a way that the time function is even (see Fig. C.1).

For even functions, the coefficients b_n are all zero, and the values of a_0 and a_n can be computed using the relations (C.4). Defining $\tau = \delta T$, we get

$$a_o = 2V_o\delta$$
$$a_n = \frac{2V_o}{T}\int_{-\tau/2}^{+\tau/2} \cos n\omega_1 t\, dt = \frac{2V_o}{\pi n}\sin\frac{n\omega_1\tau}{2} = \frac{2V_0}{\pi n}\sin n\pi\delta$$

Finally, we get:

$$s(t) = V_o\delta\left[1 + 2\sum_{n=1}^{\infty}\frac{\sin n\pi\delta}{n\pi\delta}\cos n\omega_1 t\right] \qquad \omega_1 = \frac{2\pi}{T} \tag{C.6}$$

As an example, Fig. C.1 shows two rectangular waveforms with *duty-factor* of 10 and 30 % respectively, and their spectral amplitude diagram for the first 30 harmonics.

It is worth to note that the shorter pulse has greater amplitude of high frequency components than the longer pulse.

Example 2: Rectified Sinusoidal waveform. We calculate now the spectrum of a sinusoidal signal after "cutting" its negative part. This waveform is obtained with a simple (nonlinear) circuit made by a diode and a resistor (see Sect. 7.4). The study of the rectified waveform is not purely academic as these signals are present as noise in laboratories when a large amount of direct-current power is required (up to several hundred kilowatts). The presence of multiple harmonics of the mains frequency (50 or 60 Hz corresponding respectively to angular frequency $\omega = 314\,\text{s}^{-1}$ or $377\,\text{s}^{-1}$) increases the probability to have electrical noise[6] in the circuits downgrading the quality of the power distribution.

Let us start computing the Fourier series of the output of the half-wave rectifier (see Fig. 7.7a in Chap. 7) when the input signal is $v(t) = V\sin\omega_o t$. The output signal is the positive part of the input signal as shown in Fig. C.2a. During one wave period $T = 2\pi/\omega$ the output signal is given by

$$f(t) = \begin{cases} V\sin(\omega_o t) & \text{if } 0 < t < T/2 \\ 0 & \text{if } T/2 < t < T \end{cases}$$

Choosing the form (C.1) to represent the Fourier series, with $\omega_1 = \omega_o = 2\pi/T$, the coefficients a_n and b_n are

[6] Electrical noise can be defined as all the components of the signals in the circuit not caused by the input.

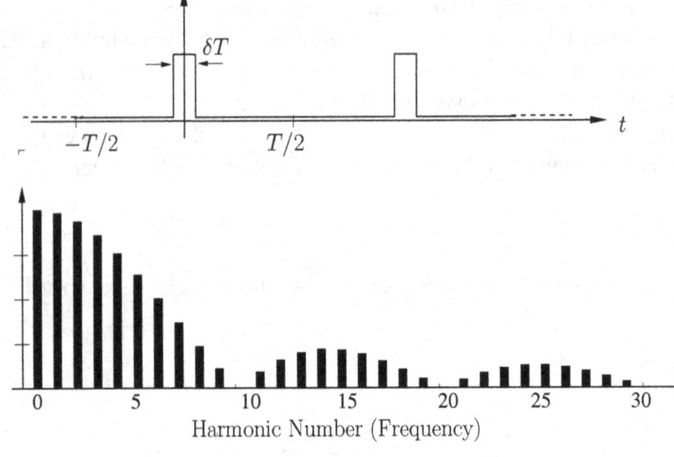

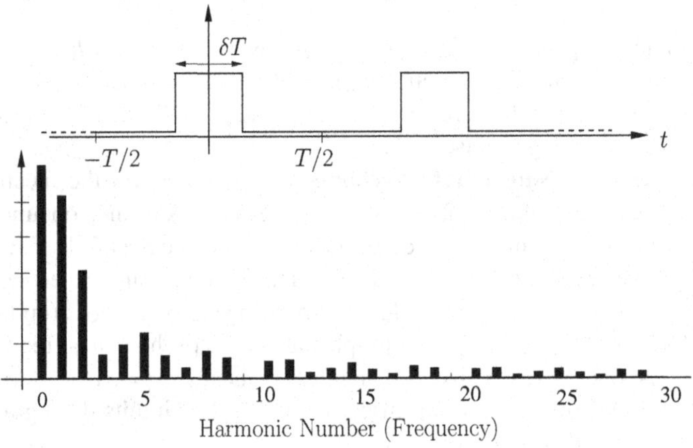

Fig. C.1 Fourier spectrum ($|A_n|$ as a function of the harmonic number n) of two rectangular waves, both with period T =10 a.u., amplitude $V_0 = 1$ a.u. and duty-factor, respectively, 10 and 30 %

$$a_0 = \frac{2V}{T} \int_0^{T/2} \sin \omega_o t \, dt = V \frac{2}{\pi} \tag{C.7}$$

$$\begin{aligned} a_n &= \frac{2V}{T} \int_0^{T/2} \sin \omega_o t \cos n\omega_o t \, dt \\ &= \frac{2V}{T} \int_0^{T/2} \sin \frac{2\pi}{T} t \, \cos \frac{2\pi}{T} nt \, dt = V \frac{1 + (-1)^n}{\pi(1 - n^2)} \end{aligned} \tag{C.8}$$

$$b_n = \frac{2V}{T} \int_0^{T/2} \sin \omega_o t \sin n\omega_o t \, dt = 0, \quad \text{for} \quad n \neq 1$$

If $n = 1$

$$b_1 = \frac{2V}{T}\int_0^{T/2} \sin^2 \frac{2\pi}{T} t\, dt = \frac{V}{2} \tag{C.9}$$

In conclusion the Fourier series of the output signal of half-wave rectifier is

$$\begin{aligned} f(t) &= \frac{V}{\pi} + \frac{V}{2}\sin\frac{2\pi}{T}t - \frac{2V}{3\pi}\cos\frac{4\pi}{T}t - \frac{2V}{15\pi}\cos\frac{8\pi}{T}t + \cdots \\ &= \frac{V}{\pi} + \frac{V}{2}\sin\omega_o t - \frac{2V}{3\pi}\cos 2\omega_o t - \frac{2V}{15\pi}\cos 4\omega_o t + \cdots \end{aligned}$$

Figure C.2a shows both the time shape of the signal and its spectral diagram.

A better way to rectify a sinusoidal signal is the "full-wave rectification," obtained with the use of a diode bridge circuit, as discussed in Sect. 7.4. This circuit is capable to invert the sign of the negative part of the input wave, doubling the efficiency with respect to the half-wave rectifier. In this case, the period of the rectified wave is equal to half of the period of the original sinusoidal waveform. If T is the period of the input signal, the fundamental harmonics is $\omega_1 = 2\omega_0 = 4\pi/T$, therefore,

$$a_0 = \frac{4V}{T}\int_0^{T/2} \sin\omega_0 t\, dt = V\frac{4}{\pi} \tag{C.10}$$

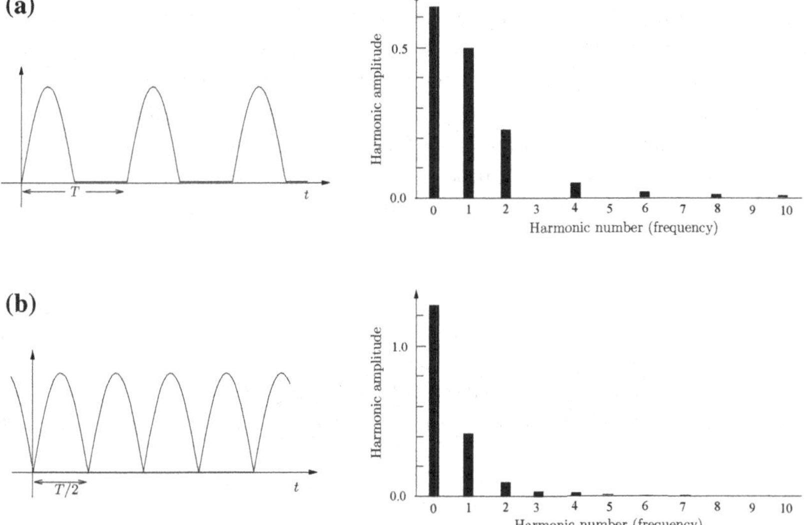

Fig. C.2 Output signals and spectral diagrams of a half-wave rectified signal panel (**a**) and of a full-wave rectifier panel (**b**)

$$a_n = \frac{4V}{T}\int_0^{T/2} \sin\omega_0 t \cos n\omega_1 t \, dt$$
$$= \frac{4V}{T}\int_0^{T/2} \sin\left(\frac{2\pi}{T}t\right)\cos\left(\frac{4\pi n}{T}t\right) dt = \frac{4V}{\pi(1-4n^2)} \tag{C.11}$$

$$b_n = \frac{4V}{T}\int_0^{T/2} \sin\omega_0 t \sin n\omega_1 t \, dt = \frac{4V}{T}\int_0^{T/2} \sin\left(\frac{2\pi}{T}t\right)\sin\left(\frac{4\pi n}{T}t\right) dt = 0 \tag{C.12}$$

In conclusion the Fourier series of the output signal of full-wave rectifier is

$$f(t) = \frac{4V}{\pi}\left(1 - \frac{1}{3}\cos\frac{4\pi}{T}t - \frac{1}{15}\cos\frac{8\pi}{T}t - \frac{1}{35}\cos\frac{12\pi}{T}t + \cdots\right)$$

Note that the mean value of the output signal of full-wave rectifier is $4V/\pi$, two times greater than the half-wave rectifier. Figure C.2b shows time shape and spectral diagram of a full rectified-wave. As can be seen from the figure, the full-wave rectification doubles the continous level component of the output, while it decreases considerably the amplitude of the AC components with respect to the half-wave rectifier.

C.1.2 The Fourier Transform

It is possible to obtain the spectral characteristic of nonperiodic signals using the *Fourier Transform* or Fourier Integral. A heuristic method to understand the Fourier transform takes into account the Fourier series in the form (C.3). Assume we compute the Fourier series for a signal in the time interval $-T, +T$. Now, leaving the signal unchanged, increase the time interval in which the signal is defined to infinite $T \to \infty$; in this limit the discrete variable $\omega_1 n$ becomes a continuous variable $\omega_1 n \to \omega$ and the sums become integrals in time.

The formal definition of the *Fourier Transform* (or *Fourier Integral*) of the function $s(t)$ is

$$S(\omega) = \int_{-\infty}^{+\infty} s(t)\, \mathrm{e}^{-j\omega t}\, dt \tag{C.13}$$

The function $S(\omega)$ is called the *spectrum* of the signal $s(t)$.

The function $s(t)$ can be obtained from the function $S(\omega)$ using the *inverse transform*

$$s(t) = \frac{1}{2\pi}\int_{-\infty}^{+\infty} S(\omega)\, \mathrm{e}^{j\omega t}\, d\omega \tag{C.14}$$

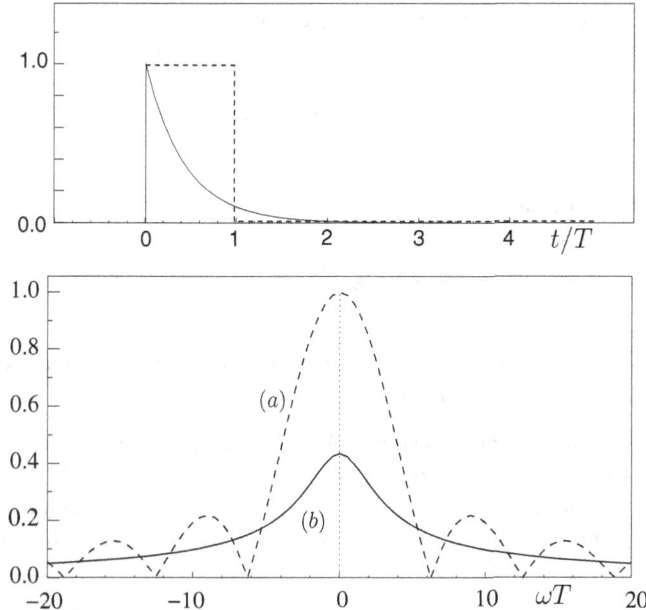

Fig. C.3 *Bottom figure* **a** Fourier transform of rectangular pulse of time width T, and **b** Fourier transform of one exponential pulse $e^{-t/\tau}$ $(t > 0)$ with $\tau = 0.43T$. The two pulses have approximately the same time length ($e^{-T/\tau} = 0.1$). *Top figure* time shape of the two signals

In order that the Fourier transform exists, the functions must satisfy certain conditions. A sufficient (but not necessary) condition is that the function $s(t)$ is *absolutely integrable* or $\int |s(t)|dt < \infty$.

Example 1: The rectangular pulse. The Fourier transform of a rectangular pulse of time width T and amplitude V is given by

$$S(\omega) = \int_{-\frac{T}{2}}^{\frac{T}{2}} V\, e^{-j\omega t}\, dt = V \frac{e^{j\omega T/2} - e^{-j\omega T/2}}{j\omega} = VT \frac{\sin(\omega T/2)}{\omega T/2} \tag{C.15}$$

The function $S(\omega)$ gives, for each ω, the weight of the corresponding frequency in the composition of the signal $s(t)$. Usually the absolute value $|S(\omega)|$ is plotted against ω to show the content of the corresponding frequency in the signal

$$|S(\omega)| = VT \left| \frac{\sin \omega T/2}{\omega T/2} \right| \tag{C.16}$$

In this case the spectrum has the absolute maximum at $\omega = 0$ and local maxima at $\omega = \pm k\pi/T$ with zeros at $\omega = \pm 2k\pi/T$. The width in angular frequency of each peak is approximatively $\Delta\omega \simeq 1/T$.

Example 2: The exponential pulse. As a second example, the Fourier transform of an exponential pulse $s(t) = s_0 e^{-t/\tau} \quad (t > 0)$ is computed as

$$S(\omega) = \int_0^{\infty} s_0 e^{-t/\tau} e^{j\omega t}\, dt = \frac{s_0 \tau}{1 + j\omega\tau} \tag{C.17}$$

The modulus and the phase of $S(\omega)$ are respectively:

$$|S(\omega)| = \frac{V_0 \tau}{\sqrt{1 + (\omega\tau)^2}} \qquad \phi = \arctan \omega\tau \tag{C.18}$$

In Fig. C.3 we show the Fourier transform of a rectangular pulse and an exponential pulse of about the same time length.

Example 3: The sinusoidal waveform of finite duration. As the last example of Fourier analysis lets us show why *it is not theoretically possible to realize a sinusoidal signal generator that contains only a single frequency. The impossibility is not due to technical limitations but to the fact that the signal starts at a given time* $t = 0$ *and ends at another time* $t = T$.

We use the notation $s(t) = s_0[\theta(t) - \theta(t - T)] \sin \omega_0 t$ for a signal of angular frequency ω_0 that starts at $t = 0$ and stops at $T = 2k\pi/\omega_0$ with k integer. The choice k integer corresponds to an integer number of wave periods and simplifies the following mathematical expressions (see Fig. C.4). The Fourier transform $S(\omega)$ of $s(t)$ is

$$\begin{aligned} S(\omega) &= \int_{-\infty}^{+\infty} s(t) e^{-j\omega t}\, dt = \int_0^T s_0 \sin \omega_0 t e^{-j\omega t}\, dt \\ &= \frac{s_0}{2j} \int_0^T s_0 \left(e^{j(\omega_0 - \omega)t} - e^{-j(\omega_0 + \omega)t}\right) dt \end{aligned}$$

$$= \frac{s_0}{2} \left(\frac{1 - e^{-j2k\pi(\omega/\omega_0 - 1)}}{\omega - \omega_0} + \frac{1 - e^{-j2k\pi(\omega/\omega_0 + 1)}}{\omega + \omega_0} \right) = \frac{s_0 \omega_0}{\omega_0^2 - \omega^2} \left(1 - e^{-j2k\pi\,\omega/\omega_0}\right) \tag{C.19}$$

where in the last relation we used that, for k integer, $\exp(\pm j2k\pi) = 1$. The spectrum of a time limited sinusoidal signal is given by the modulus of the function $S(\omega)$ given in Eq. (C.19) and is shown in Fig. C.5. As stated above, its spectral diagram shows

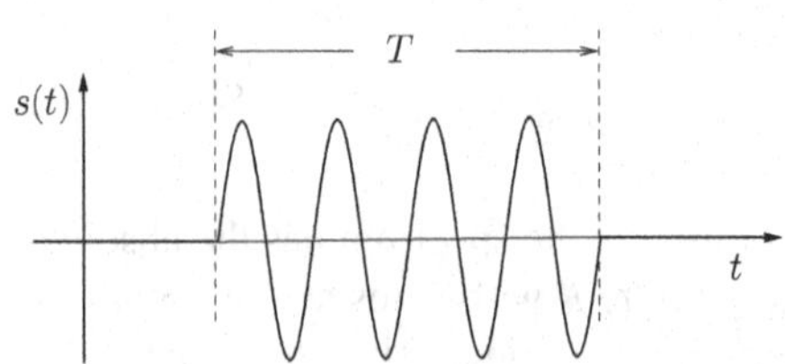

Fig. C.4 Sinusoidal signal of finite time duration

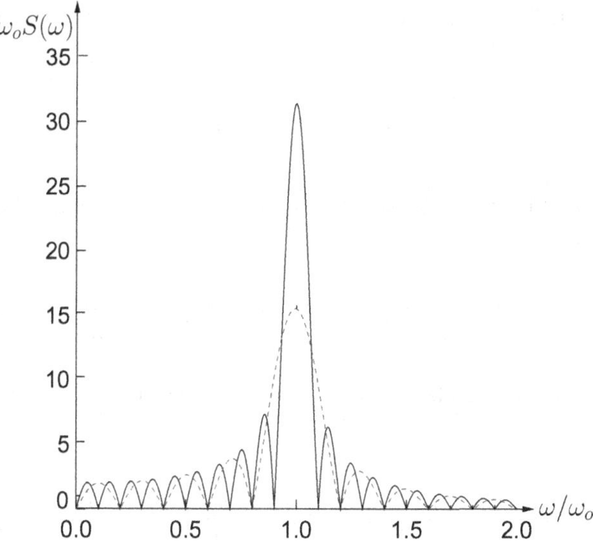

Fig. C.5 Frequency spectra of sinusoidal signals lasting 5 cycles (*dashed line*) and 10 cyles (*continous line*). By increasing the number of cycles, the width of the main peak around ω_o decreases while its amplitude increases linearly

the presence of many harmonics in addition to the fundamental one ω_0. Increasing k, the contribution to the spectrum of harmonics other than ω_0 decreases as the figure shows. The decrease in the (relative) contribution of the harmonics, different from the fundamental one, is also seen taking the limit of $S(\omega)$ for $\omega \to \omega_0$:

$$\lim_{\omega \to \omega_0} S(\omega) = k\pi s_0/\omega_0$$

that shows a linear increase with k of the spectrum around ω_0. Moreover, it can be seen that the width of the main peak of the spectrum decrases with the inverse of k.

C.1.3 Fourier Analysis and Symbolic Method

Consider again Eq. (5.7) describing the behavior of an RLC-series circuit. In this equation both the unknown variable $i(t)$ and the excitation $v(t)$ are time-dependent functions. Using their Fourier transform representations:

$$v(t) = \frac{1}{2\pi}\int_{-\infty}^{+\infty} V(\omega)\mathrm{e}^{j\omega t}\, d\omega, \qquad i(t) = \frac{1}{2\pi}\int_{-\infty}^{+\infty} I(\omega)\mathrm{e}^{j\omega t}\, d\omega \qquad \text{(C.20)}$$

we can obtain the Fourier transform of both their derivative and their integral. For the current $i(t)$ we obtain

$$\frac{di(t)}{dt} = \frac{1}{2\pi}\frac{d}{dt}\int_{-\infty}^{+\infty} I(\omega)\mathrm{e}^{j\omega t}\,d\omega = \frac{1}{2\pi}\int_{-\infty}^{+\infty} j\omega I(\omega)\mathrm{e}^{j\omega t}\,d\omega \tag{C.21}$$

where we inverted the order of the integration in ω with the differentiation respect to t, and similarly

$$\int i(t)dt = \frac{1}{2\pi}\int dt\int_{-\infty}^{+\infty} I(\omega)\mathrm{e}^{j\omega t}\,d\omega = \frac{1}{2\pi}\int_{-\infty}^{+\infty}\frac{I(\omega)}{j\omega}\mathrm{e}^{j\omega t}\,d\omega \tag{C.22}$$

Equivalent expressions can be written also for the tension $v(t)$. Using these results in (5.7) we can write

$$\frac{1}{2\pi}\left[L\int_{-\infty}^{+\infty} j\omega I(\omega)\mathrm{e}^{j\omega t}\,d\omega + R\int_{-\infty}^{+\infty} I(\omega)\mathrm{e}^{j\omega t}\,d\omega + \frac{1}{C}\int_{-\infty}^{+\infty}\frac{I(\omega)}{j\omega}\mathrm{e}^{j\omega t}\,d\omega\right]$$
$$= \frac{1}{2\pi}\int_{-\infty}^{+\infty} V(\omega)\mathrm{e}^{-j\omega t}\,d\omega$$

Since this relation is valid for all values of time t, it requires that

$$\left(j\omega L + R + \frac{1}{j\omega C}\right)I(\omega) = V(\omega) \tag{C.23}$$

This is the same equation we found in Sect. 5.5.1 where the solution of AC cicuits with the *symbolic method* was discussed. Therefore we recognize now that *the symbolic method is based on the transformation of time-dependent functions to frequency-dependent functions using formulas based on the Fourier Analysis.*

Index

R. Bartiromo and M. De Vincenzi, *Electrical Measurements in the Laboratory Practice*, Undergraduate Lecture Notes in Physics,
DOI 10.1007/978-3-319-31102-9

Printed in the United States
By Bookmasters